Sustainable Forest Management: Updated Reviews

Sustainable Forest Management: Updated Reviews

Edited by **Aduardo Hapke**

New York

Published by Callisto Reference,
106 Park Avenue, Suite 200,
New York, NY 10016, USA
www.callistoreference.com

Sustainable Forest Management: Updated Reviews
Edited by Aduardo Hapke

International Standard Book Number: 978-1-63239-584-9 (Hardback)

Contents

Preface

This research-focused book discusses the updated reviews regarding the topic of sustainable forest management. Even though the idea of forest sustainability is an age-old concept, the understanding of sustainable forest management (SFM) as a tool which balances various socio-economic and ecological concerns is relatively new. A significant change in viewpoint commenced during the 1990s as a consequence of an enhanced awareness of the deterioration of the environment, specifically regarding the alarming loss of forest resources. This book comprises of original case studies from 12 distinct countries across 4 continents (Europe, Asia, Africa and America). These case studies reflect a diversity of experiences from developed and developing countries, and should clarify the present status of SFM across the world and the problems related to its implementation.

The information shared in this book is based on empirical researches made by veterans in this field of study. The elaborative information provided in this book will help the readers further their scope of knowledge leading to advancements in this field.

Finally, I would like to thank my fellow researchers who gave constructive feedback and my family members who supported me at every step of my research.

Editor

Part 1

Africa

Obstacles to a Conceptual Framework for Sustainable Forest Management Under REDD in Central Africa: A Two-Country Analysis

Richard S. Mbatu
Environmental Sustainability
College of Science, Health and the Liberal Arts
Philadelphia University, Philadelphia
U.S.A

1. Introduction

Climate change is now an issue of concern at both national and international levels. In the past three decades efforts to address causes of climate change have focused mostly on mitigation measures of carbon emissions from conventional fossil fuels combustion – coal, oil, and natural gas. However, since 2000 after the 6th Conference of the Parties (COP) of the United Nations Intergovernmental Panel on Climate Change (IPCC) at The Hague, forests have gained increased recognition in their role in the fight against climate change. Forests are now almost at par with conventional fossil fuels at the top of the international climate change agenda.

With close to 60 gigatons of carbon (Gt C) exchanged between terrestrial ecosystems and the atmosphere every year; with the world's tropical forests estimated to contain 428 Gt C in vegetation and soils; with the loss of tropical forests as the major driver of the CO_2 flux caused by land-use changes during the past two decades (Lasco, 2010); with deforestation accounting for about 17 percent of all CO_2 emissions (Intergovernmental Panel on Climate Change [IPCC], 2007); and with little results achieved so far on mitigation measures of carbon emissions from fossil fuels combustion, the 13th COP meeting in 2007 in Bali, Indonesia, adopted a more rigorous emissions reduction mechanism on 'avoided deforestation' scheme, codenamed REDD. REDD (or REDD+)[1] is an acronym which stands for Reducing Emissions from Deforestation and Degradation. REDD provides a framework for mitigating greenhouse gas emissions from deforestation and forest degradation through market instruments and financial incentives.

Given that the majority of the world's tropical forests are located in developing countries which are generally poor, "REDD presents a tremendous opportunity to jointly address climate change and rural poverty, while sustaining ecosystem services and conserving

---

[1] Since 2008 after COP-14 in Poznań, the symbol '+' has been added to the REDD acronym in some publications in recognition of the fact that "climate benefits can arise not only from avoiding negative changes (deforestation, degradation), but also from enhancing positive changes, in the form of forest conservation and restoration" (Angelsen & Wertz-Kanounnikoff, 2008, as cited in Wertz-Kanounnikoff & Kongphan-apirak, 2009).

biodiversity" (Huberman, 2007). In this regard, sustainable forest management (SFM) in developing countries must be emphasized as an essential element for the attainment of the goals of REDD. Of the three major tropical forests regions in the world – Amazonia, Congo Basin and South-east Asia – the Congo Basin in Central Africa is the most impoverished. The implementation of REDD scheme in this region through SFM could create incentives for poverty alleviation while at the same time limiting deforestation and forest degradation thus making meaningful contribution to the fight against climate change. Although some countries in this region, since the conception of the scheme in 2007, have made significant progress in the preparedness process for a post-2012 REDD mechanism, it should be noted that mostly pilot projects have been carried out so far, with most of them marred by many difficult and controversial issues that need to be addressed before actual implementation can begin. Among the issues to be addressed are: 1) the problem of leakage. That is, the ability to control emissions beyond project and country boundaries, 2) the problem of determining the base-line. That is, how much deforestation has been avoided and how much deforestation is too much deforestation, 3) the problem of potential non-permanence. That is, how to deal with emissions resulting from natural and human causes at a later date, 4) the problem of price. That is, how the demand for carbon credits influences supply by REDD, and 5) the problem of tenure and usage rights (ownership of the land, and illegal logging control), weak economic, political and legal structures, and poor industrial practices in the forestry and agricultural sectors.

Efforts to address these problems have been largely focused on technical issues (problems 1-4), while not much attention has been given to the socio-economic and development needs (problem 5) of forest-dependent communities. With millions of people in the Congo Basin depending on the forest for their livelihoods, the importance of fully integrating the socio-economic and development needs of forest communities into REDD's agenda in this region cannot be undermined. As David Huberman of International Union for the Conservation of Nature (IUCN) observes "the success of REDD will ultimately depend on how well it contributes to the development needs of forest-dependent communities" (Huberman, 2007).

In this regard, this chapter is aim at explaining the challenges of REDD beyond the technical problem area. It focuses on the socio-economic, political and legal challenges of implementing the REDD scheme in two countries in the Congo Basin in Central Africa – Cameroon and Democratic Republic of Congo (DRC). The chapter critically reviews different factors leading to deforestation in these countries and explore potential pathways towards SFM under REDD. We argue that an architecture based on socio-economic structure that is incentive driven (financial incentive), is more likely to achieve the goals of REDD in the Congo Basin than a technical-base architecture driven by market instruments.

The methodological approach to this chapter is narrative, descriptive, and analytic review of documents and empirical data from various sources inspired by debates and events related to REDD in the Congo Basin in general and in Cameroon and DRC in particular.

2. The origins of REDD and the ongoing process

Multilateral agreements and conventions under the auspices of the United Nations are largely responsible for the resolution of many environmental problems that are of the global magnitude. This includes the convention on the Law of the sea protecting the open seas from various forms of abuses, the convention on trade on endangered species, the Montreal Protocol dealing with the elimination of substances that deplete the ozone layer,

Obstacles to a Conceptual Framework for Sustainable Forest Management Under
REDD in Central Africa: A Two-Country Analysis

5

etc. The Montreal Protocol is described by many as the most successful international environmental agreement this far. It is a fairly accurate description, especially when compared with the challenge of the global climate change problem in the wake of the failing Kyoto Protocol. The Kyoto Protocol was crafted in Kyoto, Japan in December of 1997 by the international community under the auspices of the Intergovernmental Panel on Climate Change (IPCC) in an effort to cut back on global greenhouse gas (GHG) emissions, known to be the major cause of global warming and resulting climate change. The IPCC was formed by the United Nations Environment Program (UNEP) and the World Meteorological Organization (WMO) in 1988 after the global community began noticing signs of climate change in the early 1980s. After its first and second conferences in the early 1990s, the IPCC presented a draft treaty during the second World Summit on the environment in Rio de Janeiro, Brazil in 1992, called the United Nations Framework Convention on Climate Change (UNFCCC). The UNFCCC became a binding agreement three years later in 1995 after 128 nations ratified the agreement. Every year since its ratification in 1995 the UNFCCC holds a conference of Parties of the convention (COP). It was during the 3rd COP in 1997 that the Kyoto Protocol was reached. A crucial element of the UNFCCC under the Kyoto Protocol is the undertaking by some developed countries to reduce emission of six greenhouse gases (carbon dioxide, methane, sulfur hexafluoride, nitrous oxide, hydrofluorocarbons and perfluorinated hydrocarbons) to at least 5% below 1990 levels, to be achieved by 2012. This was in recognition that the developed countries have a greater responsibility to emissions reductions than other Parties of the convention. Accordingly, all Parties of the convention are "classified [on the bases of] their levels of development and their commitments for GHG emission reduction and reporting [as follows]" (Randolph & Masters, 2008):

1. Annex I Parties: European Union (EU) member states plus other developed countries that aim to reduce emissions to pre-1990 levels.
2. Annex II Parties: The most developed countries in Annex I, which also commit to help support efforts of developing countries.
3. Countries with economies in transition (EIT): An Annex I subset mostly eastern and central Europe, and the former Soviet Union, which do not have Annex II obligations.
4. Non-Annex I Parties: All others, mostly developing countries, which have fewer obligations and should rely on external support to manage emissions.

In efforts to facilitate emissions reduction COP at its 6th meeting at The Hague in 2000, developed three flexible mechanisms to give more options to Parties of the convention in meeting their required reduction targets. These include: (1) the Clean Development Mechanism (CDM)[2], (2) the Joint Implementation (JI)[3], and (3) Emissions Trading (ET)[4]. However, in spite allowing greater flexibility in meeting emissions reduction targets, these three mechanisms considered forests only for their carbon sequestration function, leaving out their potential of reducing emissions from deforestation and degradation. The question

---

[2] The CDM mechanism allows industrialized countries to invest in clean energy projects in developing countries that are related to carbon emissions reduction and carbon sequestration in exchange for credit toward meeting their required reduction targets.
[3] The JI mechanism allows for collaborative efforts between two or more industrialized countries in meeting their respective reduction targets.
[4] The ET mechanism establishes modalities for selling and buying emissions right. It regulates the carbon market.

arises as to why reducing emissions from deforestation and degradation was excluded. Alvarado & Wertz-Kanounnikoff (2008) argue that it is not unrelated to the techno-scientific, political and methodological complexities. We will examine these obstacles later in the chapter.

Nevertheless, at The Hague meeting in 2000, the issue of avoided deforestation was brought up during deliberations on the eligibility of land use, land-use change and forestry (LULUCF) activities under the CDM. This issue was raised when Annex I Parties reported emissions from deforestation in annual GHG inventories. The concern over emissions from deforestation prompted COP to reach two compromises at The Hague as outlined in the following excerpt from Karsenty (2008):

> The compromise position proposed by President Pronk (Decision 1/CP.6) prior to the suspension of COP6 was to (1) designate avoided deforestation and combating land degradation and desertification in non-Annex I countries as adaptation activities eligible for funding through the Adaptation Fund but not through the sale of carbon credits; (2) allow only afforestation and reforestation projects in the CDM, with measures to address non-permanence, social and environmental effects, leakage, additionality and uncertainty (Karsenty, 2008).

This was evidently the first step toward the development of the REDD mechanism. This was followed by a series of behind-the-scenes meetings between some Annex I Parties and non-Annex I Parties. Collaborations between the two Parties led to the submission of a proposal by governments of Papua New Guinea (PNG) and Costa Rica (with the support of the Coalition for Rainforest Nations), to COP11 in Montreal in 2005, calling for COP to consider possible approaches to address the issue of avoided deforestation. This initiative by PNG and Costa Rica prompted COP to urge Parties to engage in a two year study of the technical, scientific, methodological, and policy and positive incentives approaches related to the issue of avoided deforestation. Hence after two deliberative workshops in Rome and Cairns, COP at its 13th meeting in 2007 in Bali, Indonesia, adopted a decision creating the REDD mechanism.

3. The concept of REDD

The concept of REDD comes from the broader ecosystems services concept developed by ecological economist Robert Costanza et al. (1987) in their pioneering work titled, *The value of the world's ecosystem services and natural capital*. Costanza et al. argue that ecosystems have economic value which must be factored into the market economy if we are to slow down or halt the global destruction of the world's natural environments. Services derived from the natural environment such as "regulating services (climate or water), provisioning services (food, fresh water), supporting services (soil conservation, nutrient cycling) and cultural services (aesthetic or traditional values)" (Alvarado & Wertz-Kanounnikoff, 2008), can be accounted for, by evaluating the direct economic value of their provision. For example, how much will it cost the U.S. state of Florida to build a water purification system that can replace the natural purification functions of the Everglades wetlands system in the coast of Florida? Although some will argued that an estimate cannot be made given that the wetlands functions of the Everglades cannot be replaced by any man-made machine, the system of 'payments for environmental services' (PES) would at least in economic terms, allow for an estimate which can then be used as incentive to protect the wetlands of the Everglades. The same principle is applied with REDD whereby through direct financial

incentives and financial flows from carbon markets, developing countries (mainly in the
tropical forests regions) are encouraged to reduce carbon emissions by adopting strategies
that would improve their capacity to reduce deforestation and forest degradation. Although
financial flows from carbon markets have the potential to motivate emissions reduction
from deforestation and forest degradation in developing countries, the greatest potential is
in direct financial incentives from the global community. According to the UN-REDD
strategy program, direct financial incentives of US$22-38 billion between 2010 and 2015
would lead to an estimated 25 percent reduction in annual global deforestation rates by 2015
(United Nations- Reducing Emissions from Deforestation and forests Degradation [UN-
REDD], 2010). Although the financial incentives approach is more likely to be favored for a
post-Kyoto implementation, "the majority of country proposals to the UNFCCC are in favor
of a mixed [market mechanisms and financial incentives mechanisms] approach" (Parker et
al., 2008).

4. Deforestation and climate change

The world's total forests area is estimated just over 4 billion hectares (Global Forest
Resource Assessment [GFRA], 2010), covering more than one-quarter (31%) of the world's
total land area. However, the world's forests cover varies in distribution with less than 2%
in the land area in some regions like North Africa, and up to 25% in others like Europe
(GFRA, 2010). Global forests also vary in cover types including the boreal forests (~ 1.3
billion hectares), the temperate forests (~ 1.0 billion hectares), and the tropical forests (~
1.7 billion hectares) (Gorte & Sheikh, 2010). These forests play an important, but unequal
role in global carbon budget as they are sinks – sequester carbon thus contribute to
climate change mitigation – and sources of carbon – emits GHG, especially CO_2 through
deforestation. While the total carbon content of the global forests in 2005 was estimated at
about 638 Gt, tropical forests store, on average, about 50% more carbon per unit area, than
temperate and boreal forests. For example, with a total area of about 1.7 billion hectares,
tropical forests store about 442 metric tons of CO_2 per hectare of plant carbon compared to
temperate and boreal forests which store only about 208 and 236 respectively (Gorte &
Sheikh, 2010). Although the current rate of tropical deforestation (2010) shows an overall
decrease of about 3 million hectares in the last ten years (GFRA, 2010), it is still
unacceptably high given the important role forests play as sinks and sources of carbon.
More so, in terms of net loss, South America and Africa which are home to two of the
world's three major tropical forests regions registered the highest net loss[5] of forests
between 2000 and 2010 – approximately 7.3 million hectares per year (GFRA, 2010). Given
that tropical forests store on average, 50% more carbon per unit area than the two other
major forests types (temperate and boreal), the global decrease of carbon stock in forests
(at an estimated 0.5 Gt per year between 2005 and 2010) can be attributed to the net loss in
tropical forests. The release of soil carbon into the earth's atmosphere is also linked to
deforestation as deforestation leads to soil exposure and disturbance (tilling), increase
dead matter, and increase soil temperature and rate of soil carbon oxidation (Gorte &
Sheikh, 2010). The boreal forests soils contain about 471 GtC per hectares of land, which
is more than twice the amount (216 GtC) in tropical forests soils (Alvarado & Wertz-
Kanounnikoff, 2008). Although tropical forests soils contain less than half the amount of

---

[5] Most of the loss was registered in the Amazonia forest in South America.

carbon in boreal forests soils, the high level of activities (agriculture, mining, ranching, road development, etc) in tropical forests soils leads to more release of soil carbon into the earth's atmosphere.

Forest Type	Area	Plant Carbon	Soil Carbon	Total Carbon
Tropical	1.76	442	450	892
Temperate	1.04	208	352	561
Boreal	1.37	236	1,260	1,496

(Area in billion hectares; carbon in metric tons of CO_2 per hectare)
Source: Adapted from Gorte & Sheikh (2010)

Table 1. Average Carbon Stocks in the World's major Forests

In all, the IPCC estimates that deforestation contributes approximately 17% of global greenhouse gas emissions (IPCC, 2007), which is equivalent to about 5.8 Gt of CO_2 per year (UN-REDD, 2010). According to the IPCC, reduced deforestation and degradation is the forest mitigation option, as it has the largest and most immediate carbon stock impact with approximately 93% of the total mitigation potential in the tropics (IPCC, 2007). This also has direct positive implications for the natural environment, notably in the area of biodiversity conservation, as well as indirect positive implications for sustainable development, notably in the area of poverty reduction in developing countries in the tropics. Therefore, the mitigation of tropical deforestation and forests degradation is crucial in the fight against climate change. It is in this regard therefore, following the IPCC decision in 2007 establishing the REDD mechanism, that the conceptual framework for managing forests in developing countries has been in progress since 2008.

5. REDD in the Congo Basin in Central Africa

The Central Africa forests region expands across the borders of six countries – Cameron, Central African Republic, the Republic of Congo, the Democratic Republic of Congo, Equatorial Guinea, and Gabon – covering an area of approximately 330 million hectares that sits largely within the geologic confines of a Basin commonly known in geographic terms as the Congo Basin. It contains the second largest area of contiguous evergreen forest in the world. The expansive forests cover of the Congo Basin presents the region with the ability to make meaningful contribution in the fight against climate change via carbon sequestration and reducing emissions from deforestation and forests degradation (REDD), as it stores an estimated 25-30 billion tons of carbon in its vegetation (Hoare, 2007). The lowland forests of the region in particular, though represents only 35% of the land area, stores more than 60% of carbon (Nasi et al., 2009, as cited in Sonwa et al., 2011). Yet the region suffers from one of the highest deforestation and forest degradation rates in the world, making it one of the biggest sources of carbon emissions (Hoare, 2007; Streck et al., 2008). Between 1990 and 2000 the region suffered a net deforestation and net degradation of 0.19 and 0.10% respectively (Congo Basin Forest Partnership [CBFP], 2007); and an annual deforestation of 0.15% during the period 2000-2005 (Hansen et al., 2008, as cited in Sonwa et al., 2011). Although the region's deforestation during the period 2005-2010 was relatively low at -0.23% per year (Food and Agricultural Organization [FAO], 2011), some studies (e.g. Zhang et al., 2002) show a projected increase of 1.3% by 2050. This will mean significant emissions from deforestation and degradation because of the region's vast area of forests.

It is difficult to separate deforestation from forest degradation as deforestation often paves
the way for other activities that degrade the forest (Kaimowitz & Angelsen, 1999). This point
is raised in the region's initial REDD proposal to the UNFCCC in 2006, in which the region's
potential contribution to emissions reduction from "deforestation and degradation" is
emphasized (UNFCCC/FCCC/ SBSTA/2007/MISC.14, 2007). The proposal stressed the
significance of including forests degradation to the reduction mechanism, arguing that
"degradation constitutes the main cause of forest cover loss, likely to affect nearly 60% of
productive lands in the Congo Basin" (Alvarado & Wertz-Kanounnikoff, 2008). Also,
because most of the countries in the Congo Basin already had or were in the process of
developing forest degradation plans at the time the proposal was presented in 2006; they
wanted to be rewarded for their early efforts (Karsenty, 2008). Even more important is that
considering forests degradation as a mechanism for emissions reduction would increase
overall emissions reduction than could be achieved with deforestation mechanism alone
(Alvarado & Wertz-Kanounnikoff, 2008).   The UNFCCC reasoned with this proposal.
Recognizing that it is difficult, if not impossible, to separate forests degradation from
deforestation in a potential reduction of forest emissions mechanism, the UNFCCC shifted
from its original position which was advocating "RED" – reducing emissions from
deforestation – to a one that included forest degradation – reducing emissions from
deforestation and degradation (REDD) – and officially endorsed the mechanism in Bali,
Indonesia in 2007.

In line with its proposal to the UNFCCC, the Congo Basin countries have made progress in
integrating forest degradation activities into the ongoing REDD pilot projects. This is shown
by the current landscape of REDD activities distinguished by three groups of activities –
demonstration activities, readiness activities, and activities without explicit carbon goals –
(Wertz-Kanounnikoff & Kongphan-apirak, 2009). Demonstration activities are those that are
designed with carbon as the 'explicit objective'. Readiness activities are designed to prepare
'an enabling framework' for adopting and implementing any REDD mechanism that the
UNFCCC might finally approve for a post-2012 Kyoto commitment. This includes preparing
documents such as R-PINs (Readiness Plan Idea Notes) which involves capacity building
through consultation with stakeholders, developing baseline projects to facilitate measuring,
monitoring, and controlling emissions in order to avoid the problem of additionality, and R-
PPs (Readiness Preparation Proposals) which entails developing a national strategy
framework. Activities without explicit carbon goals are those designed to promote the
enhancement of ecosystems management activities such as payment for ecosystem services
(PES) scheme, as well as sustainable development activities such as poverty reduction
scheme. Most countries in the region are currently engaged in readiness activities with
progress been made by Cameroon and the Democratic Republic of the Congo.

However, the region's ability to make meaningful contribution to emissions reduction could
be tampered by a number of problems including – the inability to control emissions beyond
project and country boundaries as reduced emissions in the region could be linked to
increased emissions in industrialized nations (leakage), the issue of determining how much
deforestation has been avoided and how much deforestation is too much deforestation
(base-line), the limited ability of understanding how to deal with emissions on a long-term
bases and avoiding short-lived benefits (permanence), the limited ability to respond to the
demand for carbon credits as determined by the opportunity cost and the cost of
implementing the REDD (price) and, dealing with issues of ownership of the land and
illegal logging control caused by weak economic, political and legal structures, and poor

industrial practices in the forestry and agricultural sectors (tenure and usage rights). These problems have become more evident in the past three years with the ongoing country-specific pilot project experiences with the REDD mechanism and can be explained under two categories: (1) the techno-scientific and methodological category, and (2) the socio-economic and political category.

The techno-scientific and methodological category is characterized by the following issues:

- Definitions – there are different definitions for forests and forest-related processes but currently there is no clear bottom-line definition applicable to the REDD mechanism. A crucial definitional issue is whether or not, and how to include plantations to the REDD mechanism since its already part of the climate change mitigation strategy under the clean development mechanism (CDM). But given that forest plantations have been linked to the loss of biodiversity (e.g. Butler, 2005), environmental NGOs like Greenpeace and WWF are opposed to the idea of including forest plantations in the definition. This is at odds with the already existing forests management plans in the Congo Basin which includes forest plantations, currently being pushed for acceptance by the UNFCCC. Although plantation forestry is very limited[6] in this region, it has plenty of cash-crop plantations – palm, rubber, cocoa, coffee – that the region would benefit from (though at the disadvantage of the regions rich biodiversity) if forest plantations are included in the definition.

- Estimation and monitoring of stocks carbon cycle – As is the case with forests definitions, there is currently no scientifically accepted method for accounting for stocks and flows of carbon. However, a number of tools exist which have the potential to develop internationally recognized methodologies. The accrued knowledge of remote sensing and geographic information system are proving to be useful for this purpose. Although remote sensing is proving to be useful in this regard, its application in the Congo Basin remains largely at the regional and national scales, making it inefficient for place/site specific assessment which is crucial for any eventual future REDD mechanism in the region. The importance of this element of REDD in the Congo Basin is captured best in this statement by the Central Africa Forest Commission (COMIFAC) – "No matter which final REDD mechanism is chosen, we will need to know as accurately as possible how much carbon is stored in different standing vegetation types (especially forests) and soils; released through AFOLU (agriculture, forestry and other land use) activities" (State of the Forest 2008, 2009, as cited in Central Africa Forests Commission [COMIFAC], 2010).

- Data availability – The estimation and monitoring of carbon stocks in the Congo Basin depends on quality data availability. The necessity for quality data has been emphasized by the IPCC data classification system[7]. This is very challenging for countries in the Congo Basin as they are required to use the best quality data "in order to make credible international claims for reduced degradation and forest carbon enhancement" (Skutsch et al., 2009). Although few pilot projects are currently going on

---

[6] As of 2010, only one large industrial forest plantation existed in the Congo Basin – Eucalyptus du Congo (ECO s.a.) – in Pointe Noire, Republic of Congo (Sonwa et al., 2011).

[7] The IPCC monitoring standard places data in three tiers in ascending order of quality (Skutsch et al., 2009): Tier 1 – highly generalized data that may not represent actual condition on the ground. Tier 2 – data derived from national-level activities that may be closer to on-the-ground conditions but may still not be accurate. Tier 3 – data derived from specific on-the-spot measurement with low error factor.

(e.g. Australia, Brazil, Cameroon, Guyana, Indonesia, Mexico, Tanzania) in which data from site-specific measurements are inventoried for carbon estimation analysis, the cost associated with gathering quality data (Zahabu et al., 2005; Tewari & Phartiyal, 2006; Karky, 2008) in a forest changing scenario, together with the lack of know-how by the local people is a drawback to the REDD mechanism in the region.

- Market-based instrument – A drawback to the REDD market-based instrument for the Congo Basin countries is the concept of a historical baseline. In spite of the plethora of empirical data documenting deforestation activities in the Congo Basin, most countries in the region still have relatively low deforestation rates (Dooley, 2009) compared to countries in other forest regions such as Asia (e.g. Indonesia) and the Amazon Basin (e.g. Brazil). It is a paradox that the comparably low deforestation rates in the Congo Basin works to the disadvantage of countries in the region, when it comes to the carbon markets baseline instrument under the REDD mechanism. This excerpt from Dooley (2009) explains the paradox of the historical baseline concept more clearly:

For forest credits to be traded in international carbon markets, the reductions in emissions must be measurable, and they must be over and above what would have happened otherwise. To measure this difference, a reference level must be established, which forms the baseline against which the impact of programs to reduce deforestation is measured. Most REDD proposals are based on the concept of a historical baseline: the reference scenario is determined on previous rates of deforestation, usually over a ten-year period, with the average forming the baseline. When emissions from deforestation (or any other activities included in a REDD agreement) fall below this rate, forest carbon credits are issued [...]. However, this approach favors countries with high rates of deforestation in the past. Countries with low deforestation rates, such as the Congo Basin countries – and those which have succeeded in reducing deforestation, such as Costa Rica and India – will not be able to claim emission reduction credits under this approach (Dooley, 2009).

The socio-economic and political category is characterized by issues relating to:
- Drivers of deforestation
- Sustainable forests management
- Regional and international cooperation
- Capacity building and
- Financial mechanisms.

The rest of this chapter will weave the aforementioned issues in an analyses of the socio-economic, political and legal challenges of implementing the REDD mechanism in two countries in the Congo Basin – Cameroon and the Democratic Republic of Congo (DRC).

5.1 Case analysis: Cameroon and Democratic Republic of Congo

The country Cameroon is located at 6° 0' 0" North of the Equator and 12° 0' 0" East of the Greenwich Meridian. The southern part of the country constitute part of the Congo Basin forests with an estimated 17 million hectares of tropical forests, accounting for over one-tenth of the remaining tropical forests in the Congo Basin (Sunderlin et al., 2000, as cited in Bellassen & Gitz, 2008). The forests of Cameroon are home to more than 8,300 plant species, close to 300 mammal species, and 848 bird species (International Tropical Timber Organization [ITTO], 2006), and support the livelihoods of millions of Cameroonians. The Democratic Republic of Congo lies South-east of Cameroon at 4° 31' 0" South of the Equator and 15° 32' 0" East of the Greenwich Meridian. It is the second largest country in Africa with a total area of 2,345,410 km² most of which (about 95%) is forests. The DRC has an estimated forest area of about 133 million hectares (52% of the Congo Basin forests), the largest confinement in Africa and the third largest in the world. Like in Cameroon, the forests of DRC are rich in species, and harbor some of the rarest and most endangered species of mammals in the world – gorillas, chimpanzees, elephants, etc. – birds, insects, and plants. However, in recent decades, the abundant forests in Cameroon and DRC have been disappearing at an alarming rate. In Cameroon, between 1980 and 1995 an estimated 2 million hectares of the forest was lost (Ichikawa, 2006). In 2000 the annual deforestation rate ranged between 80,000 and 200,000 square hectares (Ndoye & Kaimowitz, 2000), a trend that continued throughout 2000s, leading to the loss of about 4.4 million hectares of forest between 1990 and 2010 (Cameroon Forest Information and Data, 2010). Degradation also poses a serious threat to the country's forests. With some of the richest species of timber[8] in the world, the forests of Cameroon have attracted a lot of commercial logging enterprises over the years. Beginning in the 1970s, in an effort to bust its economy, the government of Cameroon encouraged large scale commercial logging in the country which opened up the forests for different environmental and socially damaging activities such as, road construction, slash-and-burn agriculture, hunting, illegal settlements, illegal logging, selective logging, plantation agriculture, and loss of cultural value (Mbatu, 2009). In DRC, logging practice that began in the 1920s gained momentum in the 1970s and 1980s. In spite of the devastating civil war in the 1990s logging activities continued, resulting to a 3% loss of the country's forest between 1990 and 2005 (Congo Basin Program [CBP], 2011). Although official reports indicate that 500,000 cubic meters of timber is harvested annually from the forests of DRC, many observers, including the Food and Agricultural Organization (FAO), place the figure at more than double (CBP, 2011).

The alarming rate of deforestation and degradation made Cameroon and DRC attractive choices for REDD pilot projects. The first REDD pilot project in Cameroon was initiated in 2007 by Service Element on Forest Monitoring (GSE FM) of the Global Monitoring for Environmental and Security (GMES)[9] with the aim of "integrating the application of Earth Observation (EO) technologies with the policy formulation" (COMIFAC, 2010). By integrating EO technologies with policy formulation Cameroon has made progress towards a national REDD framework as international cooperation[10], collaborative work between the country's Ministry of Environment and Nature protection, GSE FM, and other stakeholders involvement has led to the development and approval of the R-PIN, and with the R-PP nearing completion. The REDD process was launched in DRC in January 2009. Through the collaborative efforts of many bilateral and multilateral partners led by the country's Ministry of the Environment, Nature Conservation, and Tourism (MECNT) the county is making steady progress towards a post-2012 national REDD framework strategy. The

---

[8] Some of the most valued timber species in the forests of Cameroon include: Sapeli, Moabi, Azobé, Iroko, Okoumé Wengé, and Doussié.

[9] GMES is a joint venture of the European Space agency and the European Union that provides information on the global status of the environment mainly through the use of remote sensing and geographic information technologies.

[10] Cameroon's REDD initiative has benefitted from a "south-south" cooperation with the Central American nation of Bolivia which is ahead of Cameroon in meeting the REDD modalities, especially with regards to the technical aspect of the REDD scheme.

establishment of a national leadership team to coordinate the capacity building efforts led to the completion of the R-PIN with significant progress towards completing the R-PP. With significant financial and technical support from the international community both countries are on course to having a complete REDD national framework strategy in place by December 2012, just in time to begin implementing a post-Kyoto emissions reduction strategy. However, in spite of the overall progress, the emerging REDD framework for Cameroon and DRC faces a number of obstacles, especially in the area of sustainable forest management (SFM). Obstacles to SFM are generally tied to the driving forces of deforestation and forests degradation (figure 1) which many studies have attributed to a complex relationship between forest resources and socio-economic and political factors. In Cameroon and DRC the intricacies of policy and institutional factors – economic development, property rights, corruption, mismanagement, market and special dynamics, urbanization and industrialization, public attitudes, and values and beliefs (Mbatu, 2009) pose a threat to a successful adoption and implementation of SFM under REDD. The situation is even more complex in that forests provide the basic needs of a majority of people living in rural areas in these countries – fuelwood, timber, game, foodstuffs, raw materials, and other non-timber forest products. As Tieguhong (2008) notes, forest products represent up to 44% of the annual income of rural populations hence, constitute an important ecosystem for the economic wellbeing of the rural communities.

5.1.1 Poverty

Poverty is an important driver of deforestation (Angelsen & Wunder, 2003; Fischer et al., 2005; Soriaga & Walpole, 2007) and many studies in Cameroon (for example, Gbetnkom, 2008) and in DRC (for example, Iloweka, 2002) have documented the role poverty plays in forests loss and degradation in these countries. Reducing poverty in these countries, especially in forest communities could lead to lower rates of deforestation and degradation. However, Angelsen and Wunder (2003) have noted that increasing wealth of the rural poor does not generally translate to improved forest ecosystem since the accumulation of wealth, especially in the hands of a few elites could act as a springboard for attracting commercial logging enterprises.

5.1.2 Agricultural expansion

Agricultural expansion is another obstacle to the success of sustainable forest management under a potential REDD mechanism in these countries. Both subsistence and mechanized forms of agriculture play a significant role in forest loss and degradation in Cameroon and DRC. Deforestation happens in these countries partly due to agribusinesses undertaken by large multinational corporations involved in plantation agriculture. This is due largely to a growing demand for cash crops – notably robber, cocoa, palm nuts (Mitchell et al., 2007). Subsistence agriculture in forest communities in Cameroon and DRC like in most other countries in the tropics occurs through slash-and-burn, a practice that involves the cutting down of trees and burning them to open up an area for cultivation. After two or three seasons of cultivation the farmer abandons the plot and colonizes a new forest area as the old plot loses fertility due to the burning of the soil. This primitive practice, apart from its ecological downside, contributes significantly to both soil and forest carbon emissions. Bellassena & Gutz (2008) in an assessment of differential revenues that "a farmer could get from 1 ha of land out of two alternative land-uses: shifting cultivation [slash-and-burn], the traditional land-use pattern in southern Cameroon, or carbon credits as compensation for the conservation of primary forest, ...found that a break-even price of $2.85/t of carbon dioxide equivalent would level shifting cultivation with "Compensated Reduction" (Bellassena & Gitz, 2008)." The conclusion is that it is more profitable for the poor shifting cultivator farmer in Cameroon to preserve the forest under the REDD mechanism rather than to slash-and-burn it to grow crops. Unfortunately, in spite decades of educating this forest dwelling communities about the negative effects of practicing slash-and-burn agriculture, the practice still persists. This persistence can be linked to the poverty factor, a condition which the REDD mechanism must not undermined if it is to succeed in a post-2012 emissions reduction mechanism.

5.1.3 Increasing global resource consumption

Another obstacle is the growing global demand for energy, materials, and industrial processes – notably mining and petroleum production. In Cameroon, a policy decision to exploit oil and natural gas in an ecologically sensitive coastal forest area in the Southwest Region in the early 1980s set precedence for deforestation and degradation of forests areas that harbor oil and mineral resources. Following this policy the government, in the late 1990s, made a decision to allow the landlocked nation of Chad to export its oil via the port of Kribi which led to significant forest loss. The project, which involved the laying of oil pipe lines from the oil field in Doba in Southern Chad to the port of Kribi in the South of Cameroon (over 1,070 km), is responsible for the loss of over 15 km² of rainforest (Center for Environment and Development [CED], 2004). Another potential loss of forest under Cameroon's oil exploitation policy could come from the oil rich Bakassi peninsula which is home to rare species of plants and animals. Although oil resources have produced high rents and made substantial contribution to the national economy of Cameroon in the past 30 years[11], the continued falling output trends in the past 10 years has pushed the government to further open up forest areas for mining activities. For example, the discovery of a massive iron-ore deposit in the dense forest area near Mbalmayo in the East Region of the country has attracted the Australian mining company, Sundance Resources Ltd (Nyuylime, 2006). The company is forging ahead with mining operations in the area which has the potential of triggering extensive deforestation and degradation, mainly from infrastructure development such as roads and railways. Another mining prospect is the iron deposit located within the sensitive forest area at the western coast at Kribi. Though not currently under exploration, increasing demand for mineral resources from China has attracted Chinese companies in "obtaining exploration and exploitation rights to several bauxite reserves in the far northern portion of the country, at Minim-Martap and Ngouanda" (Reed & Miranda, 2007), and may also attract these Chinese companies to the Kribi iron reserve which lies within the forest zone in the South Region. Although the revision of Cameroon's mining law in 2001 led to an overhaul of the mining code, it still enables dozens of small mining operations to take place within forests areas, which leads to deforestation and forests degradation. The DRC's mineral and oil exploration policies (World Bank, 2008) are similar to those of Cameroon. However, Reed and Miranda (2007) noted that "mining in DRC has been limited largely to the fringes of the [Congo] river and

---

[11] Oil accounts for more than half of Cameroon's export earnings, around 24% of government revenue, and about 6% of GDP (Extractive Industries Transparency Initiative [EITI], 2006).

forest systems that make access difficult" (Reed & Miranda, 2007). This limited access could be attributed to the socio-economic and political instability the country went through during the last quarter of the last century. However, with growing political stability since the mid-2000s, and with the growing need of improving socio-economic conditions, it is likely that mining operations will occur in formerly less accessible forest areas. If provisions are not made in the REDD framework strategy for these countries to limit the extent and intensity of these activities within their forests areas; it could undermine the highly anticipated post-Kyoto emissions reduction contributions by these countries.

5.1.4 Market failure

A significant issue peculiar to most tropical timber exporting nations is market failure. Many services provided by tropical forests (table 2) are not traded in markets and, as a result have not established markets based on prices and therefore do not enter into the decisions of private and public sector actors (Pearce & Warford, 1993). For example, in both Cameroon and DRC the government is the country's land owner who has the responsibility of maintaining lands under forest cover, but because the government does not get the full value of social benefits provided by forests, there is little or no incentive to protect the forests and the underlying lands. The market has failed to stimulate these governments and the private sector in the direction of socially based objectives. This causes these governments to misplace their priorities, hence, ushering in inefficiencies that lead to deforestation and forests degradation and, consequently, forest loss. Lack of information regarding the value of non-market benefits of forest is the reason for market failure (Pearce, 2001; Wertz-Kanounnikoff, 2006). In Cameroon for example, Fomété (2001) notes that non-market goods such as biodiversity, watershed protection, recreation, and carbon sequestration appear to be of low importance to the government and to the local population hence, no incentives to actively engage in seeking information on the market value of forests and other environmental services. The failure of markets to capture the value of non-priced services of the forest is an indirect but meaningful cause of forest loss that shapes the non conservation attitude of forest exploiters in these countries. Therefore, assigning a price to the services provided by forests and charging the price to the beneficiaries of these services is a good strategy for the REDD mechanism for two reasons: (1) it reduces over exploitation of the forests (for timber, firewood, agriculture etc.) hence reducing emissions and (2) it generates income for the protector of these services. The concept of payment for environmental services is not new to the Congo Basin region. However, integrating it into the REDD mechanism would be challenging as a majority of the people in the region are very poor and depend on nature's free services for their livelihoods. Even with the significant funding by international organizations and financial institutions (for example, the United Nations Environment Program, the United Nations Development Program, the Food and Agricultural Organization, the African Development Bank and the World Bank) toward the implementation of the concept in the region, very little has been accomplished (Lescuyer et al., 2009). The land tenure crisis in the region will compound the problem for the REDD mechanism. This is already the case where a proposed project by Conservation International and the World Wildlife Fund to pay for biodiversity services in the Ngoyla-Mintom forest and the Bonobo forest in Cameroon and DRC respectively has been delayed due to land ownership contentions (Lescuyer et al., 2009).

Regulatory functions	Productive functions
The forest provides support to economic activities and human well-being by: - climate regulation - hydric regulation - protection against soil erosion - maintaining biodiversity - carbon sequestration - recycling organic matter and human Waste	The forest provides basic resources, notably: - building materials: wood, lianas... - energy: fuelwood... - food resources: non-timber products, game... - medicinal resources - genetic resources
Physical support functions	**Informational functions**
The forest provides the space and required substrates for: - habitat - farming zones - recreational sites - conserved natural spaces	The forest provides esthetic, cultural and scientific benefits: - source of cultural and artistic inspiration - spiritual information - historic, scientific and educational information - potential information

Source: Lescuyer et al., 2009

Table 2. Categories of Environmental Services Provided by Forests

Not only are services provided by the forests in the Congo Basin not traded in markets but charges paid by timber concessionaires in the region are relatively low compared to the market value of the resource (Fomété, 2001). These low charges are encouraging foreign timber companies especially from countries with rapidly growing economies like China to aggressively seek concessions here. In fact, the Chinese company Vicwood-Thanry which began exploring Cameroon's forest in 1997 now owns 12% of foreign-owned concessions in Cameroon (Topa et al., 2009). There is also the lack of incentive for long term management due to logging competition among the many concessionaires. High profits due to low charges are also potential drivers of corruption that has handicapped strict enforcement of concession terms (Sieböck, 2002).

5.1.5 Inefficient tax system

Poor or mistaken tax policies have created obstacles to the sustainable management of forests in Cameroon [and DRC] (Sikod et al., 1996). In an effort to resolve the issue of market failure, the government of Cameroon under the 1994 forest policy embarked on reforming the forestry taxation system, output targets, regulation, private incentives and macroeconomic management, using policy instruments such as royalty[12] for the forest area, felling tax, exit duty on logs, and factory taxes.[13] These policy interventions have instead created controversies which have contributed to more deforestation in some areas. For example, the lack of concrete modalities for handling tax revenues and rents from royalties

---

[12] Royalties are "rents" not "taxes," but are placed under tax regulations to ensure that concessionaires pay for the right to access the resource.

[13] Exit duty on sawn products and entry taxes on logs taken into factories.

has made the policy vulnerable to fraud, evasion, and misappropriation. For instance, rents from the royalties are rarely distributed as stated in the loggers' condition of contract (Fomété, 2001). Logging companies have taken advantage of the weak modalities to benefit from the system by making only partial payments to the communities and direct contributions in kind to the villagers. In some cases, powerful elites have made deals with loggers that benefit only them and not the community. In such cases, the local populations do not get the social benefits (roads, schools, hospital etc) as stipulated in the conditions of contract. This depravity fosters illicit activities as forest dwellers become engaged in clandestine logging in an effort to meet their needs (Cerutti & Tacconi, 2006). Another revenue related issue is the boundary problem. Neighboring forest villages often have quarrels related to boundary issues. This is because the larger the forest estate, the more benefit the local populations gets from it. Boundary problems are particularly challenging to authorities as lack of consensus between all stakeholders due to the absence or improper consultations with the local populations has led to vague provisions that lack legislative clarity. In DRC the situation is much the same. Klaver (2009) has noted that even though the 2002 forestry law clearly stipulates that local authorities are entitled to 15% of all forest related taxes and 25% for provincial governments "no funds have been transferred to provincial or local governments because they still lack the necessary legal authority and rights" (Klaver, 2009).

Another problem with the forest taxation system of these countries is lack of transparency. There is little or no flow of information between the different stakeholders. Most forest inhabitants in both countries are illiterate. The few who can barely read and write find it difficult to interpret policy documents. This is a huge limitation which, coupled with the lack of adequate information to estimate their expected revenue, causes inhabitants to fall prey to the greedy loggers and government official imposters. An even more serious problem with the forestry taxation system of these countries is that though it claims to be a decentralized system, both governments still place strong emphasis on centralized decision-making, ownership and control of forest resources. In Cameroon for example, the creation of community forests under the 1995 forestry law was aimed at giving complete autonomy for managing the forests to the communities themselves. However, communities are still unable to fully control the management process as local, national and regional government representatives sit in management committees and impose government directives on community leaders (Assembe, 2006; Mbatu, 2010). This has caused the local population not to cooperate with the government in managing the forests. These centralized decision-making policies have been partly responsible for failures to protect forests and for its extensive loss and degradation in Cameroon. The decentralization of forest management in Cameroon was meant to act as a model for the Central African region but in many ways has fallen short of the desired prototype. Though not a prototype of an ideal decentralized management system, the model has been adopted by some countries in the region including DRC which adopted a new forestry law in 2002 and passed a new constitution in 2006 (Government of DRC, 2006) in a move aimed at empowering communities to take control over the management of local forests. Although the 2006 constitution "stipulates that taxes are to be collected by decentralized autonomous local governments (Government of DRC, 2006), all forests-related taxes are [still] been paid directly to the national Treasury" (Klaver, 2009).

5.1.6 Lack of secure land tenure

The local populations in many communities in Cameroon and DRC have little or no incentive to protect the forest resource because they feel they have no stake in it as the governments of these countries continue to appropriate forest lands and its resources. About 97% of natural forests in Cameroon belong to the government, while all natural forests in DRC are government owned (Cotula & Mayers, 2009). This clearly shows that land tenure in both countries is unclear or insecure even though both countries allocate community forests for the benefit of the communities. In the absence of well-defined property rights (complete specification of rights, exclusive ownership, transferability, and enforceability) the forest has become a "common property" with the local populations engaging in forest clearance as a way of showing occupation and ownership. They base their actions on the customary law of *droit de hache* – axe right – whereby, ownership of land and its resources is acclaimed by putting it into productive use, by cutting down of trees. The use of common property leads to externalities ("tragedy of the common") as no one is liable for the cost of the externalities. Although the governments of Cameroon and DRC continue to exercise "authoritarian and repressive policing" in their forests sectors, the "limited government capacity to monitor compliance and sanction non-compliance" (Cotula & Mayers, 2009) has allowed the customary law practice of *droit de hache* to continue. A challenge for the REDD mechanism in Cameroon and DRC, and in all other Congo Basin countries therefore, would be to help amend current decentralization policies and encourage genuine land tenure reforms that would lead to effective local control of forest management.

5.1.7 Population pressure

Like many other natural resources, deforestation and forest degradation in Cameroon and DRC has been linked to population pressure. According to World Bank, World Development Indicators 2000, Cameroon and DRC have the highest population density in the region with a combined population of 73.85 million inhabitants. Within the past five years both countries have witnessed rapid increases in populations, with the population of Cameroon increasing from 16.30 million in 2005 to 19.7 million in 2011 (Central Intelligence Agency [CIA], 2011) while that of DRC increased from 57.5 million in 2005 to 71.7 million in 2011 (CIA, 2011). This increase in population will increase the demand for land to cultivate and fuelwood for energy. It will also lead to cheap labor and generate high profits for agribusiness owners, making them more viable and having the desire for more land to expand their business. Increase in population will also lead to increase poverty as too many people will depend on few individuals for support. According to the United Nations Development Program's Human Development Index[14] of 2006, Cameroon and DRC rate among the lowest in the region. The increase poverty will caused many families to depend totally on the forests for survival, through logging and slash-and-burn farming. The problem of population pressure is not only within these countries. The pressure is also coming from outside. Global population growth, together with economic growth mainly from industrialized countries has led to excessive foreign consumption of tropical produce like banana, cocoa, rubber, and palm nuts within the past two decades. Cameroon has responded to this increase in demand by adopting policies that support agricultural extensification hence, expanding its area of cultivation at the expense of the forests. For

---

[14] The Human Development Index is a multidimensional concept of measuring human development that goes beyond income. It includes three key measurable dimensions – life expectancy at birth, adult literacy and gross enrolment in primary, secondary and tertiary levels, and per capita GDP (PPP US$).

example, since 1991, the Cameroon Development Corporation has expanded about 1,340 hectares of its plantations into the Onge-Mokoko forests in the Sourhwest province of Cameroon (Acworth et al., 2001).

5.1.8 External debt servicing and misguided macro-economic policies

Debt service and low per capita income is another obstacle to sustainable forest management in Cameroon and DRC. External pressure from the International Monetary Fund (IMF) and World Bank to recover debts owed them by the governments of these countries has left both nations' forests vulnerable. For example, the forest policy adopted by the government of Cameroon in 1994 was strongly influenced by the World Bank and the International Monetary Fund, mainly through the Structural Adjustment Program (SAP) that favors longer durations of concessions to logging companies, and encourages large forest areas for concessions (Ekoko, 1997). Most structural adjustment policies prescribed by the World Bank and the IMF are meant to liberalize the economies of countries adopting them and open up their markets to global economy. Instead, it has presented mixed fortunes for developing countries. Since its inception only a few countries like China, Brazil and Indonesia have been able to get meaningful benefits from the program. The rest of the developing countries have instead witnessed economic downturns (Brown & Quiblier, 1994). In Cameroon, reduction in public spending has instead led to unemployment forcing people to rely on subsistence farming at the detriment of the forest (Benhin & Barbier, 1999; Gbetnkom, 2008). In DRC the SAP created a situation where high unemployment (Jauch, 1999) left many families to rely on agricultural encroachment on forests lands for survival. The IMF call for economic liberalization in recent years has also not yielded much for the people as the economic indexes of both countries remains low. According to the 2011 Heritage Foundation and Wall Street Journal's Index of Economic Freedom report, Cameroon's per capita income is $2,147. Its FDI inflow is $337.5 million (Index of Economic Freedom, 2011). However, its external debt fell from an all-time-high of $11.11 billion in 1998 to $2.94 billion in 2009 (World Bank, 2009); thanks to the increased revenues from the forest sector (Fometé & Cerutti, 2008). The situation in DRC is no better. The country's current (2011) external debt is $13.5 Billion. Its per capita income of $332 is one of the lowest in Africa. Its FDI inflow is equally low at $2.1 billion (Index of Economic Freedom, 2011). High debts, low FDI and weak economies have left the peoples of Cameroon and DRC with little option but to continue deforesting and degrading their land for a livelihood.

In all, a combination of economic pressures to accelerate harvesting, corruption, the weakness of government forest administration, inappropriate concession license allocation and timber taxation systems, and the negative impacts of macroeconomic trade, population pressure, and Structural Adjustment Program polices have led to decades of unsustainable forest management in Cameroon and DRC.

The aforementioned obstacles to sustainable forest management in Cameroon and DRC are summarized in the following conceptual framework model (figure 1 – influence diagram) of direct and indirect causes of forest loss and/or degradation. Although the influence diagram appears to show some agents as directly linked to deforestation, there is actually no clear cut separation of direct and indirect causes of deforestation. In reality, forest loss is the result of long chains of causation that are not linear, but have feedback loops at some points or stages in the causation chains (Angelsen & Culas, 1996).

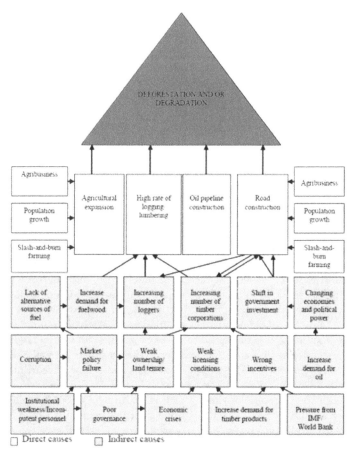

Fig. 1. Influence Diagram of Direct and Indirect Causes of Deforestation/Forest Degradation in Cameroon and the DRC

6. Conclusion

With the Kyoto protocol failing to meet its goal of reducing emissions of greenhouse gasses below the pre 1990 level by 2012, the IPCC now sees REDD as an indispensable element for the success of a post-2012 climate regime. After many decades of neglecting the importance of avoided deforestation and forest degradation in the fight against climate change, the IPCC in 2000 began integrating degradation schemes in its emissions reduction regime; thanks to the empirical evidence that deforestation and forest degradation accounts for about 17 percent of global greenhouse gas emissions, and that the world's forests sequester more carbon than it is in the atmosphere (Rogner et al., 2007; IPCC, 2007). The integration of REDD in the global emissions reduction regime does not only contribute in reducing carbon emissions and conserving biodiversity and sustaining ecosystem services, but also presents a tremendous opportunity to enhance living conditions for the world's poorest people, most of who live in forest communities in the tropics. The Congo Basin region of Central Africa in particular, has a lot to contribute and benefit from REDD. The vastness of the forests in this region means a significant amount of carbon could be sequestered here if REDD were to succeed in the region. The Democratic Republic of Congo alone, it is estimated, has a REDD potential of over 400 million tons of CO_2 per year through 2030 (Kabila,[15] 2011). In spite the tremendous opportunity REDD presents to reduce carbon emissions and foster sustainable forest management in the Congo Basin region, the obstacles discussed in this paper (and can be extrapolated to all the countries in the region) must first be overcome. Although the region faces techno-scientific problems, the biggest obstacles to meeting the goal of REDD are of socio-economic and political nature. To overcome these obstacles, a number of measures must be taken by the national REDD Coordination and Management Teams (CMTs) of various countries in the region. First, is the need for capacity building of the various CMTs. At moment, the CMTs are made up of officials from two or more government ministries and representatives from a few national and international non-governmental organizations, whose lack of coordination is, itself, an obstacle to the national coordination effort the teams are charged with. A CMT whose activities are well coordinated is more prepared to deal with socio-economic and political issues that are a potential hindrance to the success of the national REDD strategy. Second, all stakeholders must be included at every level, and at all stages of development and implementation of the REDD strategy. This will help move the process in the direction of a pro-poor approach in which, traditional farmers, hunter-gatherers, community forest managers, indigenous peoples (especially the pygmies), traditional leader, municipal councils, regional councils, etc. contribute in making decisions that directly relate to their needs. Third, countries in the region must adopt a coherent policy to address the issue of tenure and other related rights of the indigenous forest peoples to access forests and its resources. This will give REDD's investors (public and private) and its global partners a sense of security when it comes to possible tension with local groups. This will also empower the local peoples as secure tenure provide them "more leverage in relations with government and the private Sector" (Cotula & Myers, 2009). Fourth, to jointly address climate change and rural poverty under REDD would require sustainable forest management practices that confirm with the market tenets of ecosystem services. For this to be achievable, countries in the region will have to adopt measures that improve their economic and financial systems; specifically, measures that will enable them to uphold their "socio-environmental services and capital in [their] economic and political choices." Finally, REDD CMTs in the region must learn from the past successes and failures of other resources management initiatives in the agricultural and forest sectors, and the global carbon market. Specifically, attention must be given to issues related to capacity building and incentives (e.g. developing effective arrangements to channel benefits to the local level), legal framework and policy (e.g. connecting national laws and policy to relevant international principles and norms), education and information and culture (e.g. support for training workshops on resource knowledge, ecosystem management, biodiversity conservation, and cultural integration).

7. Acknowledgement

I will like to thank the anonymous reviewers for their comments on the initial draft of this chapter. Your comments have greatly shaped the outcome of this chapter.

---

[15] Kabila is the President of the Democratic Republic of Congo. He is referenced here citing a report by Mc Kinsey & Co, a consulting company for the DRC REDD initiative.

8. References

Acworth, J.; Ekwoge, H.; Mbani, J. & Ntube, G. (2001). Towards Participatory Biodiversity Conservation in the Onge-Mokoko Forests of Cameroon. *Rural Development Forestry Network Paper* No.25d, (July), pp. 1-20

Alvarado, L. & Wertz-Kanounnikoff, S. (2008). Why are we Seeing "REDD"? An Analysis of the International Debate on Reducing Emissions from Deforestation and Degradation in Developing Countries. *IDDRI Analysis,* No. 01/2008, Natural Resources, IDDRI, Paris, France

Angelsen, A. & Culas, R. (1996). Debt and Deforestation: A Tenuous Link. Working Paper No.10, Chr. Michelsen Institute, ISSN 0804-3639

Angelsen, A. & Wertz-Kanounnikoff, S. (2008). What are the key Design Issues for REDD and the Criteria for Assessing Options. In: *Moving Ahead with REDD,* A. Angelsen (Ed.), pp. 11–21, ISBN 978-979-1412-76-6, CIFOR, Bogor, Indonesia

Angelsen, A. & Wunder, S. (2003). *Exploring the Forest-Poverty Link: Key Concepts, Issues, and Research Implications.* CIFOR Occasional Paper No.40. Bogor, Indonesia.

Assembe, S. (2006). Decentralized Forest Resources and Access of Minorities to Environmental Justice: An Analysis of the Case of the Baka in Southern Cameroon. *International Journal of Environmental Studies,* Vol.63, No.5, pp. 681-689, DOI 10.1080/00207230600963825

Barbier, E. & Burgess, J. (2001). The Economics of Tropical Deforestation. *Journal of Economic Surveys,* Vol.15, No.3, (July 2001), pp. 413-433, DOI: 10.1111/1467-6419.00144

Bellassena, V. & Gitz, V. (2008). Reducing Emissions from Deforestation and Degradation in Cameroon: Assessing Costs and Benefits. *Ecological Economics,* Vol.68, No.1-2, (December 2008), pp. 336– 344, DOI 10.1016/j.ecolecon.2008.03.015

Benhin, J. & Barbier, E. (1999). *A Case Study Analysis of the Effects of Structural Adjustment on Agriculture and on Forest Cover in Cameroon.* Center for International Forestry Research and the Central African Regional Program for the Environment

Brown, N. & Quiblier, P. (1994). *Ethics and Agenda 21: Morals and Implications of a Global Consensus.* United Nations Publications, ISBN-92-1-100526-4, New York

Butler, A. (2005). *World deforestation rates and forest cover statistics, 2000-2005.* 24.08.2011, Available from http://news.mongabay.com/2005/1115-forests.html

Cameroon Forest Information and Data. (2010). 21.05.2011, Available from http://rainforests.mongabay.com/deforestation/2000/Cameroon.htm

CBFP. (2007). The Forests of the Congo Basin: The State of the Forest 2006. Congo Basin Forest Partnership, 29.08.2011, Retrieved from http://carpe.umd.edu/resources/Documents/THE_FORESTS_OF_THE_CONGO_BASIN_State_of_the_Forest_2006.pdf

CBP. (2011). Democratic Republic of Congo. 25.08.2011, Available from http://www.illegal-logging.info/approach.php?a_id=70

CED. (2004). *Inauguration: Pipeline.* CED Press Release, June 12, 2004.

Cerutti, P. & Tacconi, L. (2006). Illegal Logging and Livelihoods. *CIFOR Working Paper* No.35, Bogor, Indonesia, Retrieved from http://cameroun-foret.com/fr/system/files/18_61_48.pdf

CIA. (2011). World Fackbook. 25.08.2011, Available from https://www.cia.gov/library/publications/the-world-factbook/geos/cg.html

COMIFAC. (2010). Monitoring Forest Carbon Stocks and Fluxes in the Congo Basin. COMIFAC International Conference, 2-4 February 2010, Brazzaville, Republic of Congo, GOFC-GOLD Report No44, Available from http://www.observatoire-comifac.net/carbonCconfBrazza.php

Costanza, R.; d'Arge, R.; de Groot, R.; Faber, S.; Grasso, M.; Hannon, B.; Limburg, K.; Naeem, S.; O'Neill, R.; Paruelo, J.; Raskin, R.; Sutton, P. & van der Belt, M. (1987). The Value of the World's Ecosystems and Natural Capital. *Nature*, Vol.387, (May 1987), pp. 253–260

Cotula, L. & Mayers, J. (2009). *Tenure in REDD: Start Point or Afterthought?* International Institute for Environment and Development, United Kingdom, ISBN 978-1-84369-736-7

Dooley, K. (2009). Why Congo Basin Countries Stand to Lose out from a Market Based REDD. FERN, Briefing Note No.06, 19.08.2011, Available from http://www.scribd.com/doc/23775446/Congo-Basin-Countries-Lose-Out

EITI. (2006). Democratic Republic of Congo. 12.08.2011, Available from http://www.eitransparency.org/section/countries

Ekoko, F. (1997). *The political Economy of the 1994 Cameroon Forestry Law*. Working Paper No.3, Center for International Forestry Research ,Yaoundé, Cameroon

FAO. (2011). *The State of Forests in the Amazon Basin, Congo Basin and Southeast Asia*. A Report Prepared for the Summit of the Three Rainforest Basins, Brazzaville, Republic of Congo, 31 May–3 June, 2011

Fisher, R.; Maginnis, S.; Jackson, W.; Barrow, E. & Jeanrenaud, S. (2005). *Poverty and Conservation: Landscapes, People and Power*. IUCN. Gland, Switzerland, ISBN 2-8317-0880-x

Fometé, T. & Cerutti, P. (2008). Verification in the Forest Sector in Cameroon. Verifor Country Case, Study No.11, 25.08.2011, Available from http://www.verifor.org/RESOURCES/case-studies/cameroon_case_study_full.pdf

Fomété, T. (2001). The Forestry Taxation System and the Involvement of Local Communities in Forest Management in Cameroon. *Rural Development Forestry Network, Network Paper*, No.25b, pp. 17-28

Gbetnkom, D. (2008). Forest Depletion and Food Security of Poor Rural Populations in Africa: Evidence from Cameroon. *Journal of African Economies*, Vol.18, No.2, pp. 261–286

GFRA. (2010). Global Forest Resources Assessment 2010. *FAO Forestry Paper* No.163, 20.06.2011, Available from www.fao.org/forestry/fra/fra2010/en/

Gorle, R. & Shaikh, P. (2010). Deforestation and Climate Change, Congressional Research Service, 20.06.2011, Available from www.crs.gov

Government of DRC. (2006). *Constitution de la République Démocratique du Congo:* Cabinet du Président de la République, 2006 - *Law 18.02.2006*

Hansen, M.; Stehman, S.; Potapov, P.; Loveland, T.; Townshend, J.; DeFries, R.; Pittman, K.; Arunarwati, B.; Stolle, F.; Steininger, M.; Carroll, M. & DiMiceli, C. (2008). Humid Tropical Forest Clearing from 2000 to 2005 Quantified by Using Multitemporal and Multiresolution Remotely Sensed Data. *Proc Natl Acad Sci*, USA, Vol.105, No.27, pp. 9439–9444

Hoare, A. (2007). *Clouds on the Horizon: The Congo Basin Forests and Climate Change*. The Rainforest Foundation, ISBN: 978-1-906131-04-3, London

Huberman, D. (2007). Making REDD Work for the Poor: The Socio-economic Implications of Mechanisms for Reducing Emissions from Deforestation and Degradation. IUCN

Ichikawa, M. (2006). Problems in the Conservation of Rainforests in Cameroon. *African study monographs*, Vol.33, No.1, pp. 3-20

Iloweka, E. (2002). The Deforestation of Rural Areas in the Lower Congo Province. *Earth and Environmental Science*, Vol.99, No.1-3, pp. 245-250, DOI 10.1007/s10661-004-4028-0

Index of Economic Freedom. (2011). 26.08.2011, Available from http://www.heritage.org/index/country/

IPCC. (2007). *Fourth Assessment Report*. 12.06.2011, Retrieved from http://www1.ipcc.ch/

ITTO. (2006). *Status of Tropical Forest Management 2005*. ITTO Technical Series No24, pp. 305, Yokohama

Jauch, H. 1999. *Structural Adjustment Programs: Their Origin and International Experiences*. Labor Resource and Research Institute (LaRRI), Namibia

Kabila, J. (2011). *Inventing REDD+: Democratic Republic of Congo*. 18.08.2011, Retrieved from www.unredd.net/index.php?option=com_docman&task=doc_download&gid=408 4&Itemid=53

Kaimowitz, D. & Angelsen, A. (1999). Rethinking the Causes of Deforestation: Lessons from Economic Models. *World Bank Research Observer* Vol.14, No.1, (February 1999), pp. 73-98, DOI 10.1093/wbro/14.1.73

Karky, B. (2008). The Economics of Reducing Emissions from Community Managed Forest in Nepal Himalaya. PhD Thesis, University of Twente, Enschede, Netherlands

Karsenty, A. (2008). The Architecture of Proposed REDD Schemes after Bali: Facing Critical Choices. *International Forestry Review* Vol.10, No.3, (September 2008), pp.443-457, ISSN 1465-5489

Klaver, D. (2009). *Multi-stakeholder Design of Forest Governance and Accountability Arrangements in Equator Province, Democratic Republic of Congo*. Capacity Development and Institutional Change Programme Wageningen International, The Netherlands

Lasco, R. (2010). Facilitating Mitigation Projects in the Land Use Sector: Lessons from the CDM and REDD. World Agroforestry Centre, Los Baños, Laguna, Retrieved from http://irri.org/climatedocs/presentation_Lists/Docs/4_Lasco.pdf

Lescuyer, G.; Karsenty, A. & Eba'a Atyi, A. (2009). A New Tool for Sustainable Forest Management in Central Africa: Payments for Environmental Services, In: *The Forests of the Congo Basin : State of the Forest 2008*, C. de Wasseige, D. Devers, P. de Marcken, R. Eba'a Atyi, R. Nasi, P. Mayaux (Eds.), pp. 127-140, Luxembourg

Mbatu, R. (2009). Forest Policy Analysis Praxis: Modeling the Problem of Forest Loss in Cameroon. *Forest Policy and Economics*, Vol.11, No.1, (January 2009), pp. 26–33, DOI 10.1016/j.forpol.2008.08.001

Mbatu, R. (2010). Deforestation in the Buea-Limbe and Bertoua Regions in Southern Cameroon (1984–2000): modernization, world-systems, and neo-Malthusian outlook. *GeoJournal,*Vol.75, No.5, pp. 443-458, *DOI: 10.1007/s10708-009-9312-7*

Mitchell, A.; Secoy, K. & Mardas, N. (2007). *Forests First in the Fight against Climate Change: The VivoCarbon Initative*. Global Canopy Programme. Oxford, UK

Nasi, R.; Mayaux, P.; Devers, D.; Bayol, N.; Eba'a Atyi, R.; Mugnier, A.; Cassagne, B.; Billand, A. & Sonwa, D. (2009). Un Aperc̨u des Stocks de Carbone et Leurs Variations dans les Forêts du Bassin du Congo. In: *Les Forêts du Bassin du Congo. Etat des Forêts 2008*, C. Wasseige, D. Devers, P. de Marcen, R. Eba'a Atyi, R. Nasi, & Ph. Mayaux (Eds.), pp. 199–216, ISBN 978-92-79-132 11-7, Office des publications de l'Union Europe´enne

Ndoye, O. & Kaimowitz, D. (2000). Macro-economics, Markets and the Humid Forests of
    Cameroon, 1967–1997. *The Journal of Modern African Studies*, Vol.38, No.2, (July
    2000), pp. 225–253
Nyuylime, L. (2006). *Mbalam Iron Ore Exploration Afoot*. Cameroon Tribune, 27.10. 2006
Parker, C.; Mitchell, A.; Trivedi, M. & Mardas, N. (2008). The Little REDD Book: A Guide to
    Governmental and Non-governmental Proposals for Reducing Emissions From
    Deforestation and Degradation. Global Canopy Programme, Oxford, 25.06.2011,
    Retrieved from www.globalcanopy.org/main.php?m=4&sm=15&ssm=151
Pearce, D. & Warford, J. (1993). World without End: Economics, Environment and
    Sustainable Development. Oxford University Press, New York
Pearce, D. (2001). The Economic Value of Forest Ecosystems. *Ecosystem Health*, Vol.7, No.4
    (December 2001), pp. 284-296, DOI: 10.1046/j.1526-0992.2001.01037.x
Randolp, J. & Masters, G. (2008). *Energy for Sustainability: Technology, Planning, Policy*. Island
    Press, ISBN 13:978 – 1-59726-103-6, Washington D.C.
Reed, E. & Miranda, M. (2007). *Assessment of the Mining Sector and Infrastructure Development
    in the Congo Basin region*. WWF, Washington, D.C.
Rogner, H.; Zhou, D.; Bradley, R.; Crabbé, P.; Edenhofer, O.; Hare, B.; Kuijpers, L. &
    Yamaguchi, M. (2007). Introduction, In: *Climate Change 2007: Mitigation*, B. Metz, O.
    Davidson, P. Bosch, R. Dave, & L. Meyer, (Eds.), Contribution of Working Group
    III to the Fourth Assessment Report of the Intergovernmental Panel on Climate
    Change, University Press, Cambridge
Sieböck, G. (2002) *A Political, Legal and Economic Framework for Sustainable Forest Management
    in Cameroon: Concerted Initiatives to Save the Rainforests*. Masters Thesis, Lund
    University, Sweden, 18.08./2011, Available from
    http://www.lumes.lu.se/database/alumni/01.02/theses/sieboeck_gregor.pdf
Sikod, F.; Amin, A. & Nyamnjo, F. (1996). *Interlinkages between Trade and the Environment: A
    Case Study of Cameroon*. UNTAD/UNDP Project Report.
Skutsch, M.; van Laake, P.; Zahabu, E.; Karky, B. & Phartiyal., P. (2009). Community
    Monitoring in REED+, In: *Community Forest Management Under REED: Policy
    Conditions for Equitable Governance*, A. Angelsen, M. Brockhaus, M. Kanninen, E.
    Sills, W.D. Sunderlin & S. Wertz-Kanounnikoff) (Eds), pp. 101–112. Centre for
    International Forestry Research, Bogor, Indonesia
Sonwa, D.; Walker, S.; Nasi, R. & Kanninen, M. (2011). Potential Synergies of the main
    Current Forestry Efforts and Climate Change Mitigation in Central Africa.
    Sustainability Science, Vol.6, No.1, pp. 59–67, DOI 10.1007/s11625-010-0119-8
Soriaga, R. & Walpole, P. (2007). Forests for Poverty Reduction: Opportunities in the Asia-
    Pacific Region, In: *Forests and the Millennium Development Goals*, Mayers (Ed.), pp.47-
    48, European Tropical Forest Research Network. Wageningen, The Netherlands.
State of the forest 2008. (2009). Les Forêts du Bassin du Congo. Etat des Forêts 2008, C.
    Wasseige, D. Devers, P. de Marcen, R. Eba'a Atyi, R. Nasi, & Ph. Mayaux (eds.),
    ISBN 978-92-79-132 11-7, Office des publications de l'Union Europe´enne, Retrieved
    from http://www.observatoire-comifac.net/edf2008.php
Streck, C.; O'Sullivan, R.; Janson-Smith, T.; & Tarasofsky, R. (2008). *Climate Change and
    Forests, Emerging Policy and Market Opportunities*. Chatham House, Brookings
    Institution Press (2008), 346 pp. ISBN 978-0-8157-8192-9, London

Sunderlin, W.; Ndoye, O.; Bikié, H.; Laporte, N.; Mertens, B. Pokam, J. (2000). Economic Crisis, Small-scale Agriculture, and Forest Cover Change in Southern Cameroon. *Environmental Conservation,* Vol.27, No.3, (July 2000), pp. 284-290

Tchiofo (2008)

Tewari, A. & Phartiyal, P. (2006). The Carbon Market as an Emerging Livelihood Opportunity for Communities of the Himalayas. *ICIMOD Mountain Development,* No.49, pp. 26-27, Central Himalayan Environmental Association, Nainital, India

Tieguhong, J. (2008). *Ecotourism for Sustainable Development Economic Valuation of Recreational Potentials of Protected Areas in the Congo Basin.* PhD thesis, University of Kwazulu-Natal, South Africa

Topa, G.; Karsenty, A.; Megevand, C. & Debroux, L. (2009). *The Rainforest of Cameroon: Experience and Evidence from a Decade of Reform.* Washington, DC: World Bank

UNDP. (2006). *Human Development Index 2006: Beyond scarcity: Power, poverty and the global water crisis.* Retrieved from http://hdr.undp.org/en/media/HDR06-complete.pdf

Unfc /fc /sbsta /2007/misc.14. (2007). Views On Issues Related To Further Steps Under the Convention Related to Reducing Emissions from Deforestation in Developing Countries: Approaches to Stimulate Action: Submissions from Parties, 03.07.2011, Available from http://unfccc.int/resource/docs/2007/sbsta/eng/misc14

UN-REDD. (2010). The United Nations Collaborative Program on Reducing Emissions from Deforestation and Forest Degradation in Developing Countries: Program strategy 2011-2015. Retrieved from www.un-redd.org

Wertz-Kanounnikoff, S. & Kongphan-apirak, M. (2009). Emerging REDD+: A Preliminary Survey of Demonstration and Readiness Activities. *CIFOR Working Paper* No.46, Bogor, Indonesia

Wertz-Kanounnikoff, S. (2006). Payment for Environmental Services: A Solution for Biodiversity Conservation? *IDDRI Working Paper* No.7, Natural resources, IDDRI, Paris, France

World Bank. (2005). World's World Development Indicators. 23.08.2011, Available from http://devdata.worldbank.org

World Bank. (2006). World Bank's World Development Indicators database. 16.08.2011, Available from http://devdata.worldbank.org

World Bank. (2008). Democratic Republic of Congo: Growth with Governance in the Mining Sector. Oil/Gas, Mining and Chemicals Department AFCC2, Africa Region, Report No.43402-ZR.

World Bank. (2009). World Bank's World Development Indicators database. 25.08.2011, Available from http://devdata.worldbank.org

World Bank. (2011). World Bank's World Development Indicators database. 25.08.2011, Available from http://devdata.worldbank.org

Zahabu, E.; Malimbwi, R. & Ngaga, Y. (2005). Payments for Environmental Services as Incentive Opportunities for Catchment Forest Reserves Management in Tanzania. Paper to the Tanzania Association of Foresters Meeting. Dar es Salaam, Tanzania, 6–9 November 2005

Zhang, Q.; Justice, C. & Desanker, P. (2002). Impacts of Simulated Shifting Cultivation on Deforestation and Carbon Stocks of the Forests of Central Africa. *Agric Ecosyst Environ,* Vol.90, No.2, (July 2002), pp. 203–209, DOI 10.1016/S0167-8809(01)00332-2

2

Methodology for Forest Ecosystem Mediating Indicator – Case Mt. Kilimanjaro, Tanzania

John Eilif Hermansen

Department of Industrial Economics and Technology Management
Norwegian University of Science and Technology, NTNU, Trondheim
Norway

1. Introduction

Communication of ecological and environmental knowledge, values and concerns by means of indicators is widely accepted and adopted as a part of environmental management systems, results-oriented politics and international reporting, and benchmarking initiatives.

Application of an indicator system is a normative course of action supported by different professional perspectives and parochial interests, struggling for resource control and ownership, investigation of business opportunities, and political interests. Development and selection of indicator systems is a natural extension of questions of justice and equity regarding resources, and should accordingly be conducted in an open, transparent and consensus-based process in spirit of enlightenment and democratic traditions.

The purpose of this work is to elaborate on the asymmetrical relationship between local and indigenous people dependent on their traditional rights to tropical forest habitation and those global interests who would intervene in their traditional understanding and use of the forest resources. Forest dwellers and native forest service users in developing countries may expect a large gap between their life world and the global actors. A methodology for devising a forest ecosystem indicator system intended to balance the asymmetry and re-allocate some of the knowledge power about the forest resources back to the local community, is suggested.

A framework for mediating ecological indicators is evolved in order to keep elements of global versus local interests, nature versus society and epistemology versus ontology together in one system. This construct is referred to as the *Balanced Ecosystem Mediation Framework* (BEM-framework) (Hermansen, 2008, 2010).

The framework emerged during a case study of the catchment forest reserve at the southern slopes of Mt. Kilimanjaro. By using data from a plant ecological investigation of the forest (Hermansen et al., 2008b) an ideal typological indicator was developed to be used in the BEM-framework. The proposed indicator is generally referred to as the *Ecosystem Mediating Indicator (EMI)* and the *Forest Ecosystem Mediating Indicator (FEMI)* when applied on forest ecosystem services. Further, as an illustration of its application to the catchment forest reserve at Mt. Kilimanjaro, a special case is suggested called the *Catchment Forest Ecosystem Mediating Indicator (CFEMI)*.

CFEMI is meant to be an equitable, and ecologically acceptable, instrument for building up a reservoir of transferable knowledge. CFEMI is designed for communication and management of forest ecosystem values where there is a need for a significantly better quality communication process between the local level and global level of interests and concern.

A premise of the framework is that it should be possible to establish a negotiated understanding of tropical forest resources conveyed by a knowledge system that supports or at least evens out some of the asymmetric influence and power of the globalized community vis-à-vis the local community regarding communication of forest values.

The chapter begins with a discussion of forest management and indicators followed by a description of the Kilimanjaro case study from which the indicator and framework emerged. The framework is then described and discussed.

1.1 Local ecosystem resource governance and issues in forest management

The deterioration of tropical forests is increasing (FAO, 2007; MA, 2005; UNEP, 2007). The need for new initiatives for sustainable forest management has been raised by many authors and institutions (Studley, 2007; Van Bueren & Blom, 1996). There is a serious concern about insufficient means and instruments for a possible future sustainable use, management and governance of biodiversity and ecosystem resources (Newton & Kapos, 2002; Noss, 1990, 1999; TEEB, 2010).

Especially indigenous and poor communities are vulnerable to failed governance because of their heavy reliance on local, natural resources for subsistence and income (Lawrence, 2000; Vermeulen & Koziell, 2002; WRI, 2005). Indigenous people and communities are also on the defensive in order to protect and develop their historical rights, cultural heritage, ecosystem resources and land. UN Convention on Biodiversity (CBD) includes framework for monitoring and indicators, and new targets for biodiversity are added to the Millennium Development Goals in order to cover genetic variety, quantity of different taxon, geographic distribution and social interaction processes (CBD, 2006).

Studley (2007) states that virtually all aspects of diversity are in step decline due to the three interacting interdependent systems of indigenous knowledge, biodiversity and cultural diversity. All three are threatened with extinction. The list of threats includes rapid population growth, growth of international markets, westernised educational systems and mass media, environmental degradation, exogenous and imposed development processes, rapid modernisation, cultural homogenisation, lost language, globalisation, extreme environmentalism and eco-imperialism.

Vermeulen and Koziell (2002) give a review of biodiversity assessment and integration of global and local values including elaborating on the contrast

> *"between "global values" – the indirect values (environmental services) and non-use values (future options and intrinsic existence values) that accrue to all humanity – and "local values" held by the day-to-day managers of biological diversity, whose concerns often prioritise direct use of good that biodiversity provides. Assessments are based on values."*

Studley (2007) suggests a vision for realising the aspirations of indigenous people to ensure the enhancement of biological and cultural diversity which includes an endogenous approach dependent on building the capacity of forest development staff in acculturation, cross-cultural bridging, forest concept mapping and information technologies.

Wieler (2007) advises decision-makers that the development and implementation of an environmental monitoring system and adequate policy targets for improved environmental performance are crucial. She recommends an impact strategy that includes *relationship management* at the core to identify who are the people positioned to have influence on the changes that need to be made (Creech et al., 2006).

Especially in cases where many stakeholders and their interests pose a complex cultural and social relationship to the resources, the process to define targets for environmental improvement and performance can be difficult. The process involves negotiation and mediation between those involved. A tropical forest land where local people are directly dependent on forest resources is an example of such a case.

In order to increase the efficiency of environmental policy and management strong focus on performance is necessary and therein formulation of performance indicators. The purpose of this study is to present a deliberate and communication oriented multi-purpose forest resource indicator which may be equitable and understandable across cultural and societal borders, and also meet the requirements for *proximity- to-target* approach (Esty et al., 2006)).

1.2 Locally rooted proximity-to-target forest indicator

A wide variety of ecological indicators have been generated for the purpose of reflecting trends and needs for realising policy targets and improved nature management. The terms environmental and ecological indicators are often used as synonyms or in an arbitrary manner. Here, the notion ecological indicator is regarded as a subset under environmental indicator and use of the term ecological indicator applies directly to the ecological processes (Niemeijer & de Groot, 2006; Smeets & Wetering, 1999). Usually ecological or environmental indicators are part of a linear and hierarchical management system which includes monitoring, reporting and decision making. Van Bueren and Blom (1996) suggest a structure starting with determining goals, outlining principles and criteria with guidelines for action, which are measured and verified by indicators before they are compared with established norms and discussed. The hierarchy of the management system consists of the input (an object, capacity or intention, e.g. management plan), the process (the management process) and the output (performance and results).

The hierarchical model is systematic, logical and effective, but it is open in order to include the mediation and negotiation perspective that could increase the local people's participation and influence in local management. The model could be developed further to be more systemic and include feedback thereby reducing the asymmetry between global and local interests.

To incorporate both a systematic and a systemic forest management model it follows that a new approach to the construction of indicators is needed. Van Bueren & Blom (1996) outline very well the demand for quality in the work of designing sustainable forest indicators and they warn about incorrectly formulated criteria for management standards and indicators. However, an indicator for a forest management system that aims to increase local participation and equality regarding influence and control over local resources also must be easy to understand and use. The work for sustainable forest management rests on the assumption that local people understand how to protect the forest ecosystem services better than a scientifically constructed indicator, which fails to incorporate the knowledge of local people.

Hence, the study proposes an ecological communication model that enlarges the objectives and applications of ecological indicators. The proposed indicator framework has purposes beyond measuring ecological status, impacts or performance. The indicator should also be a tool for reflexive learning and communication including mediation and negotiation between stakeholders on the global and local scale, which includes nature itself represented by the sciences of ecology (Hermansen, 2006, 2010; Latour, 2004) as a stakeholder (Elkington, 1998).

First, ecology is addressed as a necessary knowledge system in an epistemological context for understanding the relationship and integration of natural resources to a globally recognized system, and second, the indigenous knowledge system is addressed in order to strengthen local motivation, control and proper management of community depending on a sustainable use of the ecosystem resources in an ontological context.

To make a distinction between the local context and interests and the global context and interests, two stakeholder groups, *locals* and *globals,* are introduced. The denotation of the rather new and little used term globals is not explained in dictionaries. Baumann (1998) and Strassberg (2003) refer to globals as people who are relatively free from territorial constraints, obligation, and the duty to contribute to the daily life of a community. Locals are geographically bound and they may bear the consequences of globalization. Bird and Stevens (2003) elaborate on the relationship between proximate locals and globals that may find it more difficult to work with each other because of issues of *trust.* This article attempts to enhance the understanding of locals and globals to include not only interests but also the context of the understanding of the forest ecosystem in order to make an ecosystem indicator which is ecologically founded and accepted (global perspective) and locally understood and equitable (local perspective).

Scientifically oriented assessments and validations as well as normatively oriented assessments and validations are integrated with local understanding of the forest as a source of necessary ecological goods and services to the local community. To increase the momentum of an indicator system it may be designed as a *proximity-to-target* performance indicator. The process of deciding the targets provides an opportunity for locals and globals to make reflections concerning targets, i.e. the ecological quality of the forest.

1.3 Case: Catchment forest reserve, Mt. Kilimanjaro, Tanzania

Mainland Tanzania has according to Blomley (2006) one of the most advanced community forestry jurisdictions in Africa, and Participatory Forest Management (PFM) has become the main strategy of the forest policy. He states that among the lessons learned is an increasing awareness of the importance of local forest users and managers and he espouses decentralized forest management schemes. The suggested indicator system is devised to support these efforts, and the results from an ecological study of the moist mountain forest plants at the southern slopes of Mt. Kilimanjaro are used as a case for the creation of the indicator (Hermansen et al., 2008).

The indicator is meant to be embedded in the social context of the governmental forest policy especially the Catchment Forest Project (CFP) (Hermansen et al., 1985; Katigula, 1992; Kashenge, 1995; MNRT, 1998, 2001, 2006). Creation of the indicator embeds an interpretation of possible interests and use of local ecosystem resources by the Chagga people and community (Akitanda, 1994, 2002; Bart et al., 2006; Misana, 1991, 2006; Newmark, 1991; Ngana, 2001, 2002; Soini, 2005; Stahl, 1964; Tagseth, 2006, 2008).

The Chagga people and community at the southern slopes of Kilimanjaro are included in this study as representatives for local stakeholders whose interests are then juxtaposed to the global interests. The interests of the Chagga people are presented here as an ideal typological position (space does not permit a serious and fair study of the relationship between the local community and ecosystem services). The indicators can be considered to be a measure of the interest conflicts between locals and globals, and also between ecology and people. The preparation and use of the indicator may then be a useful tool in a tool-box for the *"keepers of the forest"* (Studley, 2007) promoting interaction between the indigenous knowledge system, biodiversity and cultural diversity.

2. Case study: Construction of the catchment forest ecosystem mediating indicator

The CFEMI is pilot scheme developed on site as a specific ecological mediating indicator. CFEMI is based on experience from an ecological investigation of the plant life in a tropical moist forest at Mt. Kilimanjaro (Hermansen et al., 2008). CFEMI is a composed indicator showing how far a certain site in a specific forest deviates from norms or targets, in this case sites at different altitudes in the forest belt between 1600 and 2700 m asl on the southern slopes of Mt. Kilimanjaro (Fig. 1). The targets represent a specific defined and assumed, optimal ecological state. It is essential to point out that the purpose of CFEMI is not to be universal, but instead to be a measure for strengthening the local actors's role in defining their forest resources and sustainable forest management in the context of the catchment forest. This means that CFEMI may be regarded as a quasi-indicator (Andersen & Fagerhaug, 2002) more concerned with local and situational reality and thereby of limited value for general utilization and comparability for benchmarking with other areas.

Fig. 1. Kilimanjaro Forest Reserve and the three transects Mweka, Kilema and Marangu. The upper forest border mainly follows the Kilimanjaro National Park border. The Half Mile Forestry Strip is shaded. (Modified from Newmark, 1991)

The procedure applied for constructing the indicator includes definition of system, goals, objectives, identifying relevant ecological factors and variables, outlining methods for measurement and data collection, negotiating the construction of the index and calculation of indicators, deciding on norms and target values, and finally the presentation of the proximity-to-target performance indicator.

2.1 Management of the catchment forest

Forest reserves in Tanzania have for more than 100 years been under different forest and forestry administration and management regimes from the German colonial time to the prevailing Catchment Forestry Project (CFP) launched in 1977 and organizationally situated under the Forestry and Beekeeping Division of the Tanzanian Ministry of Natural Resources and Tourism (MNRT).

In 1941, under British colonial time, a buffer zone, *The Half Mile Forestry Strip* (HMFS), was established as a social forest zone under local management of the Chagga Council at Mt. Kilimajaro (Kivumbi & Newmark, 1991). The management worked very well the first 20 years, but after independence in 1961 the management became more centralised and the zone itself came under heavy pressure, overexploitation and encroachment from local people partly due to population growth and partly due to ineffective management. Most of the approximately 800 meter broad buffer zone along the eastern and southern part of Mt. Kilimanjaro appears even today as a seriously damaged forest far from its natural state.

Initially, the CFP did not manage the forest reserve well, and encroachment, deforestation and fragmentation of the catchment forests increased (Akitanda, 1994, 2002; Hermansen, 2008; Hermansen et al., 1985; Kashenge, 1995; Katigula, 1992; Lovett & Pocs, 1992; Mariki, 2000; Newmark, 1991; Sjaastad et al., 2003; William, 2003;). Lambrechts et al. (2002) has verified the status and the extent of encroachment of the forest by aerial survey.

New national forest polices over the last 15 years have as a goal to improve the effectiveness and promote local responsibility towards a sustainable forest management practise (MNRT, 1998, 2001, 2006) with the development of criteria and indicators for sustainable forest management in Tanzania (MNRT, 1999). Local participatory forestry (Blomley, 2006), forest management and democracy are all important issues and it not easy to find ways to transfer enough power and security to local communities and devise sustainable and effective local forest management (Wily, 2001). Global initiatives connected to fair trade strongly support the strengthening of local forest management (Macqueen, 2006).

The objectives of the CFP can be summarized to promote the utilization of the forest resources in a sustainable manner, and secure that the three key functions - production of forest goods, water generation and conservation of biodiversity of the forest - are maintained. The following interpretation of objectives forms the relationship between management purposes and ecological contents (Hermansen et al., 1985):

Water generation: Regulation and conservation of water resources and supply in the catchment area; reduction of run off and soil erosion, which is especially important in moist mountain areas.

Gene-pool conservation: Preventing extinction of rare and endemic plant and animal species in the diverse moist forest; it is essential to maintain biodiversity and keep the genetic potential for ecological and evolutionary purposes and for present and future utilisation of biological forest resources.

Production: Logging of indigenous tree species and supply of other forest products for local consumption and sale.

A number of recent studies describe, explain and discuss the forest ecosystem at Mt. Kilimanjaro, and the threats to and use of forest resources (Bart et al., 2006; Bjørndalen, 1992; Hemp, 1999, 2006a, 2006b, 2006c; Howell, 1994; Katigula, 1992; Lovett & Pocs, 1992; Lyaruu, 2002, Madoffe et al., 2005, 2006; Mariki, 2000; Misana, 1991, 2006; Misana et al., 2003; Ngana,

2001, 2002; Soini, 2005;). The arguments for understanding and supporting the conservation of plant biodiversity of the forest at Kilimanjaro are presented in many of the reference above, as well as many other articles, not referred. Burgess et al. (2007) analyse the biological importance of Eastern Arc Mountains.

Studies from Kilimanjaro and neighbouring mountain forests (eastern arc) have included inventories suitable for supporting monitoring of the forests ecosystem services and contain data which are suitable to some degree for performance indicators, but they are mainly dealing with distribution of tree species, density of trees and timber volume including regeneration (Hall, 1991; Huang et al., 2003; Jakko Pöyry, 1978; Madoffe et al., 2005, 2006; Malimbwi et al., 2001;). Water management of the Pangani river basin, which is a very important regional and national concern, is tightly connected to the management of the catchment forest at Mt. Kilimanjaro (Ngana, 2001, 2002; Røhr, 2003; Turpie et al., 2003). The river is feed from several tributaries from Kilimanjaro and other hills and mountains in the area.

2.2 Purpose and objectives of CFEMI

CFEMI offers a composite indicator of relevant ecological features that can be recognised as essential for catchment forest management; namely the conservation and protection of a specified forested area that serves local people with ecosystem services in a global perspective. Management means to keep and even enhance the forest quality within the area in order to improve water conservation and generation, to protect biodiversity and to serve local people with forest goods.

The overall goal of CFEMI is to contribute to a broad stakeholder-oriented approach (Elkington, 1998; Grimble, 1998; Grimble & Wellard, 1997) to the knowledge and under-standing of the forest and to promote an ecologically and socially wise use of the goods and services of the forest, including contributions to:

- reasonable common understanding of status and changes of the ecological conditions in the forest between globals and locals,
- motivating, learning and increasing a management oriented behaviour towards the forest resources,
- meet the requirement for local participation; application of the indicator could vary (e.g. full employment of the concept and indicator system or limited employment mainly showing the large structures in the forest).

Classes of objectives encompass:

- protection of forest ecology quality
- secure ecosystem services from the forest for the local people
- materiality for mediation and negotiation between locals and globals
- increasing local influence, control and competence regarding local resources
- provide opportunities for interactive learning loops.

The act of creating the indicator encourages mediation of the ecological aspects into a logical structure from goals to corresponding objectives, practical variables, measurement procedure and collection of relevant data.

2.3 Ecological and environmental aspects

This section will explore the variety of ecosystem assessment alternatives from the very general to the specific. Ecosystem assessment alternatives are provided from many sources.

The first group of sources are various national forest policies including the CFP (MNRT, 1998, 2001, 2006; Sjaastad et al., 2003). The second group comprises strategies and efforts from international organisations. In addition to the authoritative bodies under the UN, such as FAO and others, the new initiatives connected to Millennium Ecosystem Assessment (MA 2005) are most relevant. The third group is connected to the globalization of environmental management standards including sustainable forest management under the International Tropical Timber Organization. A fourth group is NGOs and research institutes working with tropical forest politics, management and forestry. Examples include the Forest Stewardship Council (FSC), Rainforest Alliance, Social Accountability International (SAI) and The International Social and Environmental Accreditation Labelling (ISEAL).

Macqueen et al. (2006) outline the new historical opportunities for community ownership and management of forest to realize a better position for sustainable forestry due to the alliance with a new kind of globals connected to initiatives such as fair trade and others. The World Business Council for Sustainable Development (WBCSD) in alliance with IUCN has taken the initiative in recent years to meet the requirement and opportunities connected to Millennium Ecosystem Assessment (WBCSD & IUCN, 2006).

For CFEMI, the purpose and objectives of CFP are directly relevant, as are the linkages between *Ecosystem Services and Human Well-being* of Millennium Ecosystem Assessment and the conceptual framework between biodiversity, ecosystem services, human well-being and drivers of change especially relevant.

Based on CFP and the MA framework, the ecological parameters for CFEMI can be grouped into two main categories a) forest structure and b) forest biodiversity. These categories have been chosen because maintaining these two qualities will secure that most of the other important ecological factors including microorganism and fauna and the abiotic environment, will be covered. If forest structure and biodiversity are intact on a certain level, the forest will keep its resilience potential and a number of other ecological qualities which can provide ecosystem services for human well-being in a sustainable way (Table 1).

2.4 Selection of variables and primary indicators

The case of forest management at Mt. Kilimanjaro and the Chagga people as representative stakeholders for local interests is used here as an illustration of the conceptual and practical circumstances of the indicator scheme. CFEMI is proposed as a proximity-to-target indicator meant to work in the context of negotiation and mediation between globals and locals, while strengthening the local interests, influence, control and competence regarding sustainable forest management. The distinction between globals and locals are used to underline the actor perspective of the two paramount stakeholder groups of local society and international organisations, institutions and power structure. Both globals and locals are aggregates of other more specified stakeholders.

CFEMI should support the management goals for inter alia CFP and MA in a manner that strengthens the influence of local people and mediation between locals and globals. Table 2 gives an overview of criteria for selection of ecological features that could be relevant variables or primary indicators for CFEMI. Table 3 shows the complete list and description of the measured variables, units and levels of measurement.

Composition of variables is decided based on the criteria of what are relatively easily accessible. The variables cover important features for the ecosystem services connected to biodiversity and structure where the hypothesis is that the untouched forest has the potential to provide for the demanded ecosystem services such as production of forest goods (e.g. timber, fuel wood, fodder, medical plants), conservation of biodiversity, and water regulation and supply of water of good quality.

Ecological aspect	Management goals and ecosystem services
Forest structure	Maintain a natural-like structure of trees including age/size (basal area and height of trees) and canopy cover and restore areas where forest structure is damaged.
	Main ecosystem services:
	Constructs the forest room and constitutes the system for nutrient cycling, soil formation and primary production, form the overall habitat for all organisms, regulate local climate, retain, store and purify water and moisture and makes a optimal primary production possible
	Benefit for locals:
	Secure safe water for consumption and the furrow irrigation system produce timber, fuel wood, food, cash crops, fodder and many other bio products.
	Erosion control
	Income from tourism
	Benefit for globals:
	Timber, carbon storage, climate regulation. On regional level water to irrigation, hydropower, consumption and ecosystems via Pangani River basin water system is extremely import.
	Tourism especially eco-tourism
Biodiversity	Maintain natural level of biodiversity including diversity of trees.
	Main ecosystem services:
	Provider of genetic material for large number of organism necessary for keeping the evolutionary potential intact, and provision of large number of species
	Benefit for locals:
	Secure a wide variety of organisms to be utilized by the society where some already have known benefit for people and probably many other are undiscovered useful species which will be discovered in the future.
	Income from tourism
	Benefit for globals:
	Secure biodiversity resources for future generation. Medicines
	Ecosystem resilience
	Tourism and eco-tourism. Recreation

Table 1. Main ecological aspects, goals for management and ecosystem service of the catchment forest reserve at Mt. Kilimanjaro

2.5 Measurement and calculation

Table 3 shows measured and analyzed variables and Table 4 the total average value and derived target for the nine individual variables or indicators which constitute CFEMI. Identifying variables and methods for measurement, and deciding on targets require both quantitative and qualitative approach, and are depending on local conditions.

CFEMI is proximity-to-target indicator and the target is determined for each variable as a certain value higher than the total average value for each individual variable for each site (plot). All trees within each site of 1000 m^2 along the three transects (Mweka, Kilema and Marangu) are measured and the average value for each site is then calculated. These site specific average values are then accumulated to a total average value for all sites. However,

Criterion	Description
	ECOLOGICAL ASPECTS
1	Represent important forest physiognomy and biodiversity if trees on a plant are at an ecologically acceptable level
2	Directly associated to ecosystem services (Supporting, provisioning, regulating and cultural services)
	MEDIATION AND LEARNING ASPECTS
3	Easy or intuitively understandably by local people as a relevant description of forest services and goods
4	Support learning processes
5	Supporting learning processes and local participation in selection of indicators, measurement and calculation
6	Support management efforts
	TECHNICAL ASPECTS
7	Easy to measure and calculate
8	Does not hurt the ecosystem

Table 2. Criteria for the selection of variables

among the analyzed 54 sites there are 18 sites which are too affected by human impacts and encroachment that the sites cannot be regarded as be representative for closed forest or they contain mainly dense stands of *Erica* trees. These stands are omitted from the calculation of total average value and determination of targets, but these sites are of course included in the presentation of the CFEMI score for all sites (Table 5 and Fig. 2). Hermansen et al (2008) gives a detailed description of field work and results.

Some variables are measured by using numerical data (number of tree species and stems, basal area, tree height, crown width, crown width sum and crown depth), and average value is calculated. Cover of epiphytes is variables estimated by using ordinal (categorical) data (covering of climbers and covering of vascular, lichens and bryophytes), and the average value is calculated from the ordinal values.

The score for each site is calculated as the percentage of the average value for all the nine variables for a certain site compared with the target. Hermansen (2008) contains a complete list of calculated values of variables and score for all sites.

2.6 Results

The proximity-to-target score in percentage for the sites along the three altitudinal transects from lower to upper forest borders at the southern slopes of Mt. Kilimanjaro of Mweka, Kilema and Marangu, is shown in Fig. 2. Table 5 shows average values for the sites along each transect grouped into three zones: HMFS, central part and the upper part of the forest reserve.

The HMFS shows, as expected, much lower values (average score: 60) compared with average score 99 for the central part and 92 for the upper part. Average scores for the complete transects are quite similar for Mweka (91) and Marangu (93) and lower for Kilema (80). It is the low values from HMFS (50) along the Kilema transect which draws that average down. In the Kilema transect about double as many sites were measured in the HMFS part of the transect as in the two other transects. Sites on low altitudes are over-exploited and well developed sites are situated on higher altitudes (Fig. 2).

Tree structure variables		Description	Units
A.	**Basic units**	Inventory units for identification, geo-referenced information and multivariate analysis	
	a. Tree	Individual identified and measured tree or stem	
		o Running serial number	Idnr
		o Running serial number within plot	Number
	b. Plot	Identified by transect and plot number	
B.	**Localization**		
	a. Altitude	Altitude above sea level	m asl
	b. Transect	Transect from lower to upper forest border	Nominal
	c. Exposition	Indication of exposition in 400 grades	degrees
	d. Slope	Indication of slopes in 400 grades	degrees
C.	**Stem**		
	a. Tree number	Each tree (or stem on trees divided in 2 or several stems under 1 m) is identified by transect and running number with the plot	Number
	b. Tree species	Each tree is identified	
	c. Height	Estimated height	m
	d. DBH	Measured diameter at breast height	m
	e. Basal area	Calculated basal area	m^2
	f. First branch	Height to lowest living branch	m
	g. Shape	The shape of the trunk is assessed :	Nominal
		o Straight	0
		o Leaning	1
		o Bent	2
		o Crooked	3
	h. Buttress	Each tree has been assessed if it has buttress or no	Yes or no
D.	**Canopy**		
	Crown area (total leaf area)	Estimation of the horizontal projection of the canopy of each tree. The area is calculated from estimation of the diameter of the crown along to axis through origo.	m^2
E.	**Epiphytes**		Ordinal
	a. Climbers	Estimation of the cover of climbers and lianas on each tree:	
		No climbers or lianas observed	0
		Some few / thin climbers, shorter than 2 m	1
		Some more dense / thicker climbers, more 2 m long	2
		Climbers cover the stem and some thin lianas may occur.	3
		Large and large lianas	4
		The tree is heavily affected by thick lianas	5
	b. Vascular epiphytes	Estimation of the cover of vascular epiphytes:	
		No or very few individuals observed.	0
		Less than 10 % of stem and branched cover	1
		Between 10 – 25 % of	2
		Dense mats of epiphytes may cover between 20 to 40 %	3
		Dense mats cover between 40 and 75 %. Some hanging mats.	4
		The tree is overgrown with dense and some hanging mats	5
	c. Non-vascular epiphytes	Estimation of the cover of bryophytes and lichens	
		No or very few spots or individuals observed.	0
		Less than 10 % of stem and branched cover	1
		Between 10 – 25 % of	2
		Dense mats of epiphytes may cover between 20 to 40 %	3
		Dense mats cover between 40 and 75 %. Some hanging mats.	4
		The tree is overgrown with dense and some hanging mats	5

Table 3. Measured and analyzed variables

Ecological aspects	Category	Indicators / variables	Units	Notes	Average	Target
Forest structure	Tree structure	Number of stems	no	1	40.6	50
		Basal area	m^2	2	6.0	7.5
		Tree height	m	3	19.2	24
	Leaf cover	Crown width	m^2	4	67.2	84
		Crown width sum	m^2	5	2416	3020
		Crown depth	m	6	11.8	14.7
Biodiversity and water conservation	Epiphyte cover	Covering of climbers	class	7	1.5	1.9
		Covering of vascular, lichens and bryophytes	class	7	2.3	2.9
Biodiversity	Tree species	Number of tree species	no	8	6.7	8.4

Data are based on the measurement and estimation of 1502 trees within 36 sites (plots) of 1000 m². The different targets are set close to the values for which are considered to be well developed stands (approximately 25 percentage above average values). All sites are within the forest reserve. Sites mainly containing more than 50 *Erica excelsa* trees and sites from Half Mile Forestry Strip are not included in calculation of average values and target values. Notes:

1. The number of trees per site varies between 2 and 89. Overall average number of stems is 41.
2. The sum of basal area per site varies between 0.1 and 13.2. The overall average is 6.0.
3. The tree height varies between 6 and 40 m. The overall average is 19.2 m.
4. The average crown width per site (the horizontal project of the crown for each tree) varies between from 10 to 170 m². The overall average is 67 m². The largest crown is 961 m².
5. The sum of crown width for all the trees within a site. The crowns are merged into each other and will therefore exceed 1000 m². The sum varies between 70 and 5450 m². The overall average is 2416 m²
6. The crown depth is the height between lowest living branch and top the tree and varies between 7.2 and 16.2 m as average for the different sites. The overall average is 11.8 m. The highest tree crown depth is 39 m
7. Epiphyte cover is estimated by a non-linear classification and the calculated average is the average class for the tree within the plot. Target is set to 25 % above average. Average above 3.0 implies that the average tree has a substantial cover of epiphytes and climbers, which may play an important role for water conservation and retention.
8. The number of species within the sites varies between 2 and 13. The average is 6.7.

Table 4. CFEMI variables, total average values and target values.

The most significant observation is the large range of score on the Kilema track from the lowest (30 percentage point) to the highest score (134 percentage point). Especially the sites in the HMFS are far from the target for an ideal forest composition and structure. However, this was expected and obvious from simple visual inspection of the area. The HMFS is allocated to a buffer zone. People in the adjacent home garden farm land can collect fuel wood and other goods in strip under certain rules. But for all transects, the cutting of trees degrades the forest considerably. Some sites would not be categorized as forest according to standard definition. The total area of HMFS is 8769 ha where about half of this land can be afforested (Kivumbi & Newmark, 1991) and where there is considerable potential for increasing the forest quality and hence the value of forest ecosystem services to the local people by better management. For all transects, the most well-developed and maintained sites are between 2000 to 2500 m asl as noted by the fact that many of these sites scored above 100. Based on these data, it is reasonable to conclude that the CFEMI demonstrates and represents the ecological quality of the different forest sites.

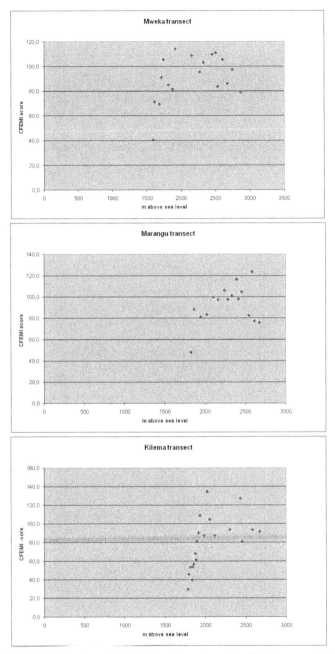

Fig. 2. CFEMI score in percentage for the sites along the altitudinal transect Mweka, Kilema and Marangu, Mt Kilimanjaro. The Half Mile Forestry Strip (HMFS) is between 1590 m asl and 1749 for Mweka, 1780 and 1880 for Kilema and 1820 and 2000 for Marunga.

	Mweka		Kilema		Marangu		Average	
HMFS	68	(4)	50	(7)	72	(3)	60	(14)
Central part	101	(8)	96	(11)	101	(9)	99	(28)
Upper part	94	(6)	93	(2)	90	(4)	92	(12)
Average	91	(18)	80	(20)	93	(16)	87	(54)

Table 5. Average CFEMI score group for the three distinct altitudinal zones of the forest along the three transects Mweka, Kilema and Marangu at the southern slopes of Mt. Kilimanjaro. Number of sites is shown within parenthesis.

3. Framework for mediating balanced ecosystem indicators

The methodology for development of EMI, FEMI and CFEMI is basically built on systems thinking and elements from systems engineering and used as tool for connecting different subsystems, such as stakeholder interests, forest ecology and management together into the larger system where the indicators are meant to work. An essential part of the methodology is the construction of the communication model Balanced Ecosystem Mediation (BEM) framework.

3.1 Construction of the BEM framework

The construction of the indicator is built on a pre-understanding of communication as an instrument for mediation and negotiation of knowledge and interests, and that these processes are integrated and accepted as fundamental for further development of the context where FEMI will contribute.

Technically, most environmental indicator systems are designed within an open system concept which includes conceptual, normative and operational elements. The notion of a system often encompasses *"a combination of interacting elements organised to achieve one or more stated purposes"* (Haskins, 2006), and could be an assemblage of elements constituting a *natural system*, a *man-made system*, an *organizational system* or a *conceptual knowledge system*.

An ecological indicator system aiming to be a management tool can be defined within all these four classes of systems and merged into an overall communication system where the indicator and the different circumstances around the indicator become elements in the system. The challenge is to design and understand how the interaction across the boundary interfaces between the elements, the subsystem and eventually the environment outside the system boundary, influence and bring the system into being. Systems thinking is an underlying concept used to assist in combining the ecological and social elements in the development of FEMI such that the indicator moves closer to a management and stakeholder approach.

Van Bueren and Blom (1996) advanced the *"Hierarchical Framework for the Formulation of Sustainable Forest Management Standards. Principles, Criteria, Indicators"* (PCI) on behalf of Tropenbos Foundation which challenges many of the aspects relevant for the FEMI indicator system. They suggest top-down oriented hierarchal framework for a forest management system with consistent standards based on the formulation of principles, criteria and indicators for sustainable forest management. In the context of development of FEMI, the PCI system appears to be an expert-oriented initiative that belongs to the sphere of influence and interests of the globals.

In order to create a structure involving the locals and strengthening their interests while supporting dialogue and continuous learning, the PCI framework has been modified. The proposed structure allocates the indicator system a more interactive role, and enlarges the system to a construct that shows an ideal typological symmetric mediation between the locals versus globals, ecology versus nature (resources or ecosystem services), and society versus culture (Hermansen, 2008). The framework is called the *Balanced Ecosystems Mediation (BEM)* Framework (Fig. 3).

The transecting lines S and V in Figure 6 represent the ideal symmetric or balanced case based on scientific and normative criteria and arguments. The vertical lines a, b, and c illustrate different constellations where the position, influence and control by the locals is more or less reduced or lost to the globals. The line **a** shows the situation where the locals are incapacitated and have lost most control over their ecosystem resources; line **b** represents the situation where the locals have managed to participate in forest management; and line **c**, the situation where the locals have substantial influence and control over local ecosystem services.

If *V* is moving upwards the ecological interests and concerns increase with stronger emphasis on protection and conservation, and if *V* is moving downwards, society utilize more of the ecosystem services with an increased ecological unsustainability impact and possibly a strong attenuation of the ecological resilience capacity.

The BEM framework should be regarded as an open system where the borders between the elements and subsystem are interfaces where mediation and negotiation can occur between the stakeholders involved. Both mediation and negotiation can take many forms depending on the question discussed or stakeholders (and subgroup of stakeholders) participating in the discourse.

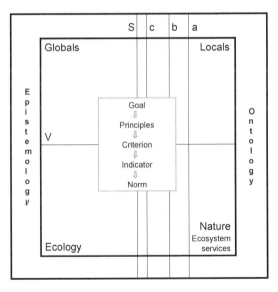

Fig. 3. The construct of the Balanced Ecosystem Mediation (BEM) Framework with the two knowledge regimes ontological and epistemological. S and V are representing the ideal typological symmetry (or balance) regarding mediation and negotiation for globals versus locals stakeholders and society versus nature (as stakeholders) respectively (Hermansen, 2010.

The corresponding influence of how the understanding of ecology (scientific) and nature, and the epistemological and ontological approach, are also illustrated in Fig. 3, and derives from the case study work in which the *indicator* was designated to be the core element in the forest management system in order to strengthen the position of the locals. The BEM framework is built on a nature versus culture model presented by Hermansen (2006, 2010). FEMI is the general and theoretical model for the indicator, while CFEMI is intended to be a specific and practical indicator reflecting the complexity of the relationship between the catchment forest ecosystem and local society.

3.2 Terms and theoretical perspectives of the communication model
The structure of the communication process based on BEM framework can be illustrated as shown in Table 6. The two knowledge regimes (empirical and methodological), which are two different ways of acquiring and constructing knowledge, are paired with the accepted viewpoints of both globals and locals about ecological issues. A normative standpoint is taken by insisting on the right of local people to understand and participate in a discursive reflection on the content and value of the indicator system. The appurtenance interests of the globals comprise the ecological area regarding empirical knowledge acquired by an epistemological methodological approach, and the appurtenance interests of the local comprise the ontological way of experience of nature and natural resources and later ecological knowledge acquired by scientific work.

	Appurtenance of ecological knowledge	
Knowledge acquisition	The globals	The locals
Experience /empirical	The ecological accepted (relevant)	The ontological regime
Scientific work/methodological	The epistemological regime	The social accepted (relevant)

Table 6. Structure of knowledge regimes and appurtenance for globals and locals for the Forest Ecosystem Mediating Indicator (FEMI) concept.

3.3 Goal and objectives of forest ecosystem mediating indicators
Objectives and practical use of the indicator are intended not only to be a measure for communication, but also for mediation and negotiation process in itself and the further understanding of forest ecosystem and management of the forest resources. The indicators are part of the process and the overall objective can be specified by separate regimes and roles into the following regimes/roles as shown in Table 6. However, goals must be stated and the context of mediation by means of indicators must be set. The paramount goal is to contribute to democratized and enlightened mediation of ecosystem knowledge, services and values between nature and society, and strengthen the locals' position in the locals versus globals relationship, and hopefully secure the wise and sustainable use of the forest. Table 7 shows the relationship between the main features of the ecological mediation.

The integrated mediation by means of indicators is dynamic and process oriented interchange and can conveniently be divided into different phases (Table 8). These phases also give an indication of the learning cycle of the activities.

Mediation is not only an end-of-project activity, but an integral part of project development. Suitable settings for mediation can be established prior to inventory (as part of planning), part of field work (inventory), part of management and part of a continuing learning and negotiation process. In a dialogue, stakeholder's interests are also maintained, represented here by local and global interests with accompanying impacts. Negotiated goals and aspects are the result of the process, integrating the consensuses of ecological content, definition of ecological service and values, and suggesting a political/management ecological regime that embraces the negotiated knowledge regime. Through genuine communicative mediation an equitable and symmetric communication process may then emerge.

Paramount objective	Democratized	Enlightened
Ecosystem knowledge	Mediation of scientific results	Mediation of scientific methods
Ecosystem services and values	Mediation of local resources	Mediation of scientific values/understanding

Table 7. Paramount goal and mediation of ecology

Mediation phases	Local interests and impact	Global interests and impact	Negotiated goals or aspects
Pre process understanding	Identify local concern and needs	Identify global concerns	Agree on concerns
Interpretation of positions	Identify local human resources	Identify scientific knowledge	Combining human resources
Designing phases	Defining need of ecological services:	Defining biodiversity and climate issues	Defining a complete description of values and resilience capacity
Pre-inventory	Practical training	Communicating support	--------
Part of inventory	Identifying and deciding	Be accepted as partner	Agree on working methods
Part of management	Control	Protect global ecological concern	Agree on management system
Part of continuously learning and negotiation process	Full access as respected partner	Move from global arrogance to universal partnership	Common interests of using communication opportunities of FEMI

Table 8. Typology of the different interests through the different mediation phases

3.4 Using FEMI to bring momentum to local management

MA (2005) is an initiative for handling the ecosystem resources under the vision of a globalized world and offers a framework both regarding ecosystem and geographic scaling. It further elaborates the relationship between the ecosystem and the human needs for ecosystem services that contribute to well-being and poverty reduction in the form of security, basic material for a good life and good social relations. This in turn necessitates requirements for freedom and choice of action. Status and quality of the forest on the global and regional scale will often be assessed in coarse categories such as area cover by forest, degree of deforestation, estimates of economic value of logs, stakeholder values etc. Application of the MA concept can easily result in a change of resource control and management away from already weak local participants to international bodies and business. FEMI is meant to adjust the management attitude in MA to facilitate a stronger local participation.

Assessments of the ecological status and trends require a set of indicator systems. *The Driving force–Pressure–State–Impact–Response* (DPSIR) framework (Smeets & Wetering, 1999)

is often used. However, Niemeijer and de Groot (2008) argue that moving the framework for environmental indicators from causal chains to causal networks could be a better tool for management decisions and they suggest an enhanced DPSIR-system that could be appropriate. FEMI can be considered as local status indicator, but based on the proximity-to-target concept for principle design of construction, the indicator is working as a performance indicator where performance (status) is compared with a defined ideal typical well developed, natural and healthy forest (the target).

Hence, the intention of FEMI is to enlarge the framework for an ecological forest indicator to include ecological integration and the potential for a larger understanding and dynamic involvement among stakeholders.

3.5 Proximity-to-target performance indicator

To measure ecological and management oriented policy categories, such as for example the wise and sustainable use of forest resources, requires a set of different measurable indicators and data. Some are easily measurable with instruments and metrics, and others by judgement, often value laden along a scale. Performance indicators on social level usually refer to different kinds of reference conditions and values, such as national or international policy targets. Especially demanding, both technically and politically, is the implementation of sustainability performance indicators. Often they are very vague and difficult to follow up and address with responsible authorities or actors. European Environmental Agency (EEA, 2007) has defined the usefulness of a proximity-to-target approach:

"… concept of environmental performance evaluation is being developed for use in an environmental management system to quantify, understand and track the relevant environmental aspects of a system. The basic idea is to identify indicators (environmental, operational and management) which can be measured and tracked to facilitate continuous improvements. Performance indicators compare actual conditions with a specific set of reference conditions. They measure the 'distance(s)' between the current environmental situation and the desired situation (target): 'distance to target' assessment."

Proximity-to-target indicators are a type of environmental performance indicator designed for ranking, benchmarking and monitoring action towards well defined and measurable objectives. The proposed CFEMI is an extension of the concepts and principles from both the macro (societal) and micro (corporate) levels including mimicry of the proximate-to-target indicator from 'Pilot 2006 EPI Environmental Performance Index' launched by Esty et al. (2006).

3.6 Reliability of measurements

To make high quality, representative measurements of forest variables, is a challenge. West (2004) gives an account of *accuracy* as the difference between a measurement or estimate of something and its true values, *bias* as the difference between the average of a set of repeated measurements or estimates of something and its true value, and *precision* as the variation in a set of repeated measurements or estimates of something.

Because much of the measurement phase of the field work is dependent on assessment of the values for the different variables, the indicator is vulnerable to the skills and experience of the observers. Within a close collaborating group of local foresters the observations can be sufficiently accurate, but comparing the results between different forests and assessment teams, the assessment could vary significantly.

4. Discussion

The scientific judgement on the feasibility of BEM framework and FEMI will depend on expectations, and many demurs and critics discussed can be raised. Concepts for ecological integrity which incorporates information from the multiple dimensions of ecosystems are, however, expected to be a useful tool for ecosystem managers and decision makers. The mediation framework and indicator are devised both to expose ecological integrity, and to be instrumental for the mediation between nature and society, and between locals and globals. This implies that the ultimate results of the application of the indicator is connected to the process of continuing improvement of genuine understanding between the globals and locals, and the continuing improvement of the management of the forest in order to secure ecosystem services for the local people as first priority and for the globals as second.

Working out the indicator system and then executing the implementation both contribute to the momentum of the learning loops and to the factual learning about the very easy accessible features of the forest ecosystem and corresponding ecosystem services.

Both selection of ecological phenomena, variables, field methods and measurements, and composition and calculation of the composed indicator are critical issues. To achieve a sufficient accuracy is difficult for many of the variables especially those depending on estimation of heights and cover. The success of the indicator will depend on how the balance of purpose, accuracy and selection of possible variables are compared with the momentum for increased local participation, increased consciousness and ecological knowledge, and increased motivation for interactive cooperation for finding wise solutions.

Local participation of sustainable management of a tropical forest requires that the knowledge about ecological status and the ecosystem services that the forest can provide, can be communicated in way that support enlightenment, democratic management processes and are environmentally sound. Hence, whole process of development and implementation using ecological indicators should be scientifically and ecologically proper (the global perspective) and locally understandable and fair (the local perspective). The case study shows that it is possible to carry out field inventory programs that encompass variables that cover main ecosystem services especially valuable for local and regional utilization, by using simple measurable ecological variables. However, many of the measured variables depend on estimations of measured values and the measurement could then be less reliable for calculation of the indicator.

The connection to the real social conditions at the slopes at Mt. Kilimanjaro in this case is rather weak due to the fact that detailed investigation of the relationship between society and ecosystems is not done. Assessing and making decisions about ecosystem resources is a normative and political action, and a challenge for an indicator system is then to make the normative dimension visible and an object for deliberative processes. To meet the requirement for local participation the indicator system has to move from a hard ecological approach with only measurable indicators to a practical and soft ecological approach and use an open, conceptual and learning oriented systems engineering approach. This movement from a hard system towards a soft system allows greater application of assessment, judgement and estimation.

5. Conclusion

The study has demonstrated and elaborated on the use of ecological indicators to support a balanced and mediating management concept in order to increase the influence of local interests on vital and ecological valuable forest resources, and to encourage knowledge insertion to achieve a proactive approach to sustainable forest management contributing to enlightenment and democratizing of ecological resource management.

Further work should explore how to develop and connect such initiatives deeper into a learning process and as a genuine measure for mediation, negotiation and decision making.

6. Acknowledgement

I am grateful to Mr Leonard Mwasumbi, Mr Frank Mbago, Mr Bernard Mponda, Mr A. Nzira, Mr Patrick Akintanda and professor Håkan Hytteborn for collaboration during field work, and to dr Martina M. Keitch and dr Cecilia Haskins for comments and support.

7. References

Akitanda, P. C. (1994). Local people participation in the management and utilization of catchment forest reserves. A case study of Kilimanjaro Catchment Forest Reserve, Tanzania. M.Sc. thesis. Agricultural University of Norway. Ås, Norway.

Akitanda, P. C. (2002). South Kilimanjaro Catchment Forestry Management Strategies – Perspectives and Constraints for Integrated Water Resources Management in Pangani Basin. In: J.O. Ngana (ed.). *Water Resources Management - The Case of the Pangani River Basin, Issues and Approaches.* University Press, Dar es Salaam.

Andersen, B. & Fagerhaug, T. (2002). Performance measurement explained: designing and implementing your state-of-the-art system. ASQ Quality Press. Milwaukee, WI, USA.

Bart, F., Mbonile, J.M. & Devenne, F. (eds.). (2006). *Kilimanjaro – Mountain, Memory and Modernity.* [*Kilimadjaro, montagne, mémoire, modernité.* (2003). Translated to English from French. Pessac.] Mkuki na Nyota Publishers Ltd. Dar es Salaam.

Bauman, Z. (1998). *Globalization: The Human Consequences.* Columbia University Press, NY.

Bird, A. & Stevens, M.J. (2003). Toward an emergent global culture and the effects of globalization on obsolescing national cultures. *Journ of Intern Management* 9: 395-407.

Bjørndalen, J.E. (1992). Tanzania's vanishing rain forests – assessment of nature conservation values, biodiversity and importance for water catchment. *Agriculture, Ecosystems and Environment* 40: 313-334.

Blomley, T. (2006). Mainstreaming participatory forestry within the local government reform process in Tanzania. *Gatekeeper Series* 128, 19 pp. IIED. International Institute for Environment and Development. London.

Burgess, N.D, Butynski, T.M., Cordeiro, N.J., Doggarts, N.H., Fjeldså, J., Howell, K.M., Kilahama, F.B., Loader, S.P.; Mbilinye, B., Menegon, M., Moyer, D.C., Nashanda, E., Perkin, A., Rovero, F., Stanley, W.T. & Stuart, S.N. (2007). The biological importance of Eastern Arc Mountains of Tanzania and Kenya. *Biological Conservation* 134: 209-231.

Creech, H., Jaeger, J., Lucas, N., Wasstol, M. & Chenje, J. (2006). *Training Module 3: Developing an impact Strategy for your Integrated Environmental Assessment.* UNEP GEO Resource Book. UNEP. Nairobi.

Esty, D.C., Levy, M.A., Srebotnjak, T., Sherbinin, A. de, Kim, C.H. & Andreson, B. (2006). *Pilot 2006 Environmental Performance Index.* Yale Center for Environmental Law and Policy, New Haven.

Elkington, J. (1998). *Cannibals with forks. The triple bottom line of 21ˢᵗ century business.* New Society Publishers. Gabriola Island BC. Cananda.

European Environmental Agency (EEA). (2007). *Multilingual environmental glossary.* Available from http://glossary.eea.europa/EEAGlossary/search_html.

FAO (2007). Global Forest Resources Assessment 2005 CD. Earthprint. Rome.

Grimble, R. (1998). Stakeholder methodologies in natural resource management. Socio-economics Methodologies. Best Practice Guidelines, 10 pp. Natural Resources Institute. Chatham, UK.

Grimble, R. & Wellard, K. (1997). Stakeholder methodologies in natural resource management: a review of principles, contexts experiences and opportunities. *Agricultural Systems.* 55(2): 173-193.

Hall, J. B. (1991). Multiple-nearest-tree sampling in an ecological survey of Afromontane catchment forest. *Forest Ecology and Management* 42: 245-266.

Haskins, C. (ed.). (2006). *Systems engineering handbook. A guide for system life processes and activities.* INCOSE-TP-2003-002-03. International Council on Systems Engineering.

Hemp, A. (1999). An ethnobotanical study of Mt. Kilimanjaro. *Ecotropica* 5: 147-165.

Hemp, A. (2006a). Continuum or zonation? Altitudinal gradients in the forest vegetation of Mt. Kilimanjaro. *Plant Ecology* 184: 27-42.

Hemp, A. (2006b). Vegetation of Kilimanjaro: hidden endemics and missing bamboo. *African Journal of Ecology* 44: 305-328.

Hemp, A. (2006c). The banana forest of Kilimanjaro: biodiversity and conservation of the Chagga homegardens. *Biodiversity and Conservation* 15: 1193-1217.

Hermansen, J.E. (2006). Industrial ecology as mediator and negotiator between ecology and industrial sustainability. *Progress in Industrial Ecology* 3. Nos ½, pp. 75-94. DOI: 10.1504/PIE.2006.010042

Hermansen, J.E. (2008). *Mediating ecological interests between locals and globals by means of indicators. A study attributed to the asymmetry between stakeholders of tropical forest at Mt. Kilimanjaro, Tanzania.* Doctoral thesis NTNU 2008: 284. Trondheim, Norway. http://ntnu.diva-portal.org/smash/record.jsf?pid=diva2:326637

Hermansen, J.E. (2010). Mediation of tropical forest interests through empowerment to locals by means of ecological indicators. *Sust. Dev.* 19(5):271-281. DOI: 10.1002/sd.478

Hermansen, J.E. Benedict, F., Corneliusen, T., Hoftsten, J. & Venvik, H. (1985). *Catchment Forestry in Tanzania. Managment and Status. Including Himo Watershed Land Use Map.* Institute of Natur Analysis and FORINDECO. Bø, Norway.

Hermansen, J.E., Hytteborn, H., Mbago, F. & Mwasumbi, L. (2008). Structural characteristics of the montane forest on the southern slope of Mt. Kilimanajaro, Tanzania. In: Hermansen, J.E. (2008). *Mediating ecological interests between locals and globals by means of indicators. A study attributed to the asymmetry between stakeholders of tropical forest at Mt. Kilimanjaro, Tanzania.* Doctoral thesis NTNU 2008: 284. Trondheim, Norway.

Howell, K. M. (1994). Selected Annotated Bibliography on Biodiversity of Catchment Forest Reserves in Arusha, Iringa, Kilimanjaro, Morogoro and Tanga Regions, Tanzania. *Catchment Forestry Report* 94.3. Dar es Salaam.

Huang, W., Pohjonen, V., Johansson, S., Nashanda, M., Katigula, M.I.L. & Luukkanen, O. (2003). Species diversity, forest structure and species composition in Tanzanian tropical forests. *Forest Ecology and Management* 173: 11-34.

Jakko Pöyry. (1978). Report on the industrial forest inventory in part of the Kilimanjaro Forest Reserve, Tanzania. TWICO. Dar es Salaam.

Kashenge, S.S. (1995). Forest Division – Catchment Plan for Catchment Forests Kilimanjaro Forest Reserve, Kilimanjaro Region 1st July 1999 – 30th June 2000. Kilimanjaro Catchment Forest Project. Moshi, Tanzania.

Katigula, M. I. L. (1992). Conservation of the Ecosystem on Mount Kilimanjaro – Reality in Experience. Kilimanjaro Catchment Forest Project. Moshi.

Kivumbi, C. O. & Newmark, W. D. (1991). The history of the half-mile forestry strip on Mount Kilimanjaro. In: W. D. Newmark (ed.). *The Conservation of Mount Kilimanjaro*, pp. 81-86. IUCN. Gland, Switzerland and Cambridge, UK.

Latour, B. (2004). Politics of Nature. How to Bring the Science into Democracy. Harward University Press. Cambridge, MA, USA.

Lambrechts, C., Woodley, B, Hemp, A, Hemp, C. & Nnytti, P. (2002). *Aerial Survey of the Threats to Mt. Kilimanjaro Forests.* UNDP. Dar es Salaam.

Lawrence, A. (ed.). (2000). Forestry, Forest Users and Research: New Ways of Learning. European Tropical Forest Research Network (ETFRN). Wageningen, Netherlands.

Lovett, J.C. and Pocs, T. (1992). *Assessment of the Condition of the Catchment Forest Reserves.* The Catchment Forestry Project, Ministry of Natural Resources and Tourism, Forestry and Beekeeping Division. Dar es Salaam.

Lyaruu, H.V. (2002). *Plant biodiversity component of the land use change, impacts and dynamics project, Mt. Kilimanjaro, Tanzania*, Land Use Change Impacts and Dynamics (LUCID). Project Working Paper 40: 1-13. Intern Livestock Research Institute, Nairobi, Kenya.

Macqueen, D.J. (2006). *Governance towards responsible forest business. Guidance on the different types of forest business and the ethics to which they gravitate.* 37 pp. IIED International Institute for Environment and Development. London.

Macqueen, D.J., Dufey, A. & Patel, B. (2006). *Exploring fair trade timber: A review of issues in current practice, institutional structures and ways of forward.* IIED Small and Medium Forestry Enterprise Series 9. IIED Edinburgh, UK.

Madoffe, S., Mwang'ombe, J., O'Connell, B., Rogers, P., Hertel, G. & Mwangi, J. (2005).*Forest Health in the Eastern Arc Mountains of Kenye and Tanzania: a baseline report on selected forest reserves.* Publised on internet, 12.30. FHM EAM Baseline Report.

Madoffe, S., Hertel, G. D., Rodgers, P., O'Connell, B. & Killenga, R. (2006). Monitoring the health of selected eastern forest in Tanzania. *African Journal of Ecology* 44: 171-177.

Malimbwi, R. E., Luoga, E. & Mwamakimbullah, R. (2001). *Inventory report of Kilimanjaro Forest Reserve in Tanzania,* 39 pp. Forest and Beekeeping Division, South Kilimanjaro Catchment Forest Project. FORCONSULT. Morogoro, Tanzania.

Mariki, S. W. (2000). Assessment of stakeholders participation in forest conservation programmes: A case of Kilimanjaro Catchment Forest Management Project – Tanzania. M. Sc. thesis. NORAGRIC, Agricultural Univ. of Norway. Ås, Norway.

Millennium Ecosystem Assessment (MA). (2005). *Ecosystems and Human Well-Being. Multiscale assessments*. Vol 4. Findings of the Sub-global Assessments Working Group. Island Press. Washington.

MNRT (Ministry of Natural Resources and Tourism. Forestry and Beekeeping Division). (1998). *National Forest Policy*. Dar es Salaam.

MNRT (Ministry of Natural Resources and Tourism, Forestry and Beekeeping Division). (1999). *Criteria and indicators for sustainable forest management in Tanzania*. Workshop Proceedings, Olmontonyi – Arusha 24 -28 May. Strategic Analyses and Planning Unit. Dar es Salaam.

MNRT (Ministry of Natural Resources and Tourism. Forestry and Beekeeping Division). 2001. *National Forest Programme in Tanzania 2001-2010*. Dar es Salaam.

MNRT (Ministry of Natural Resources and Tourism. Forestry and Beekeeping Division). (2006). *Extension and Publicity Unit - Participatory Forest Management in Tanzania. Facts and Fi*gures. Dar es Salaam.

Misana, S.B. (1991). The importance of Mount Kilimanjaro and the needs for its integrated management and conservation. In: W.D. Newmark (ed.). *The Conseravtion of Mount Kilimanjaro*. IUCN, Gland, Switzerland and Cambridge. UK.

Misana, S.B. (2006). Highlands, Community Utilization and Management of Forest Resources. In: F. Bart, M.J. Mbonile & F. Devenne (Eds.). *Mt. Kilimanjaro, Mountain, Memory, Modernity*, pp. 233-243. Mkuki na Nyota Publishers Ltd. Dar es Salaam.

Misana, S.B, Majule, A.E. & Lyaruu, H.V. (2003). *Linkages between Changes in Land Use, Biodiversity and Land Degradation on the slopes of Mount Kilimanjaro, Tanzania*, Land Use Change Impacts and Dynamics (LUCID). Project Working Paper 18: 1-30. International Livestock Research Institute. Nairobi.

Newmark, W.D. (ed.). (1991). *The Conservation of Mount Kilimanjaro*. IUCN. Gland, Switzerland and Cambridge, UK.

Newton, A.C. & Kapos, V. (2002). *Biodiversity indicators in national forest inventories*. UNEP World Conservation Centre. Expert consultation on Global Forest Resources Assessments. Kotka IV 1-5 July 2002. Kotka, Finland.

Ngana, J.O. (ed.). (2001). Water Resources Management in the Pangani River Basin, Challenges and Opportunities. University Press. Dar es Salaam.

Ngana, J.O. (ed.). (2002). Water Resources Management - The Case of the Pangani River Basin, Issues and Approaches. University Press. Dar es Salaam.

Niemeijer, D. & de Groot, R. (2008). Framing environmental indicators: moving from causal chains to causal networks. *Environ. Dev. Sustain.* 10: 89-106

Noss, R.F. (1990). Indicators for Monitoring Biodiversity: A Hierarchical Approach. *Conservation Biology* 4(4): 355-364.

Noss, R.F. (1999). Assessing and monitoring forest biodiversity: A suggested framework and indicators. *Forest Ecology and Management* 115: 136-146.

Røhr, P.C. (2003). *A hydrological study concerning the southern slopes of Mt. Kilimanjaro, Tanzania*. Doctoral thesis NTNU 2003: 39. Trondheim, Norway.

Sjaastad, E., Chamshama, S.A.O., Magnussen, K., Monela, G.C., Ngaga, Y.M & Vedeld, P. (2003). *Resource economic analysis of catchment forest reserves in Tanzania*. Ministry of Natural Resources and Tourism. Forestry and Beekeeping. Dar es Salaam.

Smeets, E. & Wetering, R. (1999). *Environmental indictors: Typology and overview*. Technical Report 25. European Environmental Agency. Copenhagen.

Soini, E. (2005). Changing livelihood on the slopes of Mt. Kilimanjaro, Tanz.: Challenges and opportunities in the Chagga home garden system. *Agroforestry Systems* 64: 157-167.

Stahl, K.M. (1964). *History of the Chagga People of Kilimanjaro*, 390 pp. Mouton & CO. London.

Strassberg, B.A. (2003). Symposium on Organ Transplants: Religion, Science, and Global Ethics. *Zygon* 38(3).

Studley, J. (2007). *Hearing a Different Drummer: a new paradigm for "keepers of the forest"*. IIED. London.

TEEB (2010). *The Economics of Ecosystems and Biodiversity: Ecological and Economic Foundations*. Edited by Kumar, P. Eartscan. London and Washington.

Tagseth, M. (2006). The Mfongo Irrigation Systems on the Slopes of Mt. Kilimanjaro. In: T. Tvedts & E. Jakobsen (eds.). *A History of Water. Vol. 1: Water Control and River Biographies*, pp 488-506.. I.B. Tauris. London.

Tagseth, M. (2008). Oral history and the development of indigenous irrigation. Methods and examples from Kilimanjaro, Tanzania. *Norwegian Journal of Geography* 62 (1): 9-22.

Turpie, J.K, Ngaga, Y.M. & Karanja, F.K. (2003). *A preliminary economic assessment of water resources of the Pangani river basin, Tanzania: Economic value, incentives for sustainable use and mechanisms for financing management*. IUCN – Eastern Africa Regionall Office and Pangani Basin Water Office. Nairobi. Kenya and Hale, Tanzania.

UNCBD (Convention on Biological Diversity). 2006. *COP Decisions VII!/15* Decisions adopted by the Conference of the Parties to the Convention on Biological Diversity at its eighth meeting. Available from http://www.cbd.int/decisions/cop/?id=11029

UNEP (United Nations Environment Programme). (2007). *Global Environment Outlook 4 – GEO-4: Environment for Development*. Earthprint. Nairobi.

Van Bueren, E.L. & Blom, E. (1996). *Hierarchical framework for the formulation of sustainable forest management standards. Principles, criteria and Indicators*. The Tropenbos Foundation. Center for International Forestry Research. Bogor, Indonesia.

Vermeulen, S. & Koziell, I. (2002). *Integrating global and local values: a review biodiversity assessments*. International Institute for Environment and Development. London.

West, P.W. (2004). *Tree and Forest Measurement*. 167 pp. Springer Verlag. Berlin.

Wieler, C. 2007. Delivery of Ecological Monitoring Information to Decision – Makers. For the Ecological Monitoring and Assessment Network, Environment Canada. International Institute for Sustainable Development. Canada.

William, C.M.P. (2003). *The implications of land use changes on forests and biodiversity: A case of the "Half Mile Strip" on Mount Kilimanjaro, Tanzania*. Land Use Change Impacts and Dynamics (LUCID) Project Working Paper 30: 1-50. International Livestock Research Institute. Nairobi.

Wily, L.A. (2001). "Forest management and democracy in East and Southern Africa: Lessons from Tanzania" *Gatekeeper Series* 95, 20 pp. IIED. London.

WBCSD (World Business Council for Sustainable Development) and IUCN (World Conservation Union). (2007). *Business for ecosystem services – New challenges and opportunities for business and environment*. WBCSD and IUCN. Geneva. Available from www.wbcsd.org

WRI (World Resources Institute) with UNDP, UNEP and World Bank. (2005). *World Resources 2005. The Wealth of the poor: Managing Ecosystems to Fight Poverty*. Washington DC.

3

Collaborative Forest Management in Uganda: Benefits, Implementation Challenges and Future Directions

Nelson Turyahabwe, Jacob Godfrey Agea,
Mnason Tweheyo and Susan Balaba Tumwebaze
College of Agricultural and Environmental Sciences, Makerere University, Kampala
Uganda

1. Introduction

In many countries including Uganda, management of forest resources has moved away from command and control system to a more participatory approach that require involvement of a broad spectrum of stakeholders. The introduction of Participatory Forest Management (PFM) was sparked by several factors: both international and local. At the international level, treaties and accords such as the Tropical Forest Action Plan (TFAP), an outgrowth of the agenda 21 framework initiated in Rio-de-Janeiro in 1992, sought to reverse the loss of forests through the involvement of stakeholders, especially adjacent communities. The Convention on Biological Diversity (CBD) (1992) highlights the importance of sustainable use and equitable sharing of benefits that arise from biodiversity resources. At the local level, the original argument for increasing community participation in the maintenance of rural conservation projects stemmed from the need to better target people's needs, incorporate local knowledge, ensure that benefits were equitably distributed and lower management costs (Wily, 1998). The inclusion of communities in the management of state-owned or formerly state-owned forest resources has become increasingly common in the last 25 years. Almost all countries in Africa, and many in Asia, are promoting the participation of rural communities in the management and utilisation of natural forests and woodlands through some form of Participatory Forest Management (PFM) (Wily & Dewees, 2001). Many countries have now developed, or are in the process of developing, changes to national policies and legislation that institutionalise PFM. PFM encompasses a wide range of different co-management arrangements with different levels of control from relatively conservative "benefit sharing" to genuine "community-based natural resource management" where local communities have full control over management of the resource and the allocation of costs and benefits (Wily, 2002).

Participatory forest management encompass processes and mechanisms that enable people who have a direct stake in forest resources to be part of decision-making in all aspects of forest management, from managing resources to formulating and implementing institutional frameworks. Notable among the participatory forestry management approaches are Joint Forest Management (JFM), Community Based Forest Management (CBFM) and Collaborative Forest Management (CFM). All these approaches tend to emphasize decentralisation or devolution of forest management rights and responsibilities to forest adjacent communities with the aim of producing positive social, economic and ecological outcomes (Carter & Grownow, 2005). Joint forest management is the type of participatory approach that allows forest adjacent communities to enter into agreements with government and other forest owners to share the costs and benefits of forest management by signing joint management agreements (Wily, 1998). Under this arrangement, local communities are co-managers of the forest owned by the central or regional government. It considers communities as "rightful beneficiaries" than as "logical source of authority and management". JFM may also be defined as a specific arrangement among different social actors around the forest sharing of rights and responsibilities in managing a specific body of resource (Borrini-Feyerabend, 1996). Community Based Forest Management (CBFM) is management of forests exclusively based on the efforts of the local communities, and at times with limited extension advice from government. In CBFM, both the ownership and user rights over the forest resource belong to the community. In CBFM, local communities declare- and ultimately, gazette-village, group or private forest reserves on village land. Under this arrangement, communities are both owners and managers of the forest Resource (Blomley *et al.*, 2010). CFM in general is loosely defined as a working partnership between the key stakeholders in the management of a given forest–the key stakeholders being local forest users and state forest departments, as well as parties such as local governments, civil groups and non-governmental organisations, and the private sector (Carter & Gronow, 2005).

In Uganda, the form of Participatory Forest Management approaches adopted for managing forest resources include Collaborative Forest Management (CFM), Community Forests (CFs) and Private Forests (PFs) (Strengthening and Empowering Civil Society For Participatory Forest Management in East Africa [EMPAFORM], 2008). CFs is the forest management approach where communities register as legal entities for purpose of seeking gazettement of a forested communal land as a Community Forest and henceforth manage it for the common good of the community. PFs is the forest management approach where local community members manage own trees on private land or participate in the management of private natural forests, private plantations, forests owned by cultural and traditional institutions. CFM is the most widely used and adopted form of participatory forest approach in Uganda today. It is a forest management approach where communities enter into agreement with the National Forestry Authority (NFA) in case of Central Forest Reserves and District Forestry Services (DFS) local governments in case of Local Forest Reserves to manage part or the whole of gazetted forest reserve. CFM is defined as a structured collaboration between governments, interested organisations and community groups, and other stakeholders to achieve sustainable forest use. It defines a local community's rights to use and/or participate in forest management and focuses on improving the livelihoods of the forest adjacent communities through mutually enforceable plans but the government does not surrender ownership of the forest to partner stakeholders (National Forestry Authority [NFA], 2003) and is the most widely used form of PFM in Uganda.

CFM is a co-management arrangement widely practiced in India, Nepal, Philippines and Latin America (Ghate, 2003; Malla, 2000) as government forest agencies and other actors recognise its potential in supporting local well-being and sustainable forest management. CFM has also gained recognition as a means of flow benefits to local people and is widely practiced in many African countries like Tanzania, Sudan, Ethiopia, Kenya, Uganda, Zimbabwe, Malawi, Cameroon, Niger, Nigeria, Gambia, Ghana, Mali and South Africa

(Willy, 2002). Many Scholars (Borrini-Feyerabend, 1997; Ghate, 2003; Malla 2000; Victor, 1996) believe that CFM provides local incentives for conservation of forest resources by sharing the costs and benefits of conservation. They further note that the implementation of CFM may result into ecological, socio-economic, institutional, infrastructural and policy impacts to both the communities and forestry sub-sector. The ecological impacts may include stabilised and/or forest resource use patterns and improved quality and or condition of forests. The economic impacts include improved livelihoods through sale of forest products, increased skills, employment and exclusion of non-CFM actors from accessing forest resources.

Other authors (Beck, 2000; Campbell et al., 2003) notes that the impact of CFM on community livelihoods directly influences people's participation or involvement. They argue that participation and commitment of communities under CFM encourages regulated legal access to socio-economic benefits. The more the community are involved in CFM, the fewer the number of illegal activities in the forest managed under CFM and the higher diameter at breast height, the basal area and density of trees. In contrast, lack of community involvement may result in high occurrence of illegal activities and lower basal area and density of trees. It is thus argued that providing socio-economic benefits to communities under CFM results into sustainable utilisation of forest resources by local communities and hence improved conditions of the forest. Improvement in the condition of the forest may also lead to increased socio-economic benefits derived by the communities and increased community participation in CFM (Ghate, 2003). If CFM provides no socio-economic benefits to communities, illegal activities may increase leading to forest degradation. Degradation of the forest may lead to loss of socioeconomic benefits to communities leading to loss of community participation in CFM. Building on experiences from India (Kothari et al., 1996; Poffenberger & McGean, 1996), collaborative forest management (CFM) was adopted in Uganda in 1993 around Bwindi Impenetrable National Park, BINP (Wild & Mutebi, 1996), and by 1996, collaborative initiatives had spread to other protected areas (national parks) such as Mt Elgon, Kibale, Mgahinga, and Murchsion falls (UWA, 2001).

In the forest sector, research on CFM began in 1996 with pilot activities in some selected Ugandan forests, for example Butto-Buvuma (Gombya-Ssembajjwe & Banana, 2000). The Forest Department, however, held consultations from 1996 to 1997, and on July 1998, the CFM programme was officially launched (Scott, 2000). Since then, pilot activities were initiated by the Forest Department (FD), emphasising equitable distribution of benefits, participation of local people at all stages, gaining consensus on the terms of management and representation; instilling the sense of ownership and authority over the resource in local management partners, ensuring flexibility on the part of the Forest Department towards the potential compromise and building mutual trust and respect as a strong foundation for future partnership. This has now been institutionalised in the 2001 Uganda Forest Policy (Ministry of Water Lands and Environment [MWLE], 2001) and in the National Forestry and Tree Planting Act, of 2003 (Government of Uganda, 2003). Guidelines have also been put in place for the implementation of CFM arrangements in the forest sector (MWLE, 2003). The CFM programme is currently being practiced in all the seven forest management ranges as designated by the NFA. A total of 27 CFM agreements have so far signed, 30 Application approved by NFA for CFM implementation and 28 initiated (Driciru, 2011).

However, the actual benefits accruing to local communities under the CFM agreement are largely unknown. Little is also known regarding the impact of CFM on the livelihoods of people. According to Scher et al., (2004) an understanding of CFM actual benefits on the peoples' livelihoods around Protected Areas (PAs) are critical in sustainable forest management. Information is also lacking to show whether CFM has improved the condition of the forest by way of controlled illegal forest access, yet this information is essential for strengthening both the CFM policy development and implementation in Uganda. Due to entrenched power structures within both government institutions and communities, it is not easy to promote social justice and sustainable livelihoods through. Overall, mechanisms of CFM are diversifying, reflecting a greater recognition of the need for partnerships in forest management.

This chapter analyses the reviews Uganda's experience in CFM to date Benefits, Strengths, Implementation Challenges and Future Directions in Uganda. It identifies a number of possible strategies and makes recommendations on how to improve CFM. The analysis done in this chapter may be used to improve or re-arrange the idea about participatory Forestry Management not only in Uganda but also in other countries with similar situations.

1.1 Status of forest resources in Uganda

Currently, there are about 4.9 million hectares of forest in Uganda (24% of the present total land area) (National Biomass Study, 2003). The forest resources comprise areas classified as savannah woodland (80.5%), natural forest (tropical high forest, 18.7%) and less than 1% of forest plantations. The existing natural forests on private land and in government reserves, together with the on-farm tree resources are the major focus of the National Forest Plan (NFP), with particular reference to decentralisation of forest management (MWLE, 2002). In terms of land ownership, 70% of the forest area is on private and customary land, while 30% is in the permanent forest estate (PFE), such as Forest Reserves (central and local), National Parks and Wildlife Reserves. Of the PFE's 1,881,000 ha, 1,145,000 ha (60.9%), is managed by the National Forestry Authority (NFA) as Central Forest Reserves (CFRs), 5,000 ha (0.3%) is controlled by District Forestry Services (DFS) of local governments as Local Forest Reserves (LFRs) and 731,000 ha (38.8%) is managed by the Uganda Wildlife Authority (UWA). The majority of private forests are woodlands, and are being depleted rapidly due to restrictions on harvesting of wood and wood products from gazetted protected areas (Jacovelli & Carvalho, 1999). A huge dependency (>90%) on fuelwood from the rapidly increasing population is clearly accelerating the problem.

1.2 The principles of CFM

According to Scott (2000), CFM must be flexible and responsive to the inputs and participation of all the parties. CFM is guided by the following principles: (a) it should be implemented by the authorities and departments responsible for forest management. CFM must be initiated and implemented by the Forest Department (FD) in partnership with other interested parties. It should be seen as an approach towards management, not as something outsiders are imposing; (b) sustainable forest management is the major objective. Sustainable forest management is the long term aim of CFM, and as much as we try to meet other aims, such as fair benefits to both partners and equity in benefit sharing within the community, this key objective is paramount; (c) the focus is not on the output but rather on how the output is arrived at. For example, the end products of negotiations are an agreement and management plan. However, the important thing is not the documents, but rather the process of negotiation that produced them. The output will be as strong or weak as the process that led to it; (d) there must be real and complete participation of all partners from the beginning. If other stakeholders are to be partners, then they must be involved in all decisions. It is not real participation if you go to the other partners and start to discuss management if a large proportion of the decisions have already been made without them; (e) the process takes time, "rapid" is never "participatory". If the process is rushed, it will not give the other party the chance to fully participate. This approach is new for the community and the FD, they will need time to adjust and feel comfortable in their new roles; (f) It should result in a fair deal for all parties. If the FD expects considerable benefits from CFM, then they should not be surprised if the other parties also expect considerable benefits. The natural tendency is to give what you get; (g) it must result in a fair distribution of benefits. In addition to ensuring a fair distribution of benefits between the community and the FD, it is vital that the benefits from CFM are shared fairly within the community; (h) flexibility is very important in CFM. The FD must be open minded, and go into the process with a clear understanding of their own objective but a fully open regarding how these objectives can be satisfied through collaboration; (i) responsibilities agreed through CFM must be appropriate. It is important during negotiations that the different partners agree to the responsibility that are appropriate. One partner will not be able to do everything. It is most beneficial if both parties take on responsibilities that maximize their capacity; (j) it should address the real issues. In order for CFM to work, it must address real issues both on the side of the FD and on the side of other partners. It should be tackled head on and a solution sought that suits both parties. Compromise on both sides will be essential; (k) it must offer long term security. In order to adopt a long-term perspective to the management of forests, both partners must be sure of their long-term security to rights and benefits; (l) all interest groups must be involved. Everybody within the community with an interest in the forest must be involved during the process of arriving at an agreement. If they are not involved in decision making that affects their lives, they are unlikely to respect and abide by the agreement and management plan; (m) agreements should be arrived at through consensus. The majority of the population must be in agreement with the decisions if they are to abide by them and be enforced by them. It is, therefore, critical to gain consensus to the greatest possible extent.

1.3 The rationale, goal, purpose and objectives of the CFM process in Uganda

Sustainable management of forest resources in Uganda has remained a challenge to forest managers and policy makers because the population is highly dependent on them for timber, agriculture, energy production and other non-timber forest products (Turyahabwe & Tweheyo, 2010). In addition, forest agencies responsible for forest management have been unsuccessfully in their effort to sustainably manage forests due to breakdown in law and order, ineffective rules and inadequate funding to manage forest resources (Banana et al., 2007). Since most of the forest reserves are small and scattered over a large area, the governmental lacks both financial and human resource to monitor the use of the resources (Buyinza & Nabalegwa, 2007). Therefore, in the current forest policy, there has been a shift of control of forest resources, especially those outside protected areas from state controlled to community level in an attempt to improve management (Kugonza et al., 2009). CFM was viewed as the one approach to achieving improved and more efficient management of the country's forest estate. The rationale behind CFM approach include: (i) a recognition that forest reserves can only be adequately managed if cooperation of forest adjacent communities is obtained; (ii) a desire to overcome conflicts with neighbouring communities; (iii) a desire to create opportunities for local people to contribute towards protection ad rehabilitation of forest resources thus reducing the costs of management; (iv) a philosophical commitment to human rights and thus to fair and equitable treatment of communities living adjacent to forest to forest reserves that they have traditionally utilised for products and services; (v) a mechanism for supporting sustainable forest based livelihoods in poor rural communities; (vi) an awareness that forest reserves are decreasing while human population is increasing; and (vii) a move towards participatory approaches and decentralised governance in natural resources management.

The goal of CFM is to contribute to the overall goal of the National Forestry Authority in sustainable forest management. The purpose is to enhance sustainable forest management through the active participation of interested parties. The specific objectives are to improve forest management through: (a) reduced costs (fairer distribution of the costs of management); (b) fairer distribution of benefits, responsibilities, decision-making authority in management; (c) reduction of conflicts over resource use; (d) creating awareness about benefits of forests; (e) creating a sense of ownership over forest resources; (f) sharing knowledge and skills (both FD and community sharing with one another; and (g) keeping abreast with trends in the rest of the world.

1.4 The larger forest sector and national context

In the larger forest sector and national context: (i) Makerere University was in the process of re-orienting its curriculum to incorporate courses in community forestry; (ii) Decentralization process aimed at creating strong local level administration; planning at the sub-county level is today nearer to the local users than before when it was done at the district level; (iii) as a result of decentralization, environmental committees at the Local Council levels were already being formed and it was envisaged that forest management issues could be addressed in these committees; (iv) the wide range of actors already operating in the forest sector provided a good opportunity for an agency like FD to play a catalytic role in integrating management efforts.

1.5 The community perspective

From the community side, concern was already being expressed about environmental degradation in the context of appreciating the value of the forest. Community members also exhibited behavioural changes, such as reporting of illegal harvesting. They also expressed willingness to take responsibility for the management of the forest so long as they can benefit from it.

2. Legal and policy framework for implementation of CFM in Uganda

2.1 National legal and policy provisions on CFM

The Constitution of the Republic of Uganda (1995), which is the supreme law, in Article 13 provides for the protection of natural resources including forestry and Article 27 provides for the sustainable management of natural resources. The traditional, protectionist approach of policing forest reserves has not been effective in reducing widespread illegal activities, has not favoured local communities in sharing the benefits from protected forest areas, and has been a source of conflict between the lead agencies and communities. The constitution thus gives ownership of resources to the people while government holds the resources in trust for all citizens. The 1995 Constitution of Uganda also incorporated decentralisation into the directives of the national policy (Government of Uganda [GOU] 1995, 1997). The government of Uganda views participation of local people and community based organisations in forest management as a practical and equitable alternative to traditional top-down approaches to forest management (MWLE 2002). It is believed that actors and agencies with grassroots experience such as NGOs and CBOs will mediate participation of local authorities and their institutions in sustainable forest management.

Following the enactment of the Resistance Councils and Committees Statute in 1987, the National Resistance Movement (NRM) government of Uganda embarked on the process of devolution of power to the district councils including the management of natural resources. In 1993, the Local Government Statute was passed and as a result some powers and responsibilities to manage forest resources were transferred from the central to local government authorities (GOU, 1993). This was further emphasised by passing of the National Environment Act (1995) and the Local Government Act of 1997 (Government of Uganda 1995, 1997). Along with other public service functions, the objectives for decentralising forestry were to: (i) enhance the role of local government with more developed responsibility to plan and implement forestry activities; (ii) reduce the burden on public finances by empowering local government to outsource financial resources and manage forestry activities; and (iii) encourage participation of local communities and farmers in the management of forest resources. The current 2001 Forestry Policy envisages that government will promote innovative approaches to community participation in forest management on both government and private forestlands and this is intended to provide a balance between the protectionist approach to forest management and open access to forest resources that may be destructive. The development of Collaborative Forest Management is intended to define the rights, roles and responsibilities of partners and provide a basis for sharing benefits from improved forest management. Therefore the CFM process is guided by principles that partners have to adhere to (Box 1).

2.2 Policy principles and opportunities specific to CFM

The 2001 National Forestry Policy for Uganda emphasizes government commitment to "promote innovative approaches to community participation in forest management on both government and private forest land" (MWLE, 2001). The Policy puts a strong emphasis on public involvement especially, forest adjacent communities, and benefit from sustainable forest management, including the application of CFM. It says in part: "Collaborative Forest Management will define the rights, roles and responsibilities of partners and the basis for sharing benefits from improved management. There will be a specific focus on wide stakeholder participation, collective responsibility and equity and on improving the livelihoods of forest dependent communities".

The National Forestry and Tree Planting Act (2003) lay out a legal framework for the development of CFM agreements for various categories of forest reserves in Uganda (GOU, 2003). Section 15 of the Act says that one or more responsible bodies may enter into a CFM arrangement with the Central or Local Government for the purpose of the management of the whole or part of a Central or Local Forest Reserve in accordance with generally acceptable principles of forest management as may be prescribed in guidelines issued by the Minister. A responsible body refers to a body appointed to manage, maintain and control a forest reserve and includes; the National Forestry Authority, a Local Council, a Local

---

Box 1. Principles guiding the CFM process in Uganda

- A process approach based on learning by doing – communities as well as forest resource managers learn from one another. This means that more time is taken to build trust and relationships.
- Meaningful participation and shared analysis – communities getting deeply involved. Stakeholders are given enough time to adjust to new roles.
- There is negotiation and consensus building – exchange of opinion, the buy-and-take approach. There is discussion of real problems that concern the parties and resources involved to fairly address local community livelihoods
- Appropriate representation and responsibilities – with due consideration of women, the elderly and the disadvantage groups.
- A supporting legal and policy framework. This involves analysing, understanding and sharing information on policy and legal provisions for CFM within the CFM Guidelines
- Building capacity for change – tolerating one another. Stakeholders are empowered to take lead and efforts to ensure good representation of all stakeholders.
- Long term perspective – forestry enterprises are long term and thus agreements must be stable and honoured by all parties.
- Transparent communication to attract marginalized stakeholders. Information is put in a format understandable to all stakeholders including women, youth and disadvantaged groups.

Community, a leady agency, a private contractor, a non-governmental organisation or stakeholders (NFA, 2003). A Responsible Body may be the National Forestry Authority (Section 52 of the Act) to manage Central Forest Reserves (CFRs) and the District Forestry Services (Section 48) to manage and Local Forest Reserves (LFRs). Section 15 of the National Forestry and Tree Planting Act (2003) mandates Responsible Bodies to enter into Collaborative Forest Management arrangements between the themselves and any forest user group(s). Section 28 of the National Forestry and Tree Planting Act (2003) also commits the Responsible Bodies to prepare management plans for all forest reserves and further guides that this "shall be in consultation with the local community", thus further emphasising the spirit of Uganda's CFM approaches. The Local Government Act (1997) assigns management of forest resources to local government and sub-county councils (Local Governments Act Part IV).The National Forestry and Tree Planting Act 2003 (Section 48) further obliges the District Local Government to establish a forestry office that is responsible for management of forest resources in the district.

To guide the step by step process of undertaking Collaborative Forest Management are CFM Guidelines (2003) that have been put in place. Part 3 of the CFM Guidelines describes the purpose for CFM as including: rehabilitation of degraded forests, maintenance of forest reserve boundaries, and regulation of access to forest products, joint law enforcement and public participation in forest management (GOU, 2003). Further to the development of CFM Guidelines are the CFM Regulations which will additionally provide for the rules and requirements for CFM and pave way for better understanding of the roles and responsibilities of concerned parties. The National Forest Plan (NFP) 2002, a sectoral plan for forestry development in Uganda that provides a framework for implementing 2001 Uganda Forestry Policy into action clarifies under makes provision for CFM by encouraging partnerships between lead forest agencies and local communities to enhance people's access to, utilisation of forest products.

2.3 Power relations and their impact on CFM

Forest Reserves were largely established during the colonial times with most of them being gazetted between the years 1920–1960 (Turyahabwe & Banana, 2008). This drew out forests from the public sector into the protective hands of the state. This also disbanded communal property ownership, access and management. This meant that forest reserves belong to government and that communities have lesser power over these resources. A power analysis of the status quo is provided examining the specific relationships relevant to proposed CFM advocacy work. Table 1 provides the management arrangements, characteristics and how they affect collaborative approaches. Based on the analysis in this Table 1, there is a strong indication for proponents of participatory forest management approaches to influence a shift in power relationship. The unequal relationships are based on the fact that forests can only be managed by a corporate entity and there are policy and legal frameworks on which to build PFM work.

2.4 Current programmes under CFM in Uganda
2.4.1 Integrated forest management planning process
In order to meet the policy demand for CFM, an integrated forest management process is being used to develop management plans for forest reserves. The purpose of adopting such a process is to ensure that local communities participate in the planning and decision making process in forest management. Basically the process involves about 10 steps: (i) formation of reserve planning team, (ii) inauguration and training of reserve planning team, (iii) resource assessment and inventories, (iv) socio-economic surveys, (v) information gathering from maps, old plans, reserve settlement agreements, logging history, etc, (vi) preparation of draft management plans, (vii) reserve planning workshop, (viii) review of draft management plan, and (ix) submission of final plan

2.4.2 Promotion of private and community forests
In accordance with the 2001 Uganda Forest policy, the government through the Forest Sector Support Department (FSSD) is encouraging the establishment of plantations and dedicated forests as a means of enriching the off reserve timber resources. A scheme is being developed to provide loans and grants for private companies and individuals to embark on forest plantations and dedicated forest as a means enriching the off reserve timber resource. With the 'right of veto' given to farmers and landowners in the procedure for felling trees off forest reserves, they are motivated to tend young indigenous trees and plant more trees.

2.4.3 Community forest committees
A major drawback to the CFM programme was the lack of a recognizable and well informed body who will liaise with the FSSD and the forest fringe communities to ensure that their aspirations, knowledge and needs from forest resources and forest management is expressed and realized. To this end, Community Forest Committees are being formed to: (a) permanently represent the forest fringe communities on forest management issues at the

Type of arrangement	Characteristics
NFA managing Central Forest Reserves	- Objectives and outputs relating to the management of CFR are clearly defined by government that is the overall trustee. - NFA may or may not enter an agreement with an interested community - Guidelines, standards, regulations binding parties are often drawn by the Responsible Body - NFA has financial resources to fund community livelihood projects under CFM - Power relationship between NFA and the communities is unequal, but can improve with increased advocacy work.
District Forest Services of Local Governments	- Local governments have powers over the Local Forest Reserves - Local governments do not have resources to manage the Local Forest Reserves and therefore not as powerful as would be expected - The power relationship with communities is unequal but expected to change in favour of communities if participatory approaches are implemented
Formal Collaborative Forest Management Agreement	- Objectives defined jointly by parties to the agreement - Roles, responsibilities, rights and benefits clearly spelled out and to some extent binding - Important stakeholders may be left out, affecting the potential for achieving management objectives - Unequal relations, not expected to be equal, but can improve
Communal Land Associations to manage Community Forests	- Objectives are clearly stated in the Land Act may become real legal entities to manage community forest reserves - Individual responsibilities may be subdued by influential members of the Association - The vulnerable groups (elderly, women, children) stand to be suppressed - Once established the power relations with Responsible Bodies will be equal with communities being the owners, managers and users of the resources
Private Forests	- Private forestry in Uganda is not well developed - Guidelines, standards, regulations for private forestry not in place - Private forest owners have no bargaining power-government continues to levy taxes and royalties without guidelines - Therefore the relationship is unequal, but can change.

Table 1. Management arrangements, characteristics and how they affect collaborative approaches in Uganda

national level and to improve upon the knowledge and capacity for collaboration at the local level; (b) enhance and encourage widespread participation in forestry matter especially those that will affect the communities; (c) mobilize wide stakeholder awareness and participation in the forest management planning process; (d) educate and assist in the development of social responsibility agreements; and (e) monitor the implementation of the social responsibility agreements. Specifically, participating communities will play important roles and responsibilities at the national, regional, district and local levels. At the national, regional and district levels the partners will: (i) participate in forest policy review and formulation; (ii) prepare proposals to promote the welfare of communities through forest resources management; and (iii) make general recommendations on forestry that will lead in to improving forest management.

2.4.4 Forest fires protection
Over the past decade most forest reserves and off reserves in Uganda have been experiencing annual forest fires. The communities are therefore expected to help in preventing and fighting forest fires in their community. This is done through: (i) planting green fire belts along the forest boundary; (ii) education of local communities on the dangers of fire and fire management especially during the dry seasons; (iii) formation of fire volunteer squads; and (iv) development and enforcement of by-laws to protect fire and sanctioning forest offenders. In addition to this collaboration, participating communities can suggest measures to conserve forest resources in their locality. They will also be responsible for encouraging and supporting the arrest and reporting of offenders to the FSSD, NFA and/or the police. In line with their protective functions, participating communities under CFM are encouraged to check the permits of people they suspect to be engaging in illegal operations.

2.4.5 Forest rehabilitation
The taungya system has been the main way in which communities were traditionally involved in forest management. A review of the past taungya system was done and this helped to inform the development of pilot programs. The review also helped the FSSD to develop new strategies and systems for forest rehabilitation called "the modified taungya system. In 2001 the government of Uganda lunched plantation activities as one of its poverty reduction strategies. In the modified taungya and plantation development programme, CFM and forest fringe communities are expected to: (a) assist in the identification of degraded portions of the forest for rehabilitation; (b) establish nurseries from which the FSSD will obtain seedlings for forest rehabilitation; (c) undertake forest rehabilitation activities such as tree planting, transplanting, tree tending operations; and (d) encourage and assist communities to plant trees on their farms.

2.4.6 Boundary cleaning and patrolling
The boundaries of the forest reserve are cleaned to ensure that farms are not extended to the reserves. In addition, it ensures that wildlife in the forest do not enter into the farms of those who share a common boundary with the forest reserve. Most often, NFA use the forest guards to patrol and clean the forest boundary at regular intervals. Currently local communities are given contracts to perform such duties.

2.4.7 Collaboration in the utilization of timber off-cuts

Uganda's Forest Policy is committed to promoting peoples' participation in resource management and a more equitable sharing of benefits from forest resources. One of the strategies of the policy is the promotion of public awareness programs as a positive community building action, to generate raw materials and income while improving the quality of the environment. Sawn timber is conveyed to the big towns and no conscious effort is made to sell lumber to the local people. Besides, huge quantities of off-cuts and sometimes logs are left behind in the forest as "waste". The communities believe that they could profitably utilize the wood and have entered into discussions with the lead forest agencies (NFA and DFS) to collect and use the wood. Their only wish is to have access to such timber to convert into merchantable and profitable products. This is likely to encourage the establishment of forest-based enterprises and generate employment. Ultimately, this will help improve upon the standard of living of forest fringe communities.

2.4.8 Community contracts jobs through boundary maintenance, seedling production, plantation development

Systems are now being implemented under which forest fringe communities enter into contracts to clean forest reserve boundaries in return for cash payments. Additionally some are also contracted to establish green-fire breaks to prevent wildfires from entering into forest reserves. The possibility of involving communities in patrolling is underway and if proved positive that system would also be adopted. Under CFM, some communities under a pilot scheme have been assisted to set up and manage their own tree nurseries to produce seedlings both for planting and sale. Apart from supporting such nurseries through the supply of inputs and offer of technical advice, NFA has been promoting the sale of seedlings from the community nurseries either through their own purchases or linking them up with tree growers to ensure their viability. It is anticipated that more of such nurseries would be set up to supply seedlings for planting in connection with the government's plantation programme.

2.4.9 Forest reserve management planning

This programme focuses on three aspects of forest management through a series of workshops and consultations including: (i) involvement of communities in forest management planning; (ii) integrated forest management process; and (iii) revenue sharing from management of forest and forest resources. To this end the issue of rights and revenues from forest management have been reviewed and recommendation made in the review of forest legislations.

2.4.10 Management of non-timber forest products (NTFPs)

This programme involves local people extraction of forest resources largely focusing on the exploitation of NTFPs for household and commercial uses. The programme also target different aspects of NTFPs exploitation, production, processing and management.

2.5 CFM operational concerns
2.5.1 Concerns of responsible bodies

There are a number of concerns for responsible bodies and these includes the following: (i) whereas it is government policy to promote community participation in forest management on government and private forest land, today participatory forest management initiatives target (rather focus) on Central Forest Reserves. This has resulted into pressure onto the NFA in terms of capacity to meet the demands by communities. It is high time these initiatives started on private forest land and Local Forest Reserves that are managed by local governments; (ii) in Uganda a nine stage process (Box 2) has been developed that has to be followed by communities applying for CFM. This process takes time and leads to anxiety. It is clear that the time and skills are inadequate in NFA for community mobilization and effective sensitization of the communities. Along the process is local political interference that favours illegal activities that applicants often prefer to indulge in illegal activities than undertake an Agreement. Subsequently there are delays in signing agreements which at times causes apathy in community; (iii) there is a general lack of capacity for implementation of CFM-inadequate staffing at NFA and DFS to monitor and give backstopping (support) to field staff. Some decision makers are still sceptical about CFM therefore need for sensitization at all levels;

(iv) Quite often communities are lured to present CFM applications (again by self seeking and self appointed leaders) that do not have good intention for genuine partnerships for collaboration with the National Forestry Authority and other responsible bodies. Some communities think that an Agreement with responsible body is a permit for undertaking unacceptable activities such as charcoal burning and cultivation of crops. Thus responsible bodies are, however, re-orienting their thinking before they undertake the CFM process; (v) good governance in CFM requires sufficient funding. So far there is inadequate funding by NFA and DFS for CFM implementation and sometimes this funding is sporadic. Such delay in funding breaks the momentum of activities in the field; (vi) there are very few Community Based Organizations with experience in facilitating the CFM process and usually these cover a relatively limited area of the country. The National Forestry Authority would be more than willing to establish working relations with such NGOs to role out CFM activities. Many NGOs have created a culture of giving handouts (food and materials) to communities and this has created a dependency syndrome with communities demanding to be given handouts. Whereas this is a positive Social Responsibility approach, it stands to demean the stigma of CFM.

Box 2. CFM process in Uganda

Step 1: Initiating the process
Step 2: Preparing an application for CFM
Step 3: Meeting between applicant and responsible body
Step 4: Participatory situation analysis
Step 5: Initial Negotiation and drafting a CFM plan
Step 6: Institutional formation and development
Step 7: Continuation of Negotiations
Step 8: Review of the Plan and Agreement by stakeholders
Step 9: Implementation of the CFM Agreement and Plan

2.5.2 Stakeholders concerns

Like responsible bodies, there are also a number of concerns for stakeholders involved in CFM operation. (i) Parties involved in CFM agreements always have a hidden agenda – forest resources managers opt for CFM to solve encroachment and not necessarily as a management tool and the communities take on CFM to legalise illegal activities in the forests; (ii) there is a lack of forest committees at district and sub county level that would hold responsible Bodies accountable. These committees would also fight for the plight of forestry extension under NAADS. The committees would also guide local political patronage to bless collaborative forest management, community forests and private forestry; (iii) funding for collaborative approaches/activities is limited. Communities argue that if government and development partners can fund private tree farmers planting trees on Central Forest Reserves, why not identify funds for implementation of forestry friendly economic activities such as apiary, harnessing of white ants, collection of rattan and mushrooms for collaborating communities and funds for community and private forests?

(iv) Lack of tangible benefits from CFM participation and lack of guidelines for benefit sharing. The burden of roles and responsibilities that are transferred to communities under the signed Agreements are not commensurate to the benefits; (v) corruption and illegal practices/activities that erode sustainable forest resource management and jeopardizes communities and members of the private sector who would otherwise have interests in private forestry. There is a general lack of professionalism in the sector with politicians wanting to get involved in the decisions that affect the electorate and this is a disappointment to committed and contending communities; (vii) the management of community forests is regulated by the National Forestry and Tree Planting Act, 2003 while the establishment of the Community Land Associations (CLAs) is regulated under the Uganda Land Act, 1998 (GOU, 1998). This has caused management delays in gazettement of Community Forests which is dependant upon an established Communal Land Associations (CLA). Establishment of CLAs has lagged behind because the implementation of the Land Act has been slow. Therefore the CLAs have not been instituted and yet Community Forests can only be gazetted if the CLAs are in place. The interest of the community is the standing forest and any barrier to communities securing their interest is dismay to participatory forest management approaches. Many communities had expressed interest in registering CLAs but in vain;

(viii) The initial lifespan of CFM agreements range between 5 and 10 years with a provision to extend them for longer periods if implemented to satisfaction. However, the benefits of some listed activities, particularly restoration of degraded forest areas would come much later. The time frame for CFM agreements with the "Responsible Bodies" is 10 years. Comparatively, the Tree Planting Permits issued by NFA to private tree planters is minimum 20 years. The 10 year period does not provide sufficient motivation for tree planting under CFM given that most trees require more than 10 years to mature. To that effect, communities are entering CFM half-heartedly. Communities would prefer a relatively longer period; (ix) communities allege that CFM is only earmarked for only degraded forest reserves or degraded compartments of forest reserves. Not only do they find nothing left to share as benefits, but also de-motivated for having been handed a degraded resource. Coupled with the 10 years margin, then this becomes totally a disincentive for communities to actively engage in CFM; (x) many of the communities are manipulated by the self seeking persons who enter into leadership and never want to retire. Someone becomes chairperson of a CFM group for as long as he lives; later becoming a proxy representative of the

Responsible Body (in this case the National Forestry Authority) with whom the community signed an agreement; (xi) Unpractical provisions within CFM agreements whereby many of the responsibilities to be executed by local communities (Box 3) and CBOs are written to be executed together with NFA. Yet going by the limited number of NFA staff, it is inconceivable that that would be practical and feasible.

Box 3: Roles and Responsibilities of communities under CFM

- Impose fines on culprits with NFA
- Arrest culprits with NFA
- Impound illegally harvested timber with NFA
- Patrol the forest with NFA

A more practical approach would entail, among others, inclusion of provisions with regard to: (a) type and limits of fines, procedures for collecting exhibit and accounting for them, and reporting about them; (b) procedures for arresting or submission of names of culprits to NFA for arrest; and (c) custody of exhibits (e.g. illegal timber) and their recording before handing over to superior authorities e.g. NFA. Police etc.

(xii) Most of the CFM agreements are signed with local community, NFA, NGOs and Local Governments. No reference is made to the private sector or business community, cultural institutions, research institutions etc, where they may have strategic roles to support CFM. The role of Uganda Wildlife Authority (UWA) in relation to wildlife within the CFM Agreements is not strong either; monitoring plan is a crucial instrument that should be annexed to the agreement. A key weakness was that there was no independent system for monitoring. Communities cannot be the "judge and party" and there is need to have an objective, independent monitoring system and most of these plans are not yet made where CFM agreements have been signed; (xiii) CFM negotiations exhibited different interests. It is the goal of NFA to maintain the wide range of values of forest resources, including both use and non-use values. To the communities the immediate attraction is to access the tangible use values. In most management plans, there are 3 types of management zones, namely: (a) buffer zone, (b) production zone and (c) strict nature reserve. Human interference is highly restricted in strict nature reserves. NFA recognizes these zones but it did not appear that communities were fully mobilized to understand and appreciate the rationale for such zoning. This message will have to be consistently communicated because in some forest reserves, the closest areas to the communities are strict nature reserves that they cannot access; (xiv) matching of benefits and costs-Communities are eager to come on board for CFM in order to gain regulated access to forest products. That eagerness overshadows the need for a basic understanding of how the potential benefits compare with costs communities incur for certain responsibilities they accept to take on e.g. joint patrols and sometimes supervision, arresting culprits, collect and provide information about illegal timber harvesting, participate in fighting wild fire, sensitizing communities about conservation to mention but a few. Owing to fewer numbers of NFA staff around protected areas and poor funding, they cannot always respond to community request for activities that are to be carried out jointly e.g. patrols.

2.6 Gaps in CFM policy and legal framework in Uganda

The practice of CFM in its present stage is yet to deliver on poverty reduction and forest conservation because of a number of issues relating to rights, privileges, roles and responsibilities. There are many implementation, conceptual as well as policy framework issues which need to be thoroughly re-examined along the key milestones. There is need to move from considering communities as forest user groups to appreciating user groups as forest managers. Finally former user groups need to be seen to become forest owners and therefore managers of the resource. The detailed account of the policy gaps are as follows: (i) the 2001 Forestry Policy provides for CFM but not community based forest management and therefore the communities are only obliged to collaborate with "responsible bodies". The law only accords them "responsible body" status when it comes to establishment of community forests and private forests on land that is communally owned by the communities or privately owned respectively; (ii) although sharing the resources is provided for in the policy, there are no guidelines for forest-benefit sharing. Quite often communities are left with low value items (mushrooms, water ponds, medicinal species etc). The high value products (reserved timber species and revenue from forestry services such as eco-tourism) are maintained by the Responsible Body (NFA). The CFM Agreement for Hanga-Kidwera community in Masindi district for example says in part: "Local inhabitants are privileged to obtain free of charge and in reasonable quantities to the discretion of the forest officer, bush firewood, bush poles, timber from unreserved tree species and sand for domestic use only. Domestic animals are allowed to visit water and salt lick points in the reserve".

(iii) Government pledges to promote CFM as indicated in Policy Statement No. 5 of the 2001 Forestry Policy, chapter 5.5 of the 2002 National Forest Plan and Section 15 and 17 of the National Forestry and Tree Planting Act 2003, but has limited institutional capacity to ably handle CFM. The process for registering Community Forests and Private Forests has stalled (The Vision 2025 of Ugandans for the 21st Century); (iv) the Forestry Inspection Division has no direct mandate for CFM, the NFA is understaffed, the District Forest Services are not fully operational. The Private Sector is not motivated to register private forest and community forests. The National Forestry Authority has limited ability to take on CFM - applications received are way above what can be handled by NFA; (v) whereas CFM seems to be taking off at a snail speed in Central Forest Reserves managed by the National Forestry Authority, there is limited attempt to introduce CFM in Local Forest Reserves managed by the DFS of local governments; (vi) the lack of guidelines for registration and declaration of Community and Private Forests. Where as there are published guidelines for CFM there are no guidelines for undertaking initiatives to declare Community Forests and guidelines for registration of Private Forests. Where as Section 17 of the National Forestry and Tree Planting provides for the declaration of the Community Forests, Sections 21–27 for Private Forests, there are no guiding principles to be followed before acceptance by the District Land Board, the District Council or even the Minister; (vii) transfer of property rights and control of resources to communities is provided for as a strategy and opportunity in the 2001 Forestry Policy but has never become a reality because forest resource managers are sceptical and pessimistic about CFM being a viable arrangement with fears that communities may indulge in illegal activities rather than implement the agreed plan.

(viii) There is a lack of information and lack of information dissemination to communities about the available opportunities in participatory forest management. Section 91 of the National Forestry and Tree Planting Act 2003 provides for access to information on forest products and services. Quite often information does not trickle down to the end users, especially local communities. CFM benefits to the communities are not well articulated. CFM information (such as its contribution to livelihoods improvement, fuel wood for more than 92% of Uganda's population, value for environmental conservation, contribution to rainfall and soil protection) is not reflected in government statistics. Therefore, the importance and contribution of CFM to the gross development strategies of the communities and the country are underestimated. This undermines the significance of CFM to rural livelihoods especially for the forest dependant poor; (ix) CFM agreements (wherever they are signed) have been drafted in English with less than 10% of the community members being able to comprehend the contents of the agreement (Organizational Capacity Assessment [OCA] Report, 2006). The issues contained therein are therefore not appropriately deciphered and has always been a major hindrance to the negotiation and implementation process. Chances of getting distorted information are high. This provides a window for manipulation of the unsuspecting communities by self seeking and opportunistic CFM leaders. Above all that, communities can not hire offices and staff; therefore their documents are kept with the Chairperson of the group and this establishes an additional barrier of access to information; (x) CFM Agreements have a life span of 10 years yet many forestry activities are of a longer gestation period. For example trees take 20 years and above to mature. Private tree farmers in Central Forest Reserves are given permits of up to 50 years. This is a disincentive for communities to undertake long term and lucrative investments under CFM restricting them to subsistence tendencies (collection of mushrooms, rattan and hunting);

(xi) Gender and equality is a mere formality under the CFM agreement. It is intended to serve the interests of the "Responsible Body" as a counterpart to the agreement. Equal participation, fairness and sharing of benefits have left a lot to be desired. Women and the elderly have fallen victims of this inequity problem with approximately 80% of the women in the agreements not being able to tell the simplest of CFM and not knowing their rights (EMPAFORM, 2006). Disadvantaged members of the community such as persons with disability, the old and vulnerable groups have little in the know about CFM. Males still dominate and take up most of the decision making positions as evidenced by the (OCA, 2006); (xii) the 2001 Forestry Policy empowers civil society organizations to be at the forefront in the management of forest resources in the country. However, there are no networks of civil societies at grass roots level fighting for CFM issues. Some of the active forestry associations, for example, the Uganda Forestry Working Group, Forestry Governance Learning Group, Mabira Forest Integrated Community Organization, Nature Conservation and Promotion Association, etc. have only expressed interest in big policy issues rather than community issues; (xiii) there is a general failure to recognize value of forest resources by decision makers (including national and local politicians) and attaching low priority to CFM. This has resulted into poor governance fuelling forest destruction, over-exploitation and encroachment to the detriment of the CFM goal. In as long as encroachment and or illegalities is seen as the easiest way to access forest resources, collaborative forest management, community forests and private forest approaches will be denied an opportunity to take root;

(xiv) Failure to recognize the value of forest resources has led to critical lack of extension and/or advisory service provision to communities. It may be argued that communities have preferred enterprises with immediate returns such as piggery, poultry, maize, beans, etc. Communities have little motivation to indulge in long-term activities that involves planting, growing and protecting trees. It may be true that high poverty levels, immediate needs like medical bills and basic household requirements are a motivation for enterprises with quick returns. However, it is also true that demand driven forestry extension service delivery has failed and communities have received little advice from the National Agricultural Advisory Services (NAADS). The National Agricultural Advisory Services (NAADS) has reluctantly adopted forestry activities as enterprises and therefore market opportunities for forest products and services remain in oblivion. No wonder that forestry does not feature in most NAADS strategic enterprises. There is also inadequate understanding of forestry based livelihood opportunities despite efforts for its inclusion in the Poverty Eradication Action Programme (PEAP); (xv) benefits from forest reserves have remained hidden. There is a lack of market information and therefore CFM communities remain attached to the traditional opportunities. This jeopardizes CFM approach to support community livelihoods initiatives. It is a fact that trees take too long to grow, but it is also a fact that many communities have not been introduced to alternatives that are of a short-term income generating nature. Eco-tourism and activities that reduce stress from the forest could be introduced as alternatives and providing a right balance between long-term and short-term investment options;

(xvi) There is political jeopardy and interference by government directly supporting encroachment in gazetted forest reserves and degazettement of reserves with preference to large scale agricultural investors (the case of Butamira Forest Reserve in Eastern Uganda and Bugala Island Forest Reserves, in Kalangala District) at the expense of the interests of the communities. Forest resources managers (NFA and DFS) are undermined. There is loss of credibility and therefore communities remain sceptical about CFM. This lessens the morale and speed of implementation of CFM; (xvii) there is a breakdown of the rule of law in the management of forests. The 2001 Forestry Policy and 2003 forest laws are defied by the civic and political leadership. This has resulted into lack of respect for professionalism on the part of government; lack of respect for the CBO/CSOs voice and opinion in this regard and therefore reduces the speed of implementing CFM approach. Recent media reports have indicated that politicians have interfered with the management of forest resources) and neglected a call by the civil society to "keep eyes on but hands off" the management of forest resources.

3. Future strategies

3.1 Awareness creation and capacity building

The need for capacity building focusing on those capacities needed for local stakeholders to adjust to changing ecological and socio-economic circumstances and institutional circumstances, including both adverse events and opportunities for livelihood improvement. The emphasis on changes means that both Resource Capacities (RCs) and Institutional Capacities (ICs) of local stakeholders need to be strengthened, where: *Resource capacities* (RCs) refer to adequacy in terms of "hardware" (funds, equipment, material and infrastructure) and "software" (information, knowledge and skills; and *Institutional Capacities* (ICs) (or governance capacities) relate to the enabling institutional environment, which allows for a cost effective use of RCs. ICs encompass several factors associated with the concept of good governance, including: (a) adequate information, Net Working and Information exchange; (b) transparency in management procedures; (c) Accountability, both upwards (to higher administrative levels) and downwards (to civil society); (d) inclusive/ participatory decision making processes and adequate representation; of local interests in decision-making fora; (e) managerial skills (in particular regarding financial matters, group dynamics); (f) cost effectiveness, business skills and management; (g) sustainability strategies and mechanisms; and (h) recourses mobilization and management skills.

3.2 Sustainability of the resource through investments in forestry
Promoting on-farm forestry and management of forests on private lands are critical. There is need for NFA to should popularise hitherto unknown species so that communities can benefit from their commercial exploitation.

3.3 Strengthening the partnership arrangements
The Community- Government partnership should: (i) include the private sector/ Civil Society Organizations (CSOs) as a partner in the assessment of CFM because private operators often: (a) replace forestry services where these have inadequate means of operation; (b) pre-finance forest related initiatives involving communities including diversification of forestry related income generating activities; and (c) easily link to other government programmes like NAADS for leverage and synergy especially in the area of capacity building; (ii) move beyond forests and consider forest management in a landscape perspective, where forests are only one amongst the possible uses of land by rural populations; (iii) mainstream CFM into Local Government Development Plans. This would open doors for capturing and mainstreaming indigenous knowledge (IK) into local planning and natural resource management initiatives. It would also avail the communities the opportunity to tap into Local Government resources for community development projects.

3.4 Development of Forestry Based Enterprises (FBEs)
Support Forest Enterprise development as alternative livelihoods. Viable forest enterprises depend not only on market demand but also on the sustainable management of resources. However, there are constraints that will affect FBEs and they should be addressed. These include: (a) inadequate information on markets; (b) weak linkages between small suppliers and large buyers; (c) limited access to credit, finance, capital and technology; and (d) shortage of business and technical skills.

3.5 Provision of incentives for community participation in forest management
Under the CFM agreements, communities have been allocated specific compartments to co manage with the NFA. However, when it comes to joint protection work (community and NFA), the community's work extends beyond the boundaries of the designated compartments. There is therefore concern among the community that the NFA should reward them for their role in the protection of forests outside their CFM compartments. Perhaps a study could be commissioned to determine how much NFA is saving through CFM. A scheme could then be worked out to pay the community a percentage of the NFA savings as an incentive. This will definitely boost community morale and interest in forestry. It will also help them to directly link their development to conservation.

3.6 Streamlining service provider activities in all NFA management ranges
According to the NFA field staff, the relationship between the service providers so far has been cordial. However, there is need to stream line the activities of the NGOs providing services to the communities so that resources and benefits are equitably shared among the

communities. For example: (a) the NGOs should reinforce and complement each other rather than seeing themselves as competitors. The most important thing in the end is delivery of services to the target group, the communities; (b) the NFA field staff is of the opinion that all the service providers and NFA should develop a joint work plan for CFM implementation; (c) the service providers can, within the period of their projects, build and shape partnerships with each other and NFA and develop a joint programme of action on common themes e.g. support to income generating activities, exchange visits, training etc.

3.7 Improving and developing internal organization which should reflect the needs of member CBOs

Some of the activities could include: maintaining, developing and strengthening partnerships and networking; securing finances to fund CFM activities; building capacity of CBOs to implement the CFM agreements; developing and delivering core conservation activities; developing and delivering livelihood activities through FBEs; investment in forestry by CBOs.

4. Conclusion

The future for participatory approaches in Uganda that includes CFM on gazetted forest reserves, community forests and private forests is bright. The struggle though remains incorporation of such initiatives in the bigger Environment and Natural Resources sector, mainstreaming it into the Poverty Eradication Action Plan (PEAP) pillar (now National Planning Authority Master Plan) on natural resources. Making sure it is in the Non-sectoral Conditional Grants, mainstreaming in the District Development Plans and making sure there are resources for implementing activities that make participatory approaches a success. It is clear that: (i) Communities need to move from positions as subordinate beneficiaries, receiving a share of access, products or other benefits, into positions where they may themselves regulate this source of livelihood and with longer-term perspectives; (ii) CFM promotes good governance and accountability in the management of gazetted forests. Collaborative Forest Management reduces the ills associated with policing, and provides for access rights and may serve as an insurance against degazettement of Forest Reserves.

5. Recommendations based on identified policy and legal gaps

It is desired by proponents of CFM that time has come for not just demanding government's support for continuing CFM as it now exist but rather to revitalize the campaign for truly benefit-oriented and equitable model of people's participation in forest management. This means restating the basic premises of collaborative forest management in Uganda and pointing to the broad directions of policy change that are required. The following recommendations are made to the corresponding policy gaps:

i.   There is need for harmonizing legislation, reviewing guidelines and finalizing regulations. The guidelines that require immediate attention include guidelines for forest benefit sharing, transfer of property rights, forest resource control by communities.

ii. There is need for the government of Uganda to build the capacity of institutions to undertake CFM. This includes building the capacity of the Forestry Inspection Division to provide guidance on CFM implementation, NFA to step up its manpower and resources for collaborative initiatives and the DFS to pilot CFM. It is also recommended that government undertakes to establish Forestry Committees provided for in the National Forestry and Tree Planting Act. These Committees will provide a forum to discuss collaborative forestry issues with due consideration of the needs of the communities.

iii. Currently, CFM is mainly undertaken by the National Forestry Authority. It is recommended that CFM is implemented by the District Forest Services and Private Forest owners. The PEAP recommends facilitation of the District Forest Services to undertake its activities (including CFM). It is recommended that this is implemented. This will not only improve forestry activities in the districts but also build their capacity to undertake collaborative forest management approach.

iv. Government should finalize guidelines for registration and declaration of Community and Private Forests to allay the anxiety arising from communities interested in registering CFs and individuals interested in registering their private forests. This will also provide the District Land Boards, the District Council and the Minister to provide guidance and speed up the process of registration of the forests.

v. There is a need for advocacy work to advocate for the transfer of property rights and control of resources to communities who would like to become owners, and users of forest resources. This means that there is need to build a strong case to demonstrate that communities are capable of managing the resources regardless of the scepticism of forest resource managers.

vi. There is sufficient information that needs to be packaged and disseminated to the communities. An effort needs to be made to translate such information into vernacular languages understood by communities in the different regions.

vii. There is need for Responsible bodies to translate CFM Agreements into vernacular languages local in the area where the Agreement is to be undertaken. This is to provide an opportunity for the community to be able to comprehend the contents of the agreement and avoid manipulation by unscrupulous CFM leaders. It is further recommended that each member should also be provided with own copy for reference from time to time.

viii. Whereas the 10 years of a CFM Agreement is tagged to the 10 years of the forest management plan, it is recommended that this period be revised to allow for projects that require a much longer time - e.g. tree planting.

ix. Affirmative action needs to be taken into account to streamline gender and equity issues in CFM Agreements. Special sensitisation meetings for women, the elderly and other vulnerable groups need to be undertaken so as to empower such groups to be able to negotiate and make informed decisions. This is the only way to allow for equal participation, fairness and sharing of benefits.

x. There is need for grassroots networks of civil societies/NGOs/CBOs that will provide a forum for to fight for participatory forest management issues at forest reserve level. This will also help develop both the capacity of communities to implement collaborative approaches, providing conducive climate for community and private forests, building the capacity of "Responsible Bodies" to implement such approaches

xi. It is important for central government to improve the budget allocation for CFM implementation under the PEAP/NAADS and within the Environment and Natural Resources Sector. It is also important that Local Governments provide necessary support to the District Forest Services to undertake participatory forest management approaches. Advocacy work needs to be stepped up at policy level to influence government on good governance issues in the forest sector, the role of both the political and civil leadership, accountability of responsible institutions, and collaborative forest management. There is need for civil society to influence government plans to incorporate collaborative forest management issues in overall government priority plans.

6. Acknowledgement

We do acknowledge National Forestry Authority (NFA), and EMPAFORM (Strengthening and Empowering Civil Society for Participatory Forest Management) for enabling us access their documents.

7. References

Banana A., Vogt N., Bahati J. & Gombya-Ssembajjwe W. (2007). Decentralized governance and ecological health. Why local institutions fail to moderate deforestation in Mpigi district, Uganda, *Science Research Essay*, 2(10):434-445.

Becker, C.D. & León, R. (2000). Indigenous forest management in Bolivian Amazon: Lessons from Yuracaré people. In: *People and forests: Communities institutions and governance.* (eds) Gibson, C.C., M.A.McKean., and E.Ostrom. Cambridge, Mass.: Institute of Technology

Blomley, T., Ramadhani H., Mkwizu, Y.& Böhringer, A. (2010). Hidden Harvest: Unlocking the Economic Potential of Community-Based Forest Management in Tanzania. In: *Governing Africa's Forests in a globalized World* eds.German, A.L., Karsenty A., Tiani, A. CIFOR, pp126-143, ISBN: 978-1-84407-756-4, Earth scan London

Borrini-Feyerabend, G. (1997). Participation in conservation: why, what, when, how? In: Borrini-Feyerabend, G. (ed) *Beyond fences: seeking social sustainability in conservation.* Vol. 2, A resource book, 26–31. IUCN, Gland, Switzerland.

Buyinza, M. & Nabalegwa M. (2007). Gender mainstreaming and community participation in plant resource conservation in Buzaya county, Kamuli District, Uganda. *African Journal of Ecology.* 45: 7-12. doi: 10.1111/j.1365-2028.2007.00730.x

Campbell, B.M. & Byron, N. (1996). Miombo woodlands and rural livelihoods: options and opportunities. In: B.M Campbell (ed.). *The Miombo in Transition: Woodlands and welfare in Africa*, 221-230. Bogor: Centre for International Forestry Research (CIFOR).

Carter, J. & Gronow, J. (2005). Recent Experience in Collaborative Forest Management. A review paper. CIFOR Occasional Paper No. 43

Clarke J., Grundy I., Kamugisha J.R., Tessema Y. & Barrow, E. (2001): Whose Power? Whose Responsibilities? An Analysis of Stakeholders in Community Involvement in Forest Management in Eastern and Southern Africa. Nairobi, IUCN-EARO.

Driciru, F. (2011). Personal Interview, National Forestry Authority, Kampala.

Strengthening and Empowering Civil Society for Participatory Forest Management in East Africa [EMPAFORM], (2006). Baseline Survey Report (EMPAFORM). Kampala Uganda

Strengthening and Empowering Civil Society for Participatory Forest Management in East Africa [EMPAFORM], (2008). The integration of participatory forest management into Uganda's policy and legal framework: an analysis of the level of Uganda's domestication, compliance and Implementation of PFM provisions in Global Legal Instruments. EMPAFORM Policy Research Paper No.1.

Ghate, R. (2003). Ensuring collective action in participatory forest management. Working Paper   No.3-03, South Asian Network for Development and Environmental Economics, Kathmandu, Nepal.

Gombya-Ssembajjwe, W.S. & Banana A.Y. (2000). Collaborative Forest Management in Uganda: The Case of Butto-Buvuma Forest Reserve. In: *Community-based Forest Resource Management in East Africa* , W. Gombya- Ssembajjwe and A.Y. Banana, eds. Makerere University Printerly.

Government of Uganda, [GOU]. (1993). *The Local Government (Resistance Councils) Statute*. Uganda Gazette No. 55, Vol. LXXXV1 of 31st December 1993. Entebbe: Government Printer.

Government of Uganda [GOU]. (1995). *The National Environment Act*, Laws of Uganda, Cap.153. Entebbe: Uganda Publishing and Printing Corporation

Government of Uganda [GOU]. (1997). *The Local Government Act*, 1997. Ministry of Local Government. Entebbe: Government Printer.

Government of Uganda [GOU]. (1998). *The Forest Reserves (Declaration) Order* 1998. Statutory Instrument Supplement No. 63 of 1998. *The Uganda Gazette* No 56, September 1998, Kampala.

Government of Uganda [GOU]. (2003). *The National Forestry and Tree Planting Act*. Acts Supplement No.5. The Uganda gazette, No. 37 Vol XCVI. Entebbe: Uganda Publishing and Printing Corporation.

Jacovelli, P. & Carvalho, J. (1999). The private forest sector in Uganda: Opportunities for greater involvement: A report for Uganda Forest Sector Review under the Forest Sector Umbrella Programme.   Uganda Forest Sector Co-ordination Secretariat, Kampala.

Kothari, A., Singh, N. & Surin, S. (1996). *People and protected areas. Towards participatory conservation in India.* New Delhi: Sage.

Kugonza A., Buyinza, M & Byakagaba P. (2009). Linking Local Community Livelihoods and Forest conservation in Masindi district, North western Uganda. *Research Journal of Applied Sciences* 4(1): 10-16.

Malla, Y.B. (2000). 'Impact of community forestry policy on rural livelihoods and food security in Nepal', *Unasylva* 51(202): 37–45

MWLE [Ministry of Water Lands and Environment]. (2001). *The Uganda Forest Policy*. Republic of Uganda, Kampala.

MWLE [Ministry of Water Lands and Environment]. (2002). The National Forest Plan. Republic of Uganda, Kampala.

MWLE [Ministry of Water Lands and Environment]. (2003). Guidelines for Implementing Collaborative Forest Management in Uganda. Produced under the EU's Forestry Programme (FRMCP). Republic of Uganda, Kampala.

National Biomass Study. (2003). Technical report for the period 1996-2002. Forest Department, Ministry of Water, Lands and Environment, Kampala.

NFA [National Forestry Authority] (2003). Guidelines for implementing Collaborative Forest Management in Uganda, Republic of Uganda, Kampala

OCA. 2006. Organizational Capacity Assessment (OCA) of CBOs under EMPAFORM Programme. Final report, Kampala Uganda.

Poffenberger, M. & McGean, B. (1996). *Village Voices, Forest Choices: Joint forest management in India*. New Delhi: Oxford University Press.

Scherl, L.M., Wilson, A., Wild, R., Blockhus, J., Franks, P., McNeely, J.A. & McShane, T.O. (2004). Can Protected Areas Contribute to Poverty Reduction? Opportunities and Limitations. Chief Scientist's Office Report, IUCN, Gland

Scott, P. (2000). Collaborative Forest Management-The Process. Paper presented at the National Workshop on Community Forestry Management, Kampala Uganda.

Turyahabwe, N. & Tweheyo, M. (2010). Does ownership influence forest vegetation characteristics? A comparative analysis of private, local and central government forest reserves in central Uganda. *International Forest Review*, 12(4): 220-238.

Turyahabwe, N. & Banana, A.Y. (2008). An overview of history and development of forest policy and legislation in Uganda. *International Forestry Review, 10 (4): 641-656.*

UWA [Uganda Wildlife Authority]. (2001). Collaborative Management within Uganda Wildlife Authority. Report of the senior staff retreat held at Ranch on Lake Hotel, Kampala 2nd June 2001.

Victor, M. (1996). Income generation through community forestry. RECOFTC Report 13. RECOFTC Kasetsart, University, Bangkok.

Wild, R.G. & Mutebi, J. (1996). Conservation Through Community Use of Plant Resources. Establishing Collaborative Management at Bwindi Impenetrable and Mgahinga National Parks, Uganda. Working paper No.5. Division of ecological sciences, People and plants. UNESCO, Paris.

Wily, L. A. (1998). Villagers as forest managers and governments "learning to let go". The case of Duru Haitemba and Mgori Forests in Tanzania. Forest Participation Series, No 9. International Institute for Environment and Development, London, UK.

Wily, L.A. & Dewees, P. (2001). From Users to Custodians: Changing Relations between People and the State in Forest Management in Tanzania. World Bank Policy Research Working Paper No. 2569. Washington DC.

Wily, L. A. (2002). *"Participatory Forest Management in Africa*: An Overview of Progress and Issues." Second International Workshop on Participatory Forestry in Africa, FAO, Rome, pp31-58.

Wily, L. A. (2003). Forest Governance Lessons from Eastern and Southern Africa. A presentation to the AFLEG Ministerial Conference, Yaounde, Cameroon, October 13-16, 2003.

Part 2

America

4

Sustainable Forest Management of Native Vegetation Remnants in Brazil

Lucas Rezende Gomide et al.*
Federal University of Lavras
Brazil

1. Introduction

A region's species diversity is an important factor, resulting as a component of social and economical development when used wisely. The correct commercialization of a region's natural resources guaranties the preservation of local culture and habitat maintenance by means of the obtained income. Hence, the idea of sustainability arises, a widespread theoretical theme which is beginning to gain force in Brazil's consumer market.

The principal conceptual shift was the erroneous notion that timber resources from forests are inexhaustible, since the processes of recomposition/restoration naturally occur after exploration. Indeed a system is capable of regeneration, but this is tied to a series of factors that are usually not respected in areas illegally explored. According to a conference realized in Melbourne by Raison et al. (2001), the concept of sustainability must encompass social and economic conditions such as: respect the forest growth rate; legislation based control; productive capacity; ecosystem's health and vitality; soil and water resource protection; carbon balance and preservation of biological diversity.

Under this scenario, Brazil presents great potential for the use of its natural resources. This is due to the country's vast territorial extension (8.5 million km^2) and high diversity of recurrent vegetation physiognomies. The country possesses about 5.2 million km^2 of forest land (60% of its territory), of this total, 98.7% consists of natural forest formation and 1.3% of planted forests. The forest types found in Brazil can be classified as Cerrado (Brazilian savanna), Amazônia (tropical rainforest), Mata Atlântica (Atlantic rainforest), Pantanal (wetlands) and Caatinga (semi-arid forest) as well of transition areas which promotes a mixture of habitats. In many cases, the deforestation of these environments is associated with illegal logging practices coupled with agriculture and cattle-raising. The damage caused by this include modifications of the carbon cycle and consequential rise of CO_2 emissions; forest fragmentation; alteration of the hydraulic cycle; species extinction; rural exodus and loss of local fauna and flora diversity.

Possibly the most logical use of these forests is the application of sustainable forest management for wood production destined for fire wood, charcoal and logs for industrial purposes. The motives for this strategy are evident, involving aspects attached to the reduction

* Fausto Weimar Acerbi Junior, José Roberto Soares Scolforo, José Márcio de Mello,
Antônio Donizette de Oliveira, Luis Marcelo Tavares de Carvalho, Natalino Calegário and
Antônio Carlos Ferraz Filho
Federal University of Lavras, Brazil

of environmental impacts during exploration; preservation of areas not economically productive; adequacy to local productive capacity by the quantification of present available stock; reduction of costs and rise of income; achievement of continued production; market expansion through forest certification; compliance of current laws; creation of employment opportunities and most importantly to legalize the activity. Forest management can be understood as the administration of a forest resource by a set of principals, techniques and norms. Its objective is to organize the actions necessary to determine production factors and to control its efficiency and productivity in order to reach pre determined objectives. According to Blaser et al. (2011) in International Tropical Timber Organization (ITTO) Brazil has presented great advance in the sustainable management of natural forests, expanding the areas subject to these practices as have other countries such as Peru, Malaysia, Gabon and Guiana. A few explications that account for this expansion are: enhancement of technologies used for forest monitoring; improvement of administration systems; forest certification and stricter demands from consumer markets concerning timber origin.

Amaral and Neto (2005) comment that sustainable forest management has conquered increasing space as an alternative for Latin Americas' rural communities, driven by governments, donors, NGOs and community organizations. According to the authors, the main problems encountered in the implementation of forest management in the Amazon are: (a) establishment of mechanisms for land regularization; (b) strengthening of local social organization; (c) credit access; (d) forestry technical assistance and (e) the need for market access mechanisms. These problems can easily be extrapolated to the other physiognomies besides the Amazon in Brazil.

Law 11.284/2006 drafted by the Brazilian Government regulates forests management in public areas; creates the Brazilian Forest Service (SFB) as the regulator agency of public forest management and at the same time the promoter of forestry development; creates the National Forest Development Fund to promote technological development, technical assistance and incentives for the development of the forestry sector. The law also defines three types of management: (a) Conservation units for forestry production; (b) Community Use and (c) Paid forest concessions in Conservation Units of Sustainable Use and in public forests. Forest concession in Brazil is still a new experience.

A sustainable cutting plan is only feasible when it respects the time needed to close the exploration cycle, i.e. the time it takes the forest to grow biomass to equal before harvest levels. The amount of biomass required determines the duration of the cycle, since it depends on the forest growth rate. Thus, an increased growth rate through silvicultural treatments promotes shorter cutting cycles, and consequently a lower demand of forest area, reducing impacts on other forest areas. For example, the cutting cycle (polycyclic) in the Amazon can be 30 years long considering a selective exploration of marketable species, removing a maximum of 30 m³/ha (IBAMA, 2007), which could represents up to 5 trees/ha (Putz et al., 2008). The determination of the cycle duration depends on the forest type and its growth rate (Braz, 2010), this information is presented in the Brazilian legislation.

2. Case studies

2.1 Savanna (Cerrado)
Cerrado is an important Brazilian biome present in the central area of Brazil (Figure 1), formed by a high number of endemic flora and fauna species, and considered a world biodiversity hotspot. According to IBAMA (2010), the biome presents more than 10,000

species of plants, 837 species of birds, 67 genera of mammals, 150 species of amphibians and 120 reptile species, of which 45 are endemic.

It is estimated that the Cerrado biome occupies an area of 203 million hectares, with about 66.4 million hectares remaining, with a population of 29 million inhabitants (SFB, 2010). The vast majority of the population lives in urban areas, the other fraction in rural areas, whose main activity is agriculture (soy, rice, wheat and livestock). Reforestations of *Eucalyptus* spp. can be found, intended for the production of coal, pulp, sawn wood among other uses.

The Brazilian Cerrado's colonization process was stimulated by the government, starting from the 60's, where the development of intensive agriculture, mineral production and extraction of native vegetation was encouraged. This process was made possible by the construction of new highways. During this period, the Cerrado became target of deforestations and fires, being gradually replaced for other uses, and in more extreme cases, areas were abandoned after intense degradation. For example from 2002 to 2008 the deforestation area was 85,074 km^2 (SFB, 2010).

Fig. 1. Spatial distribution of the Cerrado biome in Brazil (IBGE, 2004).

Despite the great interest in the use of these environments for farming, Cerrado's soil has high acidity (aluminum saturation) and low chemical fertility (phosphorus, potassium, calcium and magnesium), with average annual rainfall of 1500 mm (ranging from 750 mm to 2000 mm) and of concentrated nature. The average annual temperature fluctuates between 21°C to 25 °C. However, due to its flat topography, its lands became valuable since low water deficits in dry periods and low fertility can be controlled through modern technology. The biome's zone of occupation presents an intense river network, which contributes to the formation of the main Brazilian rivers, such as the São Francisco, Amazon, Tocantins, Paraná and Paraguay rivers. The water produced is used for human consumption, industrial ends, agricultural and electric power generation.

The Cerrado interacts with other types of Brazilian biomes, such as the Amazon rainforest, Atlantic forest, the Pantanal and Caatinga, creating diverse ecotones. However, the Cerrado has unique and peculiar features. In general terms it can be considered a transition between forests and grasslands, presenting gradual reductions in biomass and arboreal/shrub density in the landscape, possessing plants with deep root systems as a survival strategy in times of drought. In addition, it features a grass cover that receives a high intensity of light, depending on the arboreal/shrub density of each location.

Fig. 2. Examples of the variation amongst physiognomic types existent in the Cerrado in relation to tree size and density.

The vegetation's structure and size are determined by variations in soil type, climate, altitude, hydrology, anthropic actions and natural impacts such as forest fires. As a result the vegetation presents physiognomies varying from forest formations to grasslands. The forest formations are represented by the Cerradão (dense wooded savanna), Mata seca (semi-arid savanna) and Mata ciliar (riparian forest). The intermediate formations are predominantly Cerrado *Sensu Stricto* (wooded savanna), Veredas, Palmeiral (Palm forests) and Parque Cerrado (parkland). The grassland formations are composed of campo sujo, campo limpo (shrub savannas) and campo rupestre (rupestrian fields). These vegetation types are listed in the same order of biomass reduction. An illustrative example of the existing variations can be seen in Figure 2 and Table 1 presents a quantitative overview of stand characteristics for different Cerrado types. According to Scolforo et al. (2008a), the following species can be classified as the most recurrent in the Minas Gerais cerrado biome: *Astronium fraxinifolium* Schott ex Spreng.; *Caryocar brasiliense* Cambess.; *Copaifera langsdorffii* Desf.; *Qualea grandiflora* Mart.; *Qualea multiflora* Mart. and *Roupala montana* Aubl.

Due to the structural characteristics of native Cerrado trees, timber production is basically converted into charcoal, whose purpose is to supply steel and construction companies in an indirect way. According to statistics of the forestry sector, approximately 50% of the charcoal used in the Brazilian market comes from native areas, where a fraction of the production is the result of illegal deforestation. In the 80's the amount of charcoal from native areas was 85% (AMS, 2007). According to the Brazilian Silviculture Society - SBS (2008), in 2007 the Brazilian charcoal consumption of native origin was 11.88 million tons. On the other hand, the multiple use of the Cerrado can be stimulated, in view of the possibility of production of resins, barks, seeds, flowers, fruits and handicrafts, contributing to the maintenance of the biome.

Stand characteristics	Cerrado types		
	shrub savannas	Wooded savanna	Dense wooded savanna
Number of areas inventoried	5.0	57.0	5.0
Quadratic mean diameter (cm)	11.2	10.2	12.0
Mean height (m)	4.3	5.1	7.4
Average tree number (N/ha)	370.4	1168.9	1626.8
Basal area (m²/ha)	3.5	9.4	18.3
Aboveground wood volume (m³/ha)	17.7	48.5	129.0
Aboveground carbon stock (t/ha)	5.0	14.3	35.1

Table 1. Stand characteristics for different cerrado types in Minas Gerais State, where the data is provided for trees with DBH greater or equal to 5cm (Scolforo et al., 2008a).

2.1.1 Sustainable forest management of the Cerrado

In recent decades Brazil has structured and regulated the forestry sector, with the primary purpose of supplying the country's domestic needs, and subsequently the foreign market. This period was marked by government tax incentives and stimulus to reforestation. Strategic planning allowed for market advancement and expansion, driving the development of technologies to replace native wood consumption.

The era of large *Pinus* spp. and *Eucalyptus* spp. reforestation allowed for the gradual substitution of native wood use, contributing to the protection and preservation of remaining natural areas. Even with the decreased wood use from native forests, the Cerrado vegetation is still being exploited, due to low costs of wood produced. According the Brazilian Forest Service - SFB (2010) the total wood volume in the Cerrado is 870 million m^3, with an above-ground biomass of 496 million tonnes. Starting from this available stock it is necessary to draw plans and rules for its exploration, define the optimum cutting cycle without causing negative impacts to local biodiversity, and at the same time promote the increase of biomass.

In the past, and still nowadays in some regions, there were not any technical or scientific procedures to apply the efficient management of this vegetation, where the land owner established the traditional process that combined fire and deforestation for conversion into agricultural areas across the landscape (land use change).

The knowledge of the optimum cutting cycle is a predominant factor for the administration of these forests. Through its determination a detailed and exact management plan can be prepared, informing the ideal level of wood removal to ensure sustainability over time. Allied to this information, the knowledge of floristic composition correlated with environmental and spatial factors enables a holistic understanding of the system, contributing to the development of new silvicultural techniques and exploration criteria.

Forest management seems to be a good alternative in combating rampant deforestation, contributing to the reduction of native vegetation conversion in to grasslands, agriculture and degraded areas. The activities to be considered in a sustainable management should include: available forest resource inventory with a characterization of its structure and forest site; identification, analysis and minimization of environmental impacts; study of technical, economical and social viability of the project; adoption of forestry exploration procedures that minimize ecosystem damage; checking if the remaining stock is sufficient to ensure sustained production; and the adoption of an appropriate post exploration silvicultural system.

The sustainable forest management of the Cerrado case study is part of a set of experiments and research developed by the Federal University of Lavras and its Forest Science Department, dating back from the 80's until the present. The studies were conducted in the Minas Gerais State with reference to the works of Lima (1997), Mello (1999), Mendonça (2000), Oliveira (2002), Oliveira (2006a), Oliveira (2006b), and Scolforo et al. (2008a).

The main issue in vegetation management is the degree of exploration or cutting intensity to be applied, because the lack of this information leads to losses to the environment. Thus, a study was conducted in a Cerrado *Sensu Stricto* in the State of Minas Gerais (municipality of Coração de Jesus). Six intensities of basal area removal (0%, 50%, 70%, 80%, 90% and 100%) were tested and subsequent evaluation in 1986 (year of implementation of the experiment), 1996, 1998 and 2004. The experiment was installed in an area of 30 ha being applied 6 treatments/removal criteria and 5 repetitions per treatment, with sample plots of 600m². The individuals were identified botanically, and the criteria for measurement was a circumference at breast height (CBH) $\geq$ 15.7 cm. The purpose of the study was to verify the influence of the cutting intensity in recovery time of number of individuals, volume and basal area, as well as the behavior of the tree species diversity. Due to initial variation between treatments (1986), a correction factor was applied permitting comparison of the results in different periods. The results can be seen in Table 2. A gradual increase in the number of individuals occurred in all treatments after exploration. This fact was expected since Cerrado plants feature a great sprouting capacity, as noted in the works of Ferri (1960), Rizzini (1971), Thibau (1982), Toledo Filho (1988), among others.

Removals in basal area of 70% and 100% presented higher plant density in relation to the control treatment 12 years after intervention, for the other treatments this occurred after 18 years. The response in tree numbers suggests the mentioned cutting cycles ages. In contrast, basal area growth assumes a different behavior from the previous variable because it depends on the development capacity of each species. Thus, up to 1998 there was an accentuated basal area increase of all removal intensities, with variation ranging from 43.05% (80%) up to 90.62% (100%), surpassing the values presented before intervention. The control treatment presented the largest increase, followed by the clear cutting. However 18 years after installation of the experiment (in 2004), clear cutting (14.75 m^2/ha) surpassed the control treatment (12.91 m^2/ha). This demonstrates the great recovery capacity of the Cerrado after its structure has suffered intervention. Volume behaved in a manner similar to basal area. As far as a constant wood volume production is concerned, the results show that for 100% removal of basal area a silvicultural cutting cycle of 12-16 years allows the Cerrado to return to pre-intervention basal area and volume values in the studied area. In the Cerrado, stump and root sprouts are responsible for the larger part of the natural regeneration, when compared to seed rain dispersal (Durigan, 2005). This way, after exploration it is expected that the plants sprout followed by diameter growth, and then migrate to higher diameter classes.

Variable	Year	Treatments (basal area removal %)					
		0	50	70	80	90	100
Number of trees/ha	1986	1716	1716	1716	1716	1716	1716
	1996	1933 b B	1520 a A	1769 a A	1707 a A	1655 a A	2155 c A
	%	12.65	-11.43	3.09	-0.53	-3.58	25.56
	1998	1993 a B	1803 a B	2007 a B	1976 a B	1915 a B	2477 b B
	%	16.14	5.07	16.98	15.15	11.61	44.35
	2004	1820 a A	1832 a A	2044 b B	2011 b B	1853 a A	2550 c B
	%	6.06	6.76	19.11	17.17	7.97	48.62
Basal area (m^2/ha)	1986	6.72	6.72	6.72	6.72	6.72	6.72
	1996	12.94 c A	9.85 b A	9.66 b A	8.07 a A	8.15 a A	10.72 b A
	%	92.56	46.63	43.8	20.1	21.33	59.49
	1998	13.43 b A	11.01 a A	10.32 a A	9.61 a B	9.71 a B	12.81 b B
	%	99.85	63.9	53.6	43.05	44.48	90.62
	2004	12.91 b A	12.03 b B	11.26 a B	10.7 a C	10.94 a C	14.75 c C
	%	92.09	78.97	67.56	59.22	62.85	119.51
Volume (m^3/ha)	1986	21.61	21.61	21.61	21.61	21.61	21.61
	1996	44.64 c A	33.52 b A	32.81 b A	26.79 a A	27.1 a A	36.71 b A
	%	106.05	55.1	51.81	23.95	25.41	69.86
	1998	46.32 c A	37.79 b B	35.25 a A	32.62 a B	32.98 a B	44.18 b B
	%	114.34	74.87	63.13	50.94	52.6	104.44
	2004	44.53 b A	41.43 a B	38.68 a B	36.64 a B	37.54 a C	50.75 b C
	%	106.03	91.71	79	69.56	73.68	134.82

Table 2. Number of trees, basal area and volume per hectare in the different measurement dates (1986, 1996, 1998 e 2004). Equal letters in the same column (capital) or line (lower case) indicate that the values are equal, according to the Scott-Knott means grouping test at a 95% confidence level.

An economical analysis was carried out to verify the viability of the removal intensities in relation to cutting cycles studied. The method used was the Net Present Value (NPV) considering an infinite planning horizon. Thus, a greater NPV indicates a more lucrative project. Wood production costs were based on the year 2005 (USD 1.00 = R$2.165), being calculated by hectare of forest, and included depreciation costs (structures, machines, equipments and tools), administration, taxes, social charges, labor costs, alimentation, personal transport, among others. An annual discount rate of 6% was used. The future cycles' productivity was estimated through a Markov chain type prognosis system (OLIVEIRA, 2006a). Table 3 presents NPV results considering cutting cycles of 10, 12 and 18 years.

Basal area reduction	Cutting cycle		
	10	12	18
50%	-19.62	-26.72	-55.45
70%	20.04	2.15	-38.12
80%	10.23	9.22	-32.02
90%	29.16	27.12	-19.85
100%	107.88	99.53	26.39

Table 3. Net present value (US$/ha) per treatment for cutting cycles of 10, 12 and 18 years.

The smallest economic viability was obtained by removing 50% basal area and the largest in 100% removal, where longer cutting cycles (12, 16 and 18 years) made the lower intensities of basal area removal become unviable. Despite the inviability of certain treatments, charcoal prices suffer great fluctuation in short periods of time, which may make them economically viable. This variation must be closely analyzed, since a price increase in the end product can elevate the number of projects that are economically viable.

A high floristic diversity exists in the study area, with Shannon index values ranging from 3.13 to 3.23. The diversity between treatments was very similar, and the small differences found are perfectly normal in a biological system. These high values indicate that the sprouting levels are sustainable for the maintenance of local tree species diversity of the trees stratum. In summary, the family Fabaceae presented the largest number of species in the six treatments. The family Vochysiaceae presented the largest number of individuals in all treatments. The species *Qualea grandiflora*, *Qualea parviflora*, *Eugenia dysinterica*, *Qualea multiflora*, *Eriotheca pubescens*, *Magonia pubescens*, *Platymenia reticulata*, *Hymenaea stigonocarpa*, *Dimorphandra mollis*, *Caryocar brasiliense*, *Annona crassifolia*, *Buchevania tomentosa*, *Tabebuia ochracea*, *Aspidosperma macrocarpum*, *Sclerolobium aureum*, *Astronium fraxinifolium*, *Bowdichia virgilioides*, *Annona crassifolia*, *Acosmium subelegans*, *Hyptidendron cana* and *Maclinea clausseniana* are those which easily regenerate naturally, possess high basal area and are present in the lower, middle and upper stratum of the Cerrado. These characteristics indicate that these species present potential to be managed.

Figure 3 shows the cluster analysis considering a floristic matrix that included all basal area removal treatments. According to the dendrogram there is a high connection between most of treatments by 25% cut level, which basically formed 3 floristic groups. The major cluster grouped 0% and 100% basal area removal for all survey periods (1996, 1998 and 2004), which suggest a lower risk of tree species loss up to 2004. Similar tendencies were obtained by Souza et al. (2011) who related the impact of forest management in a Cerrado. The authors demonstrated that the floristic diversity increased over the years in all areas subject to vegetation removal, and did not differ statistically. They also concluded that eleven years after intervention, tree diversity change occurred in all treatments, inclusively in the unmanaged control treatment.

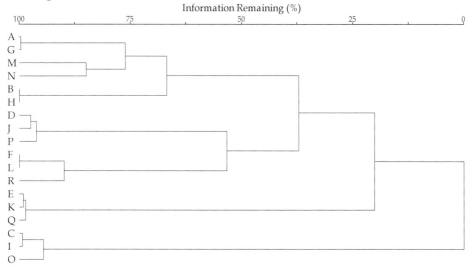

Fig. 3. Floristic dendrogram considering all treatments of percentage basal area removal and measurement dates, where: A - 0% (1996); B - 50% (1996); C - 70% (1996); D - 80% (1996); E - 90% (1996); F - 100% (1996); G - 0% (1998); H - 50% (1998); I - 70% (1998); J - 80% (1998); K - 90% (1998); L - 100% (1998); M - 0% (2004); N - 50% (2004); O - 70% (2004); P - 80% (2004); Q - 90% (2004) and R - 100% (2004).

The detreded correspondence analysis (DCA) was applied to support the discussions and interpretation of floristic recovery after intervention in 1986. The DCA was developed by Hill & Gauch (1980) being a multivariate method that contributes to analyze the connection among environment, species and other variables. Following the graphic result (Figure 4) the treatments 0% and 100% basal area removal are still close (dotted line circle), which suggest no significant loss of tree species after 1986. The rapid regeneration (sprouting) after cutting was observed and desirable for establishing the recovery of the physiognomy.

Considering only the tree stratum, it was noted in all treatments that 12 years after intervention the remaining vegetation presented diversity indexes similar to those found in Cerrado vegetation in other regions of Brazil not subject to intervention. According to Scolforo et al. (2008a) Minas Gerais state possesses potential for the sustainable timber management of the Cerrado, mainly in the North/North-West as shown in Figure 5.

Currently, the recommended option for sustainable management of the Cerrado is the clear cutting in strips, which consists in removing 100% of individuals (excepting tree species prohibited by legislation), realized in a maximum of 50% of the area destined for exploration (Figure 6). The explored and unexplored strips must be alternated, where the unexplored strips must have greater or equal dimensions in relation to the explored strips. The objective is to allocate a greater protection of the environment by preventing its degradation, as well as the possibility of seed dispersal in the explored area, therefore helping to promote natural regeneration, which as stated earlier is achieved primarily by sprouting.

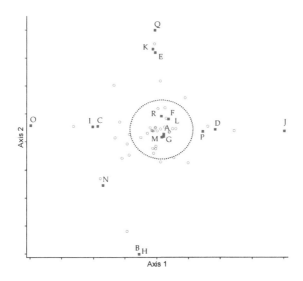

Fig. 4. The DCA analysis considering all treatments of percentage basal area removal and measurement dates, where: A - 0% (1996); B - 50% (1996); C - 70% (1996); D - 80% (1996); E - 90% (1996); F - 100% (1996); G - 0% (1998); H - 50% (1998); I - 70% (1998); J - 80% (1998); K - 90% (1998); L - 100% (1998); M - 0% (2004); N - 50% (2004); O - 70% (2004); P - 80% (2004); Q - 90% (2004) and R - 100% (2004). The blue squares represent the treatments and the open blue circles are the species.

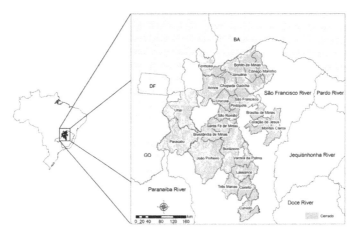

Fig. 5. Minas Gerais State's regions with potential for the application of sustainable forest management of the Cerrado.

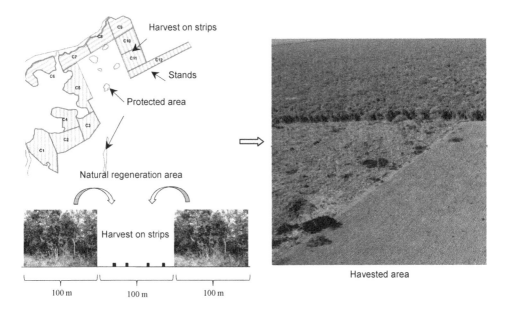

Fig. 6. Clear cutting in alternate strips scheme applied in the Brazilian Cerrado (Scolforo et al., 2008a).

The joint analysis of these studies show that the exploration of the Cerrado can be economically viable, being mainly dependent on the level of intervention, cutting cycle, productivity, land costs and market variables. The studies related here were focused in the tree stratum of the forests and as such consist of preliminary studies of the sustainability of cerrado management. Further studies on how forest management affects the fauna and other aspects of the flora (e.g. trees with DBH smaller than 5cm and herbaceous stratum) are required to provide more information on the impacts of forest management on the Cerrado. Multiple use of the Cerrado is an option, such as the already marketed species of *Dimorphandra mollis* (favela), used in the pharmaceutical industry, and the food products derived from pequi (*Caryocar brasiliensis*) and baru (*Dipteryx alata*).

2.2 Candeia tree

Formation processes of an environment over thousands of years gradually promote species selection, encouraging the development of strategic mechanisms to overcome the difficulties imposed by each habitat. The spatial distribution of species in a landscape presents a selective character, which added to between species competition, directs the occurrence and dynamics of a forest. As such, candeia (*Eremanthus* spp) predominantly occupies high altitude field areas, being quite recurrent in the State of Minas Gerais (Brazil), as shown in Figure 7.

Candeia is of the Asteraceae family, an ecotone species typical of transitional areas between wooded and grasslands. Even thought it presents several characteristics of pioneer species such as: production of large quantities of seeds, seeds dispersed by wind, high-density natural regeneration in open gaps, it must not be framed as such since its lifespan can exceed 50 years. There are several species of candeia, however *Eremanthus erythropappus*

(DC.) Macleish and *Eremanthus incanus* (Less.) Less are of greater economic importance and of higher occurrence in Minas Gerais. Candeia density ranges between 875 to 1536 trees per hectare (Scolforo et al., 2008b and Scolforo et al., 2008c), although values up to 50,000 trees per hectare (Andrade, 2009) have been reported for young candeia areas undergoing natural regeneration after exploration.

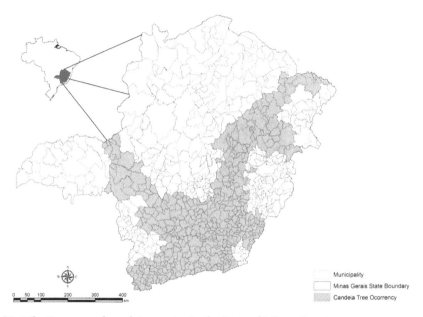

Fig. 7. Distribution map of candeia species in the State of Minas Gerais.

Eremanthus erythropappus develops predominantly in high altitude fields. Candeia is a monodominant species, such that is not uncommon to find small patches of forests formed exclusively by the species. An interesting feature of this species is its development in sites with shallow soils of low fertility. Its occurrence is heavily influenced by altitude, occurring 800 meters above sea level, where the highest abundances are found between 1,000 and 1,500 meters. Candeia develops in places where it would be difficult to employ agricultural crops or even other forest species.

Candeia possesses multiple uses, usually its wood is either used as fence posts due to its natural durability, or as a raw material from which essential oil is extracted. The essential oil's active ingredient is Alpha-bisabolol, employed in the manufacture of medicines and cosmetics such as creams, tanning lotions, sunscreens, vehicle for medicines, besides being used for prophylactic purposes and skin care of babies and adults, among others.

Native areas of candeia display decreasing diameter distribution, with trees typically reaching up to 32.5 cm. A candeia forest usually presents a average diameter of around 15 cm. However, individuals have been found that reached up to 62.5 cm. Average heights are between 6 and 7 meters. The height of the largest trees is around 9.5 to 10 meters, although an individual has been found with 16.5 meters, inside a semideciduous seasonal forest.

Candeia's trunk has thick bark presenting many fissures, newer branches have smoother bark. The characteristics of the leaves and inflorescence facilitate the identification of the species even at a distance (Figure 8). The coloration of the wood is white or grayish with a darker cross grain. Its basic density averages 675 kilograms per cubic meter of wood.

Fig. 8. Example of a representative candeia individual and its floral structures (Andrade, 2008).

The candeia *Eremanthus incanus* (Less.) Less. is an arboreal species that when adult presents average height between 5 and 7 meters, average the diameters between 10 and 12 cm, where some individuals can reach up to 20 or 25 cm. Its stem is grayish-brown, with thick bark and few branches. It occurs between 550 and 1100 m altitude, in the Cerrado, in secondary forest or in the Caatinga. Its use is basically for the production of fence posts, since it has low productivity of Alpha-bisabolol oil.

2.2.1 Sustainable forest management
Exploration of candeia populations, in the form of sustainable forest management, is only authorized by the State's environmental agency for fragments with occurrence of at least 70% of individuals of the species *Eremanthus erythropappus* or *Eremanthus incanus*. This restriction is derived from the need for restoration of the area through natural regeneration, where areas with greater dominance of the species are more likely to recover and return to the initial stage. Beyond this point, the guarantee of the sustainability of these populations is correlated with the quality of the harvest project and compliance with the State's environmental laws.

Barreira (2005) studying the genetic diversity of candeia populations (*Eremanthus erythropappus*), in which she sought to quantify and compare the intra-population genetic variation and reproductive systems of candeia before and after exploration, noted that the species is suitable for management without loss in genetic diversity, as long as 100 individuals/ha are preserved as remnants. The study also showed a strong spatial genetic structure in the population, where trees in a 200 meters radius presenting some degree of kinship, with a 95% probability.

In addition to these efforts, the owner of the area under management must present a map of his farm containing the areas to be managed as well as the areas of Legal Reserve (20% of the total farmable land) and Permanent Preservation Areas (areas adjacent to waterways, with declivity greater than 45 °, hill tops and areas 1800 meters above sea level).

2.2.2 Evaluation of the legal viability of forest use
A forest inventory is the starting point to gain knowledge about a particular forest, in which a set of sample plots are distributed to quantify the variable of interest. In the case of legal viability studies of explored forests, a systematic sampling procedure is preferred with a minimal area of 600 square meters and maximum of 1000 square meters. Instead of a forest

inventory, another option is the conduction of a census, in which all the individuals above 5 cm diameter are counted, distributed in diameter classes with amplitude of 5 cm. The operation is performed with the use of a diameter fork, which speeds up the field operation obtaining the number of individuals/ha of candeia. Sustainable exploration is legality permitted only when more than 70% of individuals present in the forest fragment are composed of candeia. In more environmentally fragile areas only the census is allowed. Local volumetric quantification can be obtained by scaling and fitting regression models, or through specific equations already adjusted to various parts of the State.

2.2.3 Silvicultural systems

A balanced silvicultural system guides the harvest and post harvest operations, in order to ensure the restocking of the location in the shortest time possible. The concern with the correct manner to intervene in a forest must be predominant and must be considered and planned before the operation. The sustainability of the project depends on the efficiency of the remaining individuals' response, as well as the quality of the natural regeneration. Among the various silvicultural systems available for tropical forest management, only those that can be applied to the management of candeia are presented. They were defined based on scientific studies, including a) Clear cutting in strips system: applied in homogeneous candeia fragments with alternating strips each 20 m wide, following the isoline curves of the land; b) Seed tree with natural regeneration system: seed trees are retained in a maximum distance of 10 meters, where the retained trees must be phenotypically superior and with the greatest crown diameter possible; c) Group selection system: gaps are opened with a maximum diameter of 15 m, within these gaps a selective exploration of individuals is conducted, preserving all the border trees (IEF, 2007). Figure 9 presents a scheme containing the spatial distribution of exploration in a candeia fragment.

In the past, farmers used selective cutting as a form of exploration. This strategy is not in favor of pure candeia fragments, since natural regeneration depends on great luminous intensity. Therefore, the strategy undertaken was to increase the canopy gap between the remaining trees in the forest according to better results for the following silvicultural systems: clear cutting in strips system, seed tree with natural regeneration system and group selection system (Scolforo et al., 2008b). To guarantee a high intensity of natural regeneration the seeds must be in contact with the ground, receiving direct sunlight and rainwater. Dispersion occurs in the months of August through October and the seeds have no dormancy problems.

After the harvest, the remaining trees must have diameter at breast height equal to or superior than 5 cm. In the Group selection and Clear cutting in strips systems a maximum of 60% of the number of candeia trees are allowed to be explored, distributed in the different diametric classes. For the Seed tree with natural regeneration system the maximum is of 70%.

A peculiar point of the exploration process is the withdrawal of epiphytes (compulsory and optional) which may exist in the area. They must be quantified and transplanted to nearby areas as similar as possible to the area under management, with the purpose of preservation. After the removal of the woody material and immediately before seed dispersal of the remaining candeias that serve as seed trees (between the months of August and October), the area explored must receive silvicultural treatments with the purpose of promoting the germination of a large contingent of candeia seeds to ensure the sustainability of production and management (IEF, 2007).

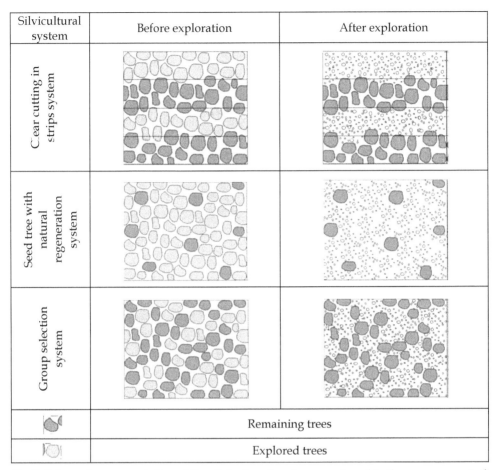

Silvicultural system	Before exploration	After exploration
Clear cutting in strips system		
Seed tree with natural regeneration system		
Group selection system		
	Remaining trees	
	Explored trees	

Fig. 9. Schematic representation of the silvicultural systems employed in the management of candeia fragments.

2.2.4 Exploration and natural regeneration

In most cases, the locations of candeia fragments are in montane regions of difficult access, usually with improper inclination for mechanization. The alternative adopted, and existent for centuries, is the withdrawal of the wood by mules to a nearby road (Figure 10). The opening of roads in the interior of the forest fragment is not recommended since it can cause erosion.

In the case of candeia (*Eremanthus erythropappus* or *Eremanthus incanus*), which colonizes areas through seed rains, a post exploration strategy is to conduct its natural regeneration, as shown in Figure 11. To this end it is necessary to evaluate its presence in the area that is being managed. If high intensity regeneration is present throughout the entire area, selective thinnings should be applied for competition reduction. However, if no natural regeneration in part of the area is detected or if its intensity is not satisfactory to promote the occupation of the site, other reforestation strategies must be adopted. This way, if the number of plants is adequate or not is not only associated to the abundance of regeneration, but also to its distribution in the area. It is important to stress that any area subject to management must be protected from domestic animals and fire, this way not compromising the natural regeneration and consequently the sustained vegetation production over time.

Fig. 10. Wood removed from the forest fragment transported by mules.

(a) soon after exploration	(b) 6 months after exploration
(c) 24 months after exploration	(d) aerial view 24 months after exploration

Fig. 11. Implementation of the group selection system (a); candeia natural regeneration after 6 months (b) and 24 months after exploration (c); aerial view of the group selection system 24 months after exploration (d).

2.2.5 Post exploration conduction

A set of treatments are carried out to stimulate candeia's natural regeneration, guarantying environmental harmony and local restoration. The operations consist of withdraw only of candeia wood, i.e. not to interfere in non candeia woody vegetation that exists in the forest fragment. The practice is linked to revolving the ground in 5 to 10 cm depths. This procedure must be done in 60 cm diameter circles or in squares with 60 x 60 cm, distant about 2.5 meters of one another, without suppressing any other shrub or tree species. Therefore, their disposition may be irregular to meet this specificity of management (Figure 12).

Practices must be adopted in the conduction of natural regeneration to increase volumetric increment, involving the elimination of invasive species such as vines that may restrict the establishment of candeia, as well as thinning when competition is above the site's capacity. Typically this thinning should be performed when the larger plants present height of around 1 m.

So that there is no risk to the genetic diversity of candeia under management, for any of the systems implemented there should be a minimum number of seeds trees equivalent to at least 100 plants per hectare. In addition, it is mandatory that the managed areas be protected against the action of domestic animals and fire.

Fig. 12. Example of post exploration practices, soil scarification using a hoe.

For the evaluation of the regeneration of managed areas, Brazilian environmental agencies require Annual Technical Reports until the 3rd year after exploration, the transplanting of epiphytes and other components that characterize sustainable management.

2.2.6 Commercialized products and oil extraction

Diametric structure of a natural forest follows a negative exponential function, with a larger abundance of individuals in the smaller diameter classes. Regarding the average production of oil/stem diameter class the production is increased in larger classes, 4.042 kg for the trees between 30 and 35 cm in diameter, contrasting to 1.585 kg for trees between 5 and 10 cm in diameter. Nonetheless, the viability of exploration in smaller classes is achieved due to the large number of individuals (Pérez, 2001). The presence of oil in plants can be found in various parts (leaves, flowers, wood, branches, twigs, fruits, rhizomes), being composed of several chemical substances such as: alcohols, aldehydes, esters, phenols, terpenes and hydrocarbons. Due to the high presence of Alpha-bisabolol (on average 66.1% of its constitution) candeia's oil odor is very characteristic and unattractive to human smell. Table 4 presents the production generated by a candeia fragment to be explored, following the principles of sustainable forest management.

The process of extracting essential oil may be done by various methods, such as hydrodistillation, maceration, solvent extraction, effleurage, supercritical gases and microwave. In Brazil there are approximately 7 companies that extract candeia oil. The oil is commercialized for the production of: astringent; liquid lipstick; sunscreen; toothpaste; baby diaper rash cream; sore cream for bedridden patients; peeling cream for cleansing and circulation stimulation; postoperative cream; toning cream; body emulsion; makeup removal wet wipes; anti-acne lotion; protective hair lotion and other goods.

Diameter class	Number of trees	Total Vob (m³)	kg of oil per m³ cc	N° of fence posts
5 ⊢ 10	530.89	9.816	85.96	1114.87
10 ⊢ 15	250.00	12.290	119.20	900.00
15 ⊢ 20	71.97	6.656	74.92	374.27
20 ⊢ 25	17.52	4.572	65.48	176.91
25 ⊢ 30	4.14	1.504	13.29	52.99
30 ⊢ 35	0.96	0.488	4.86	18.44
TOTAL	875.47	35.327	363.70	2637.48

Where: Vob - total volume over bark and m³ cc – cubic meter of wood with bark.

Table 4. Estimated production of a candeia fragment after the conduction of a census.

3. Conclusion

The joint analysis of these case studies permits one to conclude that the exploration of the Cerrado and candeia in a sustainable manner can be economically viable, being primarily dependent on the level of intervention, cutting cycle, productivity, cost of land and market variables.

Due to Brazil's territorial extension and high biological diversity, Brazilian native forest sector presents great importance on a global level, contributing to the world market supply of various products; ensuring the maintenance of biodiversity and water resources; as well as climate regulation in areas of influence. However, comparing the native and planted forest sectors in the country, the latter is more developed from a technology perspective.

The multiple use of the Cerrado is an appealing option to be pursued. In the case of candeia, advances may be focused on new product development, having as base the waste generated after oil extraction. This measure will add value to the production system, and elevate opportunities for revenue generation.

After harvesting, the cutting cycle must be respected, along with the promotion of mechanisms to maximize forest growth, accomplished through silvicultural systems. Without these cares the forest will enter into stages of degradation and fragmentation, causing permanent damage to the ecosystem. The studies related here were focused in the tree stratum of the forests and as such consist of preliminary studies of the sustainability of Cerrado and candeia management. Further studies on how forest management affects the fauna and other aspects of the flora (e.g. trees with DBH smaller than 5cm and herbaceous stratum) are required to provide more information on the impacts of forest management on the Cerrado and candeia trees' fragment.

4. Acknowledgement

Many thanks to all academic students and professionals involved in the collection of the experiments data across the years, and Mr. Joseph S. Catalano Ph.D. for kindly looking over the English of the manuscript. We thank also MMA (Ministry of Environmental) and IEF (Forestry State Institute).

5. References

Amaral, P.; Neto, M. A. (2005). *Manejo florestal comunitário: processos e aprendizagens na Amazônia brasileira e na América Latina*. IEB: IMAZON, Belém, Brazil.

AMS (2007). *Anuário estatístico de 2007*. Belo Horizonte.

Andrade, I. S. (2009). *Avaliação técnica e econômica de sistemas de manejo de candeais nativos*. UFLA, Tese Doutorado, Lavras, Brazil.

Barreira, S. (2005). *Diversidade genética em população natural de Eremanthus erytropappus (DC.) MacLeish como base para o manejo florestal* USP/ESALQ. Tese doutorado, Piracicaba, Brazil.

Blaser, J.; Sarre, A. & Poore, D. & Johnson, S. (2011). *Status of Tropical Forest Management 2011*, Technical Series nº 38, ITTO, Yokohama, Japan.

Braz, E. M. (2010). *Subsídios para o planejamento do manejo de florestas tropicais da Amazônia*. UFSM. Tese doutorado, Santa Maria, Brazil.

Durigan, G. (2005). Restauração da cobertura vegetal em região de domínio do cerrado. In: Galvão, A. P. M.; Silva, V. P. da (Ed.). *Restauração florestal: fundamentos e estudos de caso*. pp.103-118. Colombo: Embrapa Florestas.

Ferri, M. G. (1960), Boletim da Faculdade de Filosofia, Ciências e Letras. *Nota preliminar sobre a vegetação do Cerrado em Campo do Mourão – PR*. No.17, pp.109-115, ISBN 0372-4832.

Hill, M. O.; Gauch, H. G. (1980). Detrendet correspondence analysis, an improved ordination tecnique. *Vegetatio*. Vol.42.

IBAMA, (2007). *Normas Florestais Federais para a Amazônia* – Brasília: IBAMA/Diretoria de Uso Sustentável da Biodiversidade e Florestas, pp.417.

IBAMA, (2010). Cerrado, 25.05.2011, Available from http://www.ibama.gov.br/ecossistemas/Cerrado.htm

IBGE, (2004). Mapa de Biomas do Brasil e o Mapa de vegetação do Brasil, 25.05.2011, Available from http://www.ibge.gov.br/home/presidencia/noticias/noticia_visualiza.php?id_no ticia=169

IEF, (2007). *Portaria nº 01, de 5 janeiro de 2007. Dispõe sobre as normas para elaboração e execução do Plano de Manejo para Produção Sustentada da Candeia (Eremanthus erythropappus e Eremanthus incanus) no Estado de Minas Gerais*. 25.05.2011, Available from http://www.siam.mg.gov.br/sla/download.pdf?idNorma=6692

Lima, C. S. A. (1997). *Desenvolvimento de um modelo para manejo sustentável do cerrado*. UFLA, Dissertação de Mestrado, Lavras, Brazil.

Mello, A. A. (1999). *Estudo silvicultural e da viabilidade econômica do manejo da vegetação do Cerrado*. UFLA, Dissertação de Mestrado, Lavras, Brazil.

Mendonça, A. V. R. (2000). *Diagnostico dos planos de manejo e o potencial de exploração da vegetação do Cerrado e Mata Seca no Estado de Minas Gerais*. UFLA, Dissertação de Mestrado, Lavras, Brazil.

Oliveira, A. D.; Mello, A. A. & Scolforo, J. R. S. & Resende, J.L.P. & Melo, J. I. F. (2002). Árvore. *Avaliação econômica da regeneração da vegetação de Cerrado, sob diferentes regimes de manejo*. Vol.26, No.6, pp.715-726, ISBN 0100-6762.

Oliveira, S. L. (2006 a). *Análise técnica e econômica do manejo sustentado do Cerrado sob diferentes níveis de intervenção*. UFLA. Dissertação de Mestrado, Lavras, Brazil.

Oliveira, M. C. (2006 b). *Avaliação dos impactos de sistemas de manejo sustentável na diversidade e estrutura da flora de um Cerrado sensu stricto.*UFLA. Dissertação de Mestrado, Lavras, Brazil.

Pérez, J. F. M. (2001). *Sistema de manejo para a candeia (Eremanthus erythropappus (DC.) MacLeish.* Dissertação de Mestrado, Lavras, Brazil.

Putz, F.E.; Sist, P. & Fredericksen, T. & Dykstra, D. (2008). Reduced-impact logging: Challenges and opportunities. *Forest Ecology and Management*, No.256, pp.1427–1433.

Raison, J. R.; Brown, A. G.; Flinn, D. W. (2001). Criteria and indicators for sustainable forest management. IUFRO Research series 7. p.462.

Rizzini, C. T. (1971). *Aspectos ecológicos da regeneração em algumas plantas do Cerrado.* In: Simpósio sobre o Cerrado, p.61-64, EDUSP, São Paulo, Brazil.

SBS. (2008). *Fatos e Números do Brasil Florestal*, Sociedade Brasileira de Silvicultura, 08.06.2011, Available from http://www.sbs.org.br/FatoseNumerosdoBrasilFlorestalpdf

Scolforo, J. R. S; Mello, J. M. & Oliveira, A. D. (2008a). *Inventário Florestal de Minas Gerais: Cerrado.* UFLA, Lavras, Brazil.

Scolforo, J. R. S.; Oliveira, A. D. & Silva, C. P. C. & Andrade, I. S. & Camolesi, J. F. & Borges, L. F. R. & Pavan, V. M. (2008b). *O manejo da candeia nativa.* UFLA, Lavras, Brazil.

Scolforo, J. R. S; Mello, J. M. & Silva, C. P. C. (2008c). *Inventário Florestal de Minas Gerais: Floresta Estacional Semidecidual e Ombrófila.* UFLA, Lavras, Brazil.

Scolforo, J. R. S.; Oliveira, A. D. & Davide, A. C. (2002). Manejo sustentado das candeias *Eremanthus erythropappus* (DC.) McLeisch e *Eremanthus incanus* (Less.) Less. Relatório Técnico Científico. Lavras. UFLA-FAEPE.

SFB. (2010). *Florestas do Brasil em resumo - 2010: dados de 2005-2010*, Brasília: Serviço Florestal Brasileiro, Brazil.

Souza, F. N.; Scolforo, J. R. S. & Santos, R. M. & Castro, C. P. S. (2011). Assessment of different management systems in an area of cerrado sensu strict. *Cerne*, Vol.17, No. 1., pp.85-93, ISSN 0104-7760.

Thibau, C. E. (1982). *Produção sustentada em florestas, conceitos metodológicos.* In: Produção e utilização de carvão vegetal, v.1, p.10-57, CETEC, Belo Horizonte, Brazil.

Toledo Filho, D.V. (1988). Competição de espécies arbóreas de Cerrado. *Boletim Técnico do Instituto Florestal*, Vol.42, pp.54-60.

5

Sustainable Forest Management in a Disturbance Context: A Case Study of Canadian Sub-Boreal Forests

X. Wei[1] and J. P. Kimmins[2]
[1]Department of Earth and Environmental Sciences
University of British Columbia (Okanagan), Kelowna, British Columbia
[2]Department of Forest Sciences
University of British Columbia, Vancouver, British Columbia
Canada

1. Introduction

In many forests, timber harvesting is displacing natural disturbance (e.g., wildfire, wind, insects and disease) as the major agent of ecosystem disturbance. There is a growing concern over the impacts of intensive timber harvesting on the long-term site productivity (Nambiar et al., 1990; Nambiar and Sands, 1993; Johnson, 1994). The significant yield decline of Chinese fir (*Cunninghamia lanceolata* [Lamb] Hook) in southern China (Yu, 1988; Sheng and Xue, 1992) and of radiata pine (*Pinus radiata* D. Don) in southeastern Australia (Keeves, 1966; Squire, 1983) and New Zealand (Whyte, 1973) after several forest-harvest rotations exemplifies this concern, and the issue has gained renewed attention as interest has grown in forest certification, biodiversity, protection, forest carbon management and sustainability.

A key issue in the discussion of sustainability is the comparability in ecological impacts between timber harvesting and natural disturbance (e.g., wildfire, insects, and disease). Much of the focus in this discussion has been on the characteristics that clear-cutting and natural disturbance have in common (Hammond, 1991; Keenan and Kimmins, 1993). However, the debate has frequently been frustrated by the lack of an adequate description of the range of ecological effects of both natural disturbance and forest harvesting. A variety of recent initiatives in forest policy in both the United States and Canada have emphasized natural disturbance processes and their structural consequences as models of forest management (Lertzman et al., 1997). However, implementing this approach is often limited by our incomplete understanding of natural disturbance regimes (Lertzman and Fall, 1998; Perera and Buse, 2004). Many studies have demonstrated importance of wildfire disturbance in forest ecosystems (Attiwill, 1994; Lertzman and Fall, 1998). If natural disturbance is fundamental to the development of forest ecosystems, then our management of natural areas should be based on an understanding of disturbance processes (Attiwill, 1994; Poff et al., 1997; Richter et al., 1997; Andison, 2000; Johnson et al., 2003)). This also highlights that ecological impacts of harvesting must be evaluated within a broad disturbance context.

In the British Columbia (BC) interior forest ecosystems that are described as having a natural disturbance type maintained by frequent stand-initiating fires, forest managers are particularly interested in understanding natural wildfire disturbances in order to maintain the ecological patterns and functions of these forests (Parminter, 1998; Kimmins, 2000; Wei et al. 2003). In these areas, wildfire disturbance is part of natural ecosystem processes, and its functions are, from a long-term perspective, part of natural variability. Lodgepole pine forest (*Pinus contorta ssp. latifolia* Engelm. ex S. Wats.) is a major type of forests in the central interior of BC, and concerns have been expressed over potential impacts of intensive timber harvesting on long-term site productivity (Kimmins, 1993; Wei et al., 1997).

Both timber harvesting and wildfire disturbances can vary substantially in size, intensity, severity, frequency and internal heterogeneity, greatly complicating comparisons of the effects of different disturbance types. The differences relate to differences in forest type, topography, timing of the disturbance, local management methods, and management objectives (Lertzman and Fall, 1998; Parminter, 1998). From a nutrient perspective, a major difference between timber harvesting and wildfire disturbance is the biomass of woody debris (WD) left in the ecosystem, and the quantity of nutrients removed.

WD, particularly coarse woody debris (CWD), has been shown to be an important structural and functional element in many forested ecosystems (Lambert et al., 1980; Sollins, 1982; Harmon et al., 1986; Spies et al., 1988). It provides a key habitat component (especially large logs) for many forms of wildlife (Reynolds et al. 1992). Studies by Harvey et al. (1981) and Harvey et al. (1987) showed that organic materials, especially humus and buried residue in the advanced stage of decay, are excellent sites for the formation of ectomycorrhizal root tips. Graham et al. (1994) used ecotomycorrhizal activity as a primary indicator of a healthy forest soil. Further, CWD may play a significant role in long-term nutrient cycling; it can be an important site for asymbiotic nitrogen fixation, and it acts as a source of slow nutrient release during its long period of decay.

This paper summarizes a series of our research and publications in lodgepole pine forests in the sub-boreal regions of British Columbia, Canada. The objectives of this study are: (1) to quantify the difference in the mass and nutrients of woody debris remaining immediately following harvesting and wildfire disturbances; (2) to evaluate long-term implications of those differences in site productivity; and (3) to determine the management strategies for achieving sustainability of long-term site productivity in lodgepole pine forests in the BC interior.

2. Study area

The study area is located west of Williams Lake in the Chilcotin plateau of interior British Columbia (52°-53°N, 123°-124°W) in the very dry and cold subzone of the Sub-Boreal Pine Spruce biogeoclimatic zone (SBPSxc). Average monthly temperatures range from -13.8°C in January to +11.6°C in July. The mean daily temperature is below 0°C from November to March. Average annual precipitation is 464 mm, of which 195 mm is snow. Soils are well drained brunisols and luvisols of sandy or sandy loam texture. Soil parent material is primarily glaciofluvial or morainal. The forest is relatively pure lodgepole pine with trembling aspen (*Populus tremuloides* Michx) in some of the newly disturbed stands. White spruce (*Picea glauca* (Moench) Voss) is the theoretical climax tree species over most of the SBPS. In the very dry and cold SBPSxc subzone area, however, the abundance of pine regeneration and the virtual absence of spruce regeneration on zonal sites suggests that lodgepole pine is the climatic climax tree species (Steen and Demarchi, 1991). Lodgepole pine grows relatively slow under such a dry and cold subzone. Average diameter and height are 17.8 cm and 14.4 m in mature stands, respectively.

Wildfire is a common, natural disturbance which cycles this type of forest about every 100 - 125 years (mean fire return interval in the SBPS Zone, J. Parminter, Ministry of Forests, Victoria, British Columbia). Other natural disturbance agents such as mountain pine beetles (*Dendroctonus ponderosae* Hopk) and dwarf mistletoe (*Arceuthobium americanum* Nutt) are common in the study area. The former can cause extensive tree mortality, and much of the harvesting in the study area was in stands where some (about 5-25%) of the dominant trees had been killed by mountain pine beetles. Dwarf mistletoe does not usually kill lodgepole pine but can cause severe losses of volume production.

There are two types of timber harvesting including stem-only harvesting (SOH) and whole-tree harvesting (WTH) applied in lodgepole pine forests in the central interior of British Columbia. WTH removes most of the above-ground woody biomass including crown materials while SOH removes most of the above-ground woody biomass but leaves crown materials on the site. Concern has focused on the ecological impacts of the removal of nutrient-rich crown materials, and in WTH such removal may result in considerably more site nutrient depletion. For this study, we included both harvesting methods.

3. Methods

A combination of field investigation with ecosystem modeling was used for this study. The purpose of the field survey is to quantify the differences immediately following wildfire disturbance and harvesting, while the ecosystem modeling is to evaluate the long-term implication of those differences in site productivity. The ecosystem model FORECAST, or its forerunner FORCYTE, has been used as a management evaluation tool in several types of forest ecosystems (Sachs and Sollins, 1986; Kellomäki and Seppälä, 1987; Wang et al., 1995; Wei and Kimmins, 1995; Morris et al., 1997; Wei et al., 2000; Seely et al., 2002; Welham et al., 2002). The model was specifically designed to examine the impacts of different management strategies or natural disturbance regimes on long-term site productivity. A brief description of the FORECAST model approach is presented in the next section; details are found in Kimmins (1993); Seely et al., (1999); and Kimmins et al., (1999).

3.1 Field investigation
WD in this study includes CWD and fine woody debris (FWD). We define CWD as woody stems ≥ 2.5 cm diameter and FWD as woody stems < 2.5 cm diameter. We investigated FWD as well as CWD as the former should account for the major difference in WD loading between SOH and WTH sites.

3.1.1 Measurements of mass of above-ground WD
Thirteen plots (five for WTH and four each for wildfire disturbance and SOH) with records of the time of disturbance were located, interspersed across the study area. Most selected plots were harvested or burned within less than 20 years ago, with an exception of the plot burned in 1961. It was not possible to locate more fire-killed plots because no records are available for the fires occurred 30 years ago, and because recent fire protection in the area has limited the number of fire disturbances.

The line intersect sampling method was employed to quantify WD volume (McRae et al., 1979). This involves 3 lines (each 30 m in length) laid out in an equilateral triangle. The triangular layout is used to minimize bias in situations where the logs are not randomly oriented and to cover the variation in WD distribution. Five triangles were randomly set up in each plot. The more details on measurements and calculations were described by Wei et al. (1997).

3.1.2 Measurements of decay of above-ground WD

Changes of wood density in WD for early and medium decay classes on both fire-killed and harvested sites were used to estimate WD decay coefficients. We assumed that WD decay follows a single-exponential decay equation (Fahey, 1983; Busse, 1994):

$$y_t = y_0 * e^{-k*t}$$

where y_t is wood density in WD after decay of t years, y_0 is initial wood density at year 0, and k is the decay coefficient. Using this equation, we estimated k values for early and medium decay classes on both fire-killed and harvested sites, which enable us to compare differences in WD decay between these two types of disturbance. No attempt was made to measure k values for advanced WD because of lack of data on years of decay in advanced WD on the study sites.

WD in the plot burned in 1961 can obviously be classified as medium and advanced decay classes after 33 years of decay, depending on the degree of contact with ground. The mean decay rate of these advanced decaying woody materials in this plot was estimated based on the above equation. This rate was assumed to be the rate for advanced decaying WD carried over from pre-disturbance forests in order to estimate the advanced WD loading in the year right after disturbances in all selected plots.

3.1.3 Estimates of below-ground WD

The stump-breast height diameter table from Omule and Kozak (1989) was used to estimate DBH values from the measured stump diameters for the SOH and WTH sites. A DBH - height equation was established in adjacent uncut stands to estimate heights of stands prior to disturbances. The validated equations and biomass component ratios from Comeau and Kimmins (1989), along with estimated H and DBH were then used to estimate above- and below-ground biomass of each plot before harvesting for harvested sites. For the fire-killed sites, these biomass parameters were estimated using the decomposition model developed in this study and the biomass component ratios from Comeau and Kimmins (1989). The below-ground biomass component includes fine and coarse roots.

3.1.4 Quantification of nutrient removals

In order to quantify the amount of nutrients removed by different disturbances, samples of the different decay and size classes of CWD and WD (from two plots of each disturbance), fresh woody materials, living needles and fresh litter (from neighboring uncut stands of above sampling plots) were collected to analyze N and P nutrient concentrations.

Total N and P losses through harvesting were estimated from data on the biomass removed and nutrient concentrations therein. The biomass removed by harvesting was calculated from the estimated total biomass before disturbances and total WD mass left after harvesting. Total N and P losses through wildfire were estimated using assumed % loss of forest floor and foliage due to burning. Because wildfires in the study area are very variable in intensity and severity of impact, a range of severities was assumed. Based on our observations, we assumed a range of wildfire severities from 25% to 85% loss of both forest floor and foliage. It was also assumed that, on average, 10% of branch mass was lost during the fire. The average loading of the forest floor (L and F layers) biomass in mature lodgepole pine forests in the study area was 14.5 ($\pm$3.3) Mg.ha^{-1}.

3.1.5 Measurement of asymbiotic nitrogen fixation rates in woody debris and forest floor

Measures of nitrogenase activity were used as an index of nitrogen fixation, based on an adaptation of the acetylene reduction assay (Hardy et al., 1973). Sampling was carried out on five occasions (August 22 and September 25, 1994, and May 15, June 18 and July 20, 1995) at the three study sites. From an analysis of historical weather data and preliminary sample tests, we believe that no nitrogen fixation activities occur from November to April due to low temperature. The more details on measurements and calculations were described by Wei et al. (1998).

3.1.6 Statistical analyses

Homogeneity of variances and normality of distributions of data sets were checked. Data that were not homogeneous (CWD, total WD and nitrogen fixation rates) were logarithmically transformed prior to analysis. Using SYSTAT version 5.0 (Wilkinson, 1990), analyses of variance (ANOV) were performed on WD variables (e.g., above-ground CWD and total WD), nitrogen fixation rates and moisture contents (%). Where there was a significant difference, means of measured variables were compared between WTH, SOH and wildfire disturbances using the Turkey test.

3.2 Ecosystem modeling
3.2.1 A brief description of the ecosystem model FORECAST

The ecosystem management simulation model FORECAST uses the hybrid simulation approach. The model employs empirical data, from sites of different nutritional quality, which describe tree and plant biomass accumulation over time and plant tissue nutrient concentrations. These data form the basis from which the rates of key processes are estimated, such as canopy function (photosynthesis), carbon allocation responses to changing resource availability (nutrients), competition-related mortality (largely competition for light), and rates of nutrient cycling. FORECAST, which accounts explicitly for changes in nutritional site quality over time caused by various simulated autogenic successional processes and types (allogenic and biogenic) of disturbance, was designed for the evaluation of forest management strategies in forests where potential net primary production is limited by nutrient availability, and in which nutrient availability is altered by management or natural disturbance events. The more details on the FORECAST model and model calibration were described by Kimmins et al. (1999) and Wei et al. (2003).

3.2.2 Defining the disturbance scenarios and sustainability indicators

Characteristics of disturbance can be best described by the frequency (return intervals for wildfire and rotation lengths for harvesting) and intensity (severity for wildfire and utilization levels for harvesting) of disturbance. Based on input from local ecologists and soil scientists, we defined three severity categories (low, medium, high) and three fire return intervals (40, 80, 120 years) for wildfire simulations, and two utilization levels (SOH, WTH) and three rotation lengths (40, 80, 120 years) for harvesting simulations. A description of those severity and utilization categories is given in Table 1. Each scenario was then simulated commencing with the initial ECOSTATE file described in Wei et al. (2003). The unrealistically short 40-year rotation length was included in the study in order to compare harvesting at this frequency with fire, and because it helps to define a response curve (Powers et al., 1994).

Four output parameters (production, mass of decomposing litter, total available soil nitrogen and nitrogen removal) were used in the assessment of the sustainability of site productivity. Total production is a direct indicator of achieved productivity, while decomposing litter, total available soil nitrogen and nitrogen removal are indirect indicators of site productivity potential. Woody debris, as part of decomposing litter, is also a source of asymbiotic nitrogen fixation (Wei and Kimmins, 1998), and can protect soil from erosion and play an important role in maintaining some aspects of biodiversity (Harmon et al., 1986; Hunter, 1990).

Disturbance	Severity*	Biomass burned or removed for each component (%)							
		Stemwood	Stembark	Branch	Foliage	Large root	Medium root	Small root	Cones
Wildfire	Low (Fire-L)	0	10	10	20	0	0	10	10
	Medium (Fire-M)	15	60	50	60	20	20	20	60
	High (Fire-H)	50	95	95	100	30	30	60	95
Harvesting	SOH	90	90	0	0	0	0	0	0
	WTH	90	90	90	90	0	0	0	90

Table 1. Definition of disturbance severity for both wildfire and harvesting for simulations of lodgepole pine forests (*Fire-L: low severity fire; Fire-M: medium severity fire; Fire-H: high severity fire; SOH: stem-only harvesting; WTH: whole-tree harvesting)

4. Results and discussion

4.1 Differences immediately following disturbances
4.1.1 Difference in WD mass between SOH, WTH and wildfire disturbances
There were significant differences in above-ground CWD ($p < 0.001$) and total WD mass ($p < 0.001$) between the fire-killed and the harvested sites that we sampled. The greatest CWD and WD mass was created by wildfire while the lowest was left on the WTH sites; wildfire left about 3 to 5 fold more above-ground WD on the sites than did clearcutting. No significant differences in above-ground CWD and total WD mass were detected between SOH and WTH (Table 2), except the difference in above-ground FWD ($p < 0.001$) (Table 2). However, the means of CWD and WD mass were greater on the SOH sites than on the WTH sites.

Timber harvesting removed far more above-ground WD than did the wildfire disturbance that was investigated. In the case of a severe wildfire, the difference would be much less. However, harvesting did leave between 18% (WTH) and 24% (SOH) of total above-ground WD on the sites. These above-ground WD retention percents on the harvested sites were higher than we expected. This is because a relatively high proportion of the living stems in the stands that were studied were smaller than the utilization criteria; more than 92% of above-ground WD on the harvested sites was contributed by woody material with a diameter less than 15 cm, which was the lower limit for utilization on these sites. The remaining WD was contributed by small quantities of logs in advanced decay stages that were a carryover from previous natural disturbances, and unharvested low-quality trees.

There were no significant differences in below-ground WD mass between the three types of disturbance (Tables 2). The amount of below-ground WD was similar to the amount of above-ground WD on the harvested sites, and was about 30% of the above-ground mass on the fire-killed sites. This suggests that below-ground WD makes an important contribution to total WD loading on the disturbed sites in the study area.

Diameter size class (cm)	Mass of WD $(Mg.ha^{-1})$		
	WTH	SOH	Wildfire
0.0 -- 0.49	0.25(0.0)	0.63(0.0)	0.49(0.0)
0.5 -- 0.99	0.75(0.1)	1.78(0.2)	1.37(0.1)
1.0 --2.49	1.61(0.1)	3.60(0.5)	2.77(0.3)
2.5 -- 4.99	1.28(0.1)	2.67(0.2)	2.53(0.1)
5.0 -- 6.99	1.73(0.2)	3.83(0.6)	4.34(0.4)
≥ 7.0	11.71(1.3)	15.51(2.3)	91.09(7.3)
advanced decay	1.88(0.3)	3.68(1.4)	1.29(0.3)
stump	1.62(0.1)	1.96(0.4)	0.00
above-ground CWD	18.22(1.2)[a]	27.65(1.6)[a]	99.25(4.2)[b]
above-ground FWD	2.61(0.1)[a]	6.01(0.4)[b]	4.63(0.2)[b]
total above-ground WD	20.83(1.1)[a]	33.66(2.2)[a]	103.88(4.3)[b]
total below-ground WD	31.98(1.9) [a]	31.53(2.2) [a]	37.12(1.6) [a]
total WD	52.81(3.0)[a]	65.19(4.1)[a]	141.00(5.8)[b]

Table 2. Comparison of woody debris (WD) mass $(Mg.ha^{-1})$ in the year immediately following disturbance between stem-only harvested (SOH), whole-tree harvested (WTH) and wildfire-killed sites, based on modification of the field data to account for mass losses due to decomposition since disturbance (Note: FWD: fine woody debris; CWD: coarse woody debris; standard error of the mean is in parentheses; the sample size (n) is 25 for WTH, and 20 for others; means with the same letter within a row are not significantly different ($p > 0.05$) from each other (the Tukey test))

Timber harvesting removed most of the above-ground biomass, but like natural disturbances it left all below-ground biomass on site. The below-ground biomass (total roots) in lodgepole pine stands accounts for an important portion of total tree biomass, ranging from 20% on mesic sites to 28% on xeric sites (Comeau and Kimmins, 1989). This below-ground biomass may play an important role in long-term site productivity on these disturbed sites because our data show that asymbiotic nitrogen fixation rates in below-ground WD are significantly higher than in above-ground WD (see the section on asymbiotic nitrogenase fixation). Most WD studies have ignored below-ground WD, which may result in an incomplete understanding of the post-disturbance role of WD in forest ecosystems.

4.1.2 Above-ground CWD decay

The single-exponential decay coefficient (k) is an indicator of the rate of the decay process (Swift et al., 1979). The higher the k value, the faster CWD decays. Table 3 showed that, when CWD is in early or medium decay stages (data limited to about <30 years of decay since disturbance in this study), k values associated with above-ground CWD on the harvested sites were significantly higher ($p < 0.001$) than those on the fire-killed sites. These differences were attributed to the degree of contact between CWD and ground. CWD after harvesting generally has full contact with the ground, while CWD after wildfire disturbance experiences a long period of time before fully reaching the ground (from snag to partial suspended CWD and finally to fully ground-contact CWD). Decomposition is accelerated when logs are in contact with the ground, probably as a result of higher moisture content and increased interaction with the soil fauna and microflora.

Decay class	Years of decay*	Treatment		Sample size (n)
		Harvested	Fire-killed	
Early	5 - 10	0.021 (0.004)a	0.004 (0.001)b	16
Medium	20 - 30	0.018 (0.005)a	0.009 (0.002)b	12

Table 3. Single-exponential decay coefficients (k) for 10-20 cm diameter above-ground coarse woody debris (CWD) of different decay classes on both harvested and fire-killed sites (Wei et al. 1997) (Note: Means with the same letter within a row are not significantly different ($p > 0.05$) from each other (t-tests); standard error of the mean is shown in parentheses; *data on various years of wood decay was grouped into two time classes as shown in the table)

4.1.3 Nutrient losses through harvest or wildfire

The differences in nutrient removal between the three disturbances are presented in Table 4. The nutrient removals associated with timber harvesting generally were within the range of estimated wildfire removals; the P removal by WTH was equal to the largest value estimated for wildfire losses. WTH removed more N and P nutrients (about 2-fold) than SOH harvesting did, confirming that SOH is the more nutrient-conserving of the two harvesting methods.

Wildfire removes nutrient-rich crown material and forest floor, but leaves most large woody material in the ecosystem. In contrast, timber harvesting removes most large woody material, but leaves the nutrient-rich forest floor and part of the crown materials, depending on harvesting technique (e.g., SOH vs. WTH). Our estimates on N loss by wildfire, based on consumption of woody and litter materials, are reasonable due to volatilization and particulate convection. However, P loss by wildfire may be overestimated because P in burned material may be lost as fly-ash, or simply added to the ground as ash. No attempt was made in this study to compare the nutrient losses through soil leaching process following the SOH, WTH and fire disturbances. However, nitrogen leaching losses after harvesting or wildfire disturbances are believed to be low in these dry interior ecosystems.

Nutrient variables	Disturbance types		
	WTH	SOH	Wildfire
N			
total removed[1]	97.8 (3.9)	57.0 (4.6)	50.0 (1.5) ~~ 166.7 (4.8)[3]
removal ratio[2]	0.31 (0.01)	0.18 (0.01)	0.15(0.01) ~~ 0.49(0.02)
P			
total removed	8.8 (0.4)	5.2 (0.4)	2.8 (0.1) ~~ 9.5 (0.5)
removal ratio	0.44 (0.01)	0.26 (0.01)	0.13(0.01) ~~ 0.44(0.02)

Table 4. Estimated nutrient losses (kg.ha^{-1}) caused by stem-only harvesting (SOH), whole tree harvesting (WTH) and wildfire (from Wei et al. 1997)(Note: Standard error of the mean is shown in parentheses; [1]. total removed is the amount of nutrients removed during disturbance; [2]. the removal ratio is the amount of nutrients removed during disturbance divided by total amount of nutrients in biomass (above and below-ground biomass and forest floor) prior to disturbance; [3]. the range given here is based on an assumed range of fire severities: see text for explanation)

4.1.4 Asymbiotic nitrogenase activity in decaying wood, forest floor and soil

The nitrogen fixation rate in dead root was significantly higher than in the other substrates examined (Table 5), and the rate in mineral soil was the lowest. Table 5 also showed that nitrogen fixation rates in more advanced decay wood (medium or advanced decay classes) were significantly higher than in early decay wood. The difference in nitrogen fixation rates between medium and advanced decay wood was not significant. The nitrogen fixation rate in May 15, 1995 was the lowest, due to low temperature, among all other sampling dates.

The differences between early and medium decay classes of WD for both wildfire-killed and harvested sites indicate that nitrogen fixation activity increases as wood decay progresses (Table 5). However, there was little change in activity between medium and advanced decay stages on the wildfire-killed sites, suggesting that the increase in activity may only occur during the early stages of decay and then reach a steady level, depending on moisture content. The nitrogen fixation activity associated with dead roots was the highest among all substrates we studied (Table 5) probably due to the high moisture content of this material. The lowest activity occurred in mineral soil probably because of insufficient carbon substrates and low moisture content and also because soil weights more per volume. The nitrogen fixation activity in stumps and litter was higher than that in soil and early decaying stems.

Rates of nitrogen fixation in our study are generally consistent with other measures in northern forest ecosystems (Jurgensen et al., 1987 and 1989; Hendrickson, 1988; Harvey et al., 1989; Sollins, 1982; Roskoski, 1980). Cushon and Feller (1989) found much lower values; aerobic conditions during incubation in their assay might be responsible for this deviation.

Substrate	Nitrogen fixation rates (nm C2H4g⁻¹day⁻¹)					
	Aug. 22, 94	Sept. 25, 94	May 15, 95	June 18, 95	July 20, 95	mean
Fire-killed sites						
decaying wood						
early	2.60 (0.10)	1.27 (0.32)	0.00	1.86 (0.91)	0.92 (0.30)	1.33 (0.36) [c]
medium	3.91 (0.92)	3.85 (0.71)	0.14 (0.10)	10.94 (2.66)	9.00 (3.32)	5.58 (1.14) [b]
advanced	6.32 (1.72)	8.69 (1.49)	0.33 (0.20)	12.05 (6.40)	2.81 (0.71)	5.64 (1.61) [b]
dead roots	11.40 (2.82)	14.87 (4.56)	1.23 (0.46)	37.15 (16.1)	10.94 (3.11)	15.1 (4.75) [a]
branches	4.70 (2.20)	4.96 (1.98)	0.00	0.29 (0.24)	0.00	1.99 (0.75) [c]
floor litter	6.92 (3.11)	7.10 (2.02)	0.00	8.75 (4.11)	4.23 (1.52)	5.40 (1.43) [b]
humus	1.33 (0.32)	0.10 (0.03)	0.21 (0.20)	7.05 (3.47)	1.68 (1.30)	2.07 (0.87) [c]
mineral soil	0.00	0.05 (0.03)	0.00	0.50 (0.10)	0.00	0.11 (0.04) [d]
Harvested sites						
decaying wood						
early	3.22 (0.30)	3.62 (1.14)	0.00	3.49 (0.88)	0.72 (0.24)	2.21 (0.74) [c]
medium	3.90 (1.01)	3.60 (1.20)	0.00	10.70 (2.56)	7.37 (2.10)	5.20 (1.10) [b]
advanced	10.02 (3.07)	1.76 (0.52)	0.60 (0.40)	10.11 (2.72)	9.11 (3.82)	6.32 (1.24) [b]
dead roots	11.90 (3.01)	10.38 (1.34)	6.20 (0.31)	11.34 (1.29)	13.52 (5.94)	10.7 (2.01) [a]
branches	3.01 (0.80)	5.57 (1.67)	1.41 (2.01)	1.06 (0.14)	2.57 (1.01)	2.72 (0.98) [c]
floor litter	12.1 (2.05)	8.30 (2.89)	2.30(2.21)	1.20 (0.22)	4.30 (1.12)	5.64 (1.47) [b]
humus	1.13 (0.24)	0.34 (0.15)	0.49 (0.28)	2.28 (0.89)	0.41 (0.31)	1.16 (0.37) [c]
mineral soil	0.10 (0.05)	0.15 (0.07)	0.00	0.00	0.10 (0.10)	0.07 (0.03) [d]

Table 5. Estimated nitrogenase activity for each substrate and sampling time and the mean nitrogenase activities for all sampling dates in 1994-1995 (from Wei et al. 1998) (Note: Each value is the mean and (the standard error) of 4 samples; means with the same letter within a column are not significantly different ($p > 0.05$) from each other (the Tukey test))

4.2 Long-term implications of these differences in sustainability of productivity
4.2.1 Impacts of disturbance frequencies (rotation length or intervals)

As expected, the total productivity over a 240-year simulation increased with the length of the interval for all disturbance types (Figure 1a). This is clearly related to lower nitrogen losses over the 240-year simulation period (Figure 1b) and consequently more nitrogen and forest floor accumulation (Figures 1c,d, respectively). This indicates, as expected, that the sites we studied would be more productive under less frequent disturbance by the regimes defined in this study.

The rate of increase in productivity between disturbance scenarios varies, with a sharp increase from intervals of 40 years to intervals of 80 years, but only a modest increase from 80 years to 120 years. This reflects not only the difference in percentage change in interval length between these two scenarios, but a decrease in stem mass accumulation at stands ages greater than 80 years. The combined effects of genetically-determined, age-related decline in growth rates and the altered geochemical balance at longer disturbance intervals results in a declining sensitivity of total productivity to disturbance frequency at intervals longer than 80 years.

Figure 1a also shows that the difference in total productivity between the five disturbance types becomes progressively smaller as the disturbance interval increases, suggesting that these lodgepole pine ecosystems are fairly resilient in the face of a disturbance interval of 120 years. This is particularly evident for timber harvesting (SOH and WTH) and low-severity wildfire disturbance. However, rotation lengths of longer than 120 years may not be suitable from a timber management perspective because (1): they lead to little gain of productivity within a rotation, and a decline of total productivity over the 240-year simulation period; and (2) they increase problems with mistletoe. Therefore, we conclude that 120 years would be the upper limit for rotation length in terms of maximization of site productivity for the medium quality site.

The trend of site productivity over multiple consecutive rotations is a useful indication of sustainability. Figure 2 shows that with a harvest (WTH, SOH) or low-severity wildfire interval of 80 years or longer, site productivity is sustainable over a 240-year simulation. In contrast, site productivity at 40-year frequency is only sustainable with SOH; the other four scenarios were not shown to be sustainable (Figure 1a and Table 6). Therefore, 80 years appears to be the lower limit of sustainable rotation lengths of the three examined for the management system that we simulated, and 80 to 120 years would probably be the range of suitable rotation lengths for medium quality sites in the study area. Simulations at intermediate rotation lengths would be needed to define sustainable rotation length more accurately.

Lodgepole pine forests in the study area are thought to have been recycled for thousands of years under natural wildfire return intervals of about 100-125 years. This is similar to the disturbance interval that was estimated by this simulation to be sustainable, and suggests that the study of natural disturbance regimes can be helpful in designing sustainable management strategies. However, although the average wildfire return interval is 100 –125 years in the study sites, its variability is very high, ranging from 40 to 200 years (Pojar, 1985). This is much different from human-caused disturbance such as timber harvesting which tends to apply roughly equal harvest frequencies in a specific type of forest. The variability in frequency of natural disturbance may be important for the maintenance of certain ecosystem values because it affects the dynamics of WD loading and stand structures. Our study has demonstrated that both above-ground and below-ground WD plays an important role in the nitrogen economy in these lodgepole pine forests. The implications of this natural disturbance variability for other ecological attributes, such as wildlife habitat, remain unknown and are beyond the scope of this study, as are the implications of imposing a more uniform disturbance frequency.

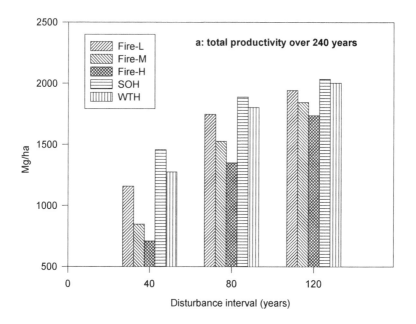

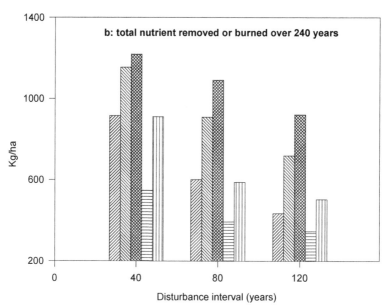

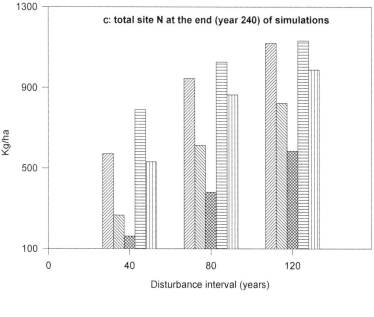

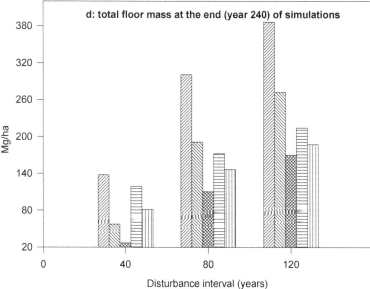

Fig. 1. (a-d). Four simulation output indicators (total productivity, total nitrogen removal, available soil nitrogen and forest floor mass) under five disturbance scenarios on a site of medium quality over a period of 240-year simulation (from Wei et al. 2003)

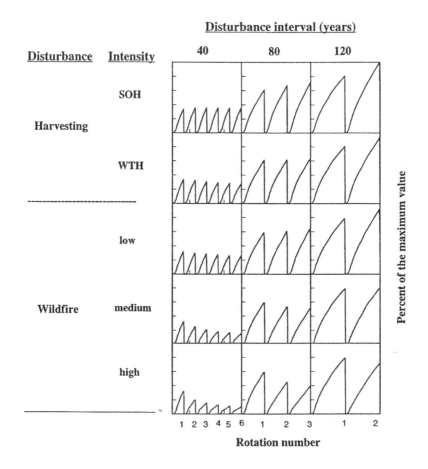

Fig. 2. Dynamics of site productivity under fiver disturbance scenarios on a site of medium quality over a simulation period of 240 years (from Wei et al. 2003)

In some forest types, and depending on how it is done, clear-cut harvesting reverts the ecosystem to an earlier stage of the sere (defined as the sequence of plant and animal communities which successively occupy a site over a period of time). However, in some forests, or with some techniques, clear-cutting may simply recycle the existing seral stage, promptly replacing the mature trees with young trees of the same species with little or no change in the understory. In other forests, clear-cutting in the absence of fire and soil disturbance may accelerate succession by facilitating earlier development of the subsequent seral stage. This generally involves the release of shade-tolerant seedling of the next seral stage. Clear-cutting in pure lodgepole pine forests in the study area generally tend to recycle the existing seral stage.

Disturbance Interval (years)	Disturbance Severity				
	Fire-L	Fire-M	Fire-H	SOH	WTH
40	1157	847	709	1457	1275
80	1745	1525	1349	1887	1801
120	1942	1843	1734	2034	2002

Table 6. Difference in total tree productivity (Mg/ha) between disturbance severity scenarios and between disturbance interval scenarios over a period of 240-year simulation (from Wei et al. 2003) (Note: Abbreviations are given in Table 1)

4.2.2 Impacts of disturbance severities

Figures 1a-d show that medium and high-severity wildfire had the largest impact on most of the simulated indicators. Total productivity (Figure 1a) and total soil nitrogen (Figure 1c) are much less for the moderate and severe fire simulations than for the harvested or low-severity wildfire simulations. This reflects the greater loss of N in the medium and high-severity wildfire simulations (Figure 1b).

The simulations suggest that timber harvesting (SOH, WTH) is a relatively nutrient conservative disturbance compared to wildfire. All simulated indicators for the harvesting treatments are within the range of the wildfire treatment (Figures 1a-d), and close to those for the low-severity wildfire treatment. These simulation results support one of the conclusions from our field investigation of the differences between harvested and fire-killed stands in the study area: the nutrient removals caused by harvesting were within the estimated range of nutrient removals caused by wildfire.

The difference in all indicators between SOH and WTH treatments declined as the rotation lengths increased (Figures 1a-d). There is only a minor difference in total productivity over the 240-year simulation at a rotation of 120 years. This suggests that both WTH and SOH are acceptable harvesting methods for the maintenance of long-term site productivity in these lodgepole pine forests if a rotation of 120 years is used. However, the simulations suggest that WTH could reduce productivity by up to 20% compared with SOH if the rotation length was as short as 40 years. SOH is a more nutrient conservative harvest method because it leaves more of the relatively nutrient-rich crown materials on the ground, and should be used instead of WTH for rotations less than 80 years. Our results from both simulation (Figure 1d) and field studies demonstrate that the total mass of decomposing organic matter on wildfire-killed sites, particularly for long-interval, lower-severity wildfires, would be much higher than on harvested sites. This reflects the larger accumulation of aboveground CWD on the wildfire-killed sites, and slower decomposition of this material on fire sites than on harvested sites because much of it is suspended above the ground on branch-stubs on the burned sites. The WD left on the harvested sites is smaller in diameter and in closer contact with the ground, resulting in faster decomposition and, therefore, lower persistence. The lower level of decomposing litter on high-severity wildfire sites (Figure 1d) is attributed to much larger loss of forest floor and crown materials compared with lower severity fire. Because of these differences, harvesting conserves nutrients more than wildfire does at time of disturbance. However, because decomposing litter on wildfire sites consists largely of persistent WD that supports asymbiotic nitrogen fixation, even severely burned sites eventually recover from nutrient losses caused by fire. WD may also play an important role in providing wildlife habitat and microclimatic shelter for regeneration.

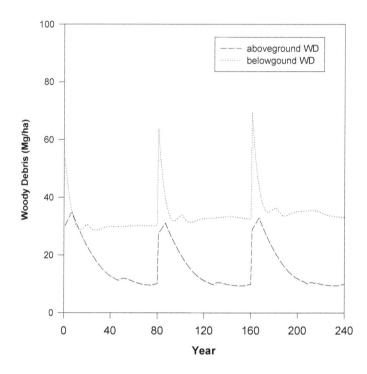

Fig. 3. Simulation of the mass of decomposing woody debris (WD) in above-ground and below-ground over three consecutive 80-year rotations in a stem-only harvested lodgepole pine forest (from Wei et al. 2003)

WTH has been a common harvesting method in the study area. SOH has been applied recently as a result of concern over nutrient removal caused by WTH. While SOH leaves much more of the fine woody debris (<2.5 cm) and crown materials on the ground, both SOH and WTH leave the same mass of stump and root systems (Wei et al., 1997). In harvested lodgepole pine forests, the mass of this largely unseen below-ground WD is normally much greater than the mass of the visible above-ground WD (Figure 3), and asymbiotic nitrogen fixation associated with the former is much higher than for the latter due to its higher moisture content. From a nitrogen perspective, below-ground WD in these forests is more important than above-ground WD, and consequently, the N-fixation associated with below-ground WD reduces the nutritional significance of differences in aboveground debris loading between SOH and WTH sites. The importance of this belowground tree biomass also suggests that complete tree harvesting, which includes removal of stumps, major roots, and all above-ground biomass, would require much longer rotations for it to be sustainable.

4.2.3 Interactive effects of disturbance severity and frequency:
The ecological rotation concept

Sustainability in stand level forestry involves non-declining patterns of change. Such non-declining patterns require a balance between the frequency and severity of disturbance, and the resilience of the ecosystem in question. An ecological rotation is defined as the period required for a given site managed under a specific disturbance regime to return to an ecological state comparable to that found in pre-disturbance condition, or to some new desired condition that is then sustained in a non-declining pattern of change. Too short a recovery period for a given disturbance and ecosystem recovery rate, or too large a disturbance for a given frequency and recovery rate, can cause reductions in future forest productivity and other forest values (Kimmins, 1974). Stand level sustainability can thus be achieved by using the design of management on ecological rotations. However, estimating the length of ecological rotations is difficult, and will generally require the use of an ecosystem management simulation model.

The ecological rotation concept asserts that stand level sustainability can be achieved under several different combinations of disturbance severity and disturbance frequency, for a given level of ecosystem resilience. Figure 2 shows that stemwood production is sustained over successive rotations for SOH-40 year and WTH-80 year harvest disturbances, and sustained production for low fire-80 year and medium fire-120 year combinations. The low fire-40 year, medium fire-80 year and the high fires-120 year combinations all show about the same degree of decline in productivity. These results support the concept of ecological rotations, and suggest the ecological rotation could be a useful template for the design of sustainable stand level forestry. Managers may either choose harvest frequency based on economic, technical or other social/managerial considerations and then limit the degree of ecosystem disturbance required by ecological rotation for the site in question with the chosen frequency. Alternatively, they may choose the level of ecosystem disturbance (type of harvest system; severity of post harvest site treatment), but then be constrained in terms of how frequently this can be applied (e.g. the rotation length). If neither of these alternatives is acceptable, they can increase ecosystem resilience by means of silvicultural interventions.

Our FORECAST simulation results suggest that, from a nitrogen-related productivity perspective, the ecological rotation of the medium quality site would average about 100 years, with a possible range of 80 –120 years. However, the ecological rotation should be evaluated in a broader context than soil fertility and site productivity. It should also be calculated for attributes such as understory vegetation, wildlife habitat, and soil physical and chemical conditions.

The ecological rotation for soil fertility is site-specific (Kimmins, 1974). For example, on a site receiving nutrients in seepage water or having large reserves of readily weatherable soil minerals, even substantial losses of nutrients may be replaced relatively rapidly. Similarly, on a site with very slow replacement and/or poorly developed nutrient accumulation mechanisms, even a small loss of nutrients may require a substantial period for replacement. The concept of ecological rotation is defined and applied at the stand level. As noted above, sustainable management at this spatial scale implies non-declining patterns of change, rather than unchanging conditions. As a consequence, no single value or ecological process can be supplied from any one stand continuously. The objectives of sustainable management in terms of even-flow of values must, therefore, be evaluated at the landscape level. This requires that models such as FORECAST be incorporated into landscape level models such as HORIZON (Kimmins et al., 1999).

5. Conclusion

Timber harvesting has substantially reduced the mass of above-ground CWD compared with natural wildfire disturbance in lodgepole pine forests in the central interior of British Columbia. There were no significant differences in above-ground CWD and total WD between SOH and WTH sites, but there was a significant difference in the mass of FWD between these two types of harvesting. WD on the harvested sites decays more rapidly and persists for less time due to its smaller size and closer contact with the ground, compared with those on the fire-killed sites. Below-ground WD is an important component of total WD in terms of its mass loadings and associated nitrogen fixation activities. It should be given greater consideration in future investigations of the role of WD, particularly in the dry interior forests of British Columbia.

Both measured and simulated nutritional impacts of timber harvesting were within the simulated range of impacts caused by the wildfire defined in this study. They were similar to the simulated long-interval, low-severity wildfire regimes. Either of the current timber harvesting methods (SOH or WTH) can maintain long-term site productivity in the study area if rotations of 80-120 years are used. Shorter rotations should use SOH. Because of lack of validation, application of these simulation results must be cautious and adaptive.

This research also demonstrates that the combination of field measurement with ecosystem simulation is a useful and effective approach for evaluating sustainability of long-term forest productivity.

6. Acknowledgment

Funding for this study was provided by a collaboration grant of NSERC (Natural Sciences and Engineering Research Council of Canada), Ministry of Forests, British Columbia and Timber Supply Association (Williams Lake, British Columbia).

7. References

Andison, D. W. (2000). The natural disturbance program for the Foothills Model Forest: a foundation for growth. In R. G., D'Eon, K. Johnson, and E. Alex Ferguson (Eds), *Ecosystem management of forested landscape--directions and implementation.* ISBN-10:096865410x. Vancouver: University of British Columbia Press.

Attiwill, P. M. (1994). The disturbance of forest ecosystems: the ecological basis for conservative management. *Forest Ecology and management*, 63, 247-300.

Busse, M. D. (1994). Downed bole-wood decomposition in lodgepole pine forests of central Oregon. *Soil Sci. Soc. Am. J.*, 58, 221-227.

Comeau, P. G., and Kimmins, J. P. (1989). Above and below-ground biomass and production of lodgepole pine on sites with differing soil moisture regimes. *Can. J. For. Res.*, 19, 447-454.

Cushon, G. H., and Feller, M. C. (1989). Asymbiotic nitrogen fixation and denitrification in a mature forest in coastal British Columbia. *Can. J. For. Res.*, 19, 1194-1200.

Fahey, T. J. (1983). Nutrient dynamics of above ground detritus in lodgepole pine (*Pinus contorta ssp. latifolia*) ecosystems, southeastern Wyoming. *Ecol. Monogr.*, 53, 89-98.

Graham, R. T., Harvey, A. E., Jurgensen, M. F., Jain, T. B., Tonn, J. R., and Page-Dumroese, D. S. (1994). Managing Coarse Woody Debris in Forests of the Rocky Mountains. U.S. Department of Agriculture and Forest Service. Intermountain Research Station. *Research Paper* INT-RP-477.

Hammond, H. (1991). *Seeing the Forest Among the Trees: the Case for Wholistic Forest Use.* ISBN: 0-919591-58-2. Vancouver: Polestar Press.

Hardy, R. W. F., Burns, R. C., and Holsten, R. D. (1973). Applications of the acetylene-ethylene assay for measurement of nitrogen fixation, *Soil Biol. Biochem.*, 5, 133-302.

Harmon, M. E., Franklin, J. F., Swanson, F. J., Sollins, P., Gregory, S. V., Lattin, J. D., Anderson, N. H., Cline, S. P., et al. (1986). Ecology of coarse woody debris in temperate ecosystems. *Adv. Ecol. Res.*, 15, 133-302.

Harvey, A. E., Jurgensen, M. F., and Larsen, M. J. (1981). Organic reserves: importance to ectomycorrhizae in forest soils in western Montana. *For. Sci.*, 27, 442-445.

Harvey, A. E., Jurgensen, M. F., Larsen, M. J., and Graham, R. T. (1987). Decaying organic materials and soil quality in the Inland Northwest: a management opportunity. Gen. Tech. Rep. INT-225. Ogden, UT: U.S. Department of Agriculture, *Forest Service, Intermountain Research Station.* 15p.

Harvey, A. E., Larsen, M. J., and Jurgensen, M. F. (1989). Nitrogenase activity associated with decayed wood of living Northern Idaho conifers. *Mycologia*, 5, 765-771.

Hendrickson, O. Q. (1989). Implication of natural ethylene cycling processes for forest soil acetylene reduction assays. *Can. J. Microbiol.*, 35, 713-718.

Hendrickson, O. Q. (1988). Use of acetylene reduction for estimating nitrogen fixation in woody debris. *Soil Sci. Soc. Am. J.*, 52, 840-844.

Hunter, M. L. JR. (1990). Wildlife, Forests, and Forestry: Principles of Managing Forests for Biological Diversity. ISBN: 0139594795 9780139594793. *Englewood Cliffs*, N. J.: Prentice Hall.

Johnson, E.A., Morin, H., Miyanishi, K., Gagnon, R., and Greene, D.F. (2003). A process approach to understanding disturbance and forest dynamics for sustainable forestry. In P.J. Burton, C. Messier, D.W. Smith and W.L. Adamowicz (Eds). Towards Sustainable Management of the Boreal Forest. NRC-CNRC, NRC Research Press, Ottawa, Canada

Johnson, D. W. (1994). Reasons for concern over impacts of harvesting. In: W. J. Dyck, D. W. Cole and N. B. Comerford (Eds). *Impacts of Forest Harvesting on Long-Term Site Productivity.* ISBN 10: 0412583909. ISBN-13: 978-0412583902. London: Chapman and Hall.

Jurgensen, M. F., Larsen, M. J., and Harvey, A. E. (1989). A comparison of dinitrogen fixation rates in wood litter decayed by white-rot and brown-rot fungi. *Plant Soil*, 115, 117-122.

Jurgensen, M. F., Larsen, M. J., Graham, R. T., and Harvey, A. E. (1987). Nitrogen fixation in woody residue of Northern Rocky Mountain conifer Forests. *Can. J. For. Res.*, 17, 1283-1288.

Keane, M. G. (1985). Aspects of needle morphology, biomass allocation and foliar nutrient composition in a young fertilized stand of repressed lodgepole pine. *Ph.D thesis*. Faculty of Forestry, University of British Columbia. Vancouver, British Columbia.

Keenan, R. J. and Kimmins, J. P. (1993). The ecological effects of clear-cutting. *Environ. Rev.*, 1, 121-144.

Keeves, A. (1966). Some evidence of loss of productivity with successive rotations of Pinus radiata in the south-east of South Australia. *Australian Forestry*, 30, 51-63.

Kellomäki, S., and Seppälä, M. (1987). Simulations on the effects of timber harvesting and forest management on the nutrient cycle and productivity of Scots pine stands. *Silva Fennica*, 21, 203-236.

Kimmins, J. P. (2000). Respecting for nature: an essential foundation for sustainable forest management. In Robert G. D'Eon, K. Johnson, E. Alex Ferguson (Eds), *Ecosystem management of forested landscape--directions and implementation*. ISBN: 0139594795 9780139594793. Vancouver: University of British Columbia.

Kimmins, J. P., Scoullar, K. A., Seely, B., Andison, D. W., Bradley, R., Mailly, D., and Tsze, K. M. (1999). FORCEEing and FORECASTing the HORIZON: Hybrid Simulation--Modeling of Forest Ecosystem Sustainability. *In: A. Amaro and M. Tome (Editors), Empirical and process based models for forest tree and stand growth simulation*. Lisboa, Portugal: Edicoes Salamandra.

Kimmins, J. P., Mailly, D. and Seely, B. (1999). Modelling forest ecosystem net primary production: the hybrid simulation approach used in FORECAST. *Ecol. Modelling*, 122,195-224.

Kimmins, J. P. (1993). Scientific foundations for the simulation of ecosystem function and management in FORCYTE-11. CFS. *Information Report. NOR-X-328*. 88p.

Kimmins, J. P. (1974). Sustained yield, timber mining, and the concept of ecological rotation: a British Columbia view. *For. Chron.*, 50, 27-31.

Lambert, R. C., Lang, G. E., and Reiners, W. A. 1980. Loss of mass and chemical change in decaying boles of subalpine balsam fir forest. *Ecology* 61: 1460-1473.

Lertzman, K., and Fall, J. (1998). From forest stands to landscapes: spatial scales and the roles of disturbances. In: D. Peterson and V. T. Thomas (Eds), *Ecological Scales--Theory and Applications*. ISBN:-10: 0231105037. New York: Columbia University Press.

Lertzman, K., Spies, T., and Swanson, F. (1997). From ecosystem dynamics to ecosystem management. In: P.K. Schoonmaker, B.von Hagen and E.C. Wolf (Eds), *The Rain Forests of Home: Profile of a North American Bioregion*. ISBN: 1-55963-470-0. Washington, D.C.: Island Press.

McRae, D. J., Alexander, M. E., and Stocks, B. J. (1979). Measurement and Description of Fules and Fire Behavior on Prescribed Burns: a Handbook. Canadian Forestry Service. Department of the Environment. Report O-X-287. *Great Lakes Forest Research Centre. Sault Ste.* Marie, Ontario.

Morris, D. M, Kimmins, J. P. and Duckert, D. R. (1997). The use of soil organic matter as a criterion of the sustainability of forest management alternatives: A modeling approach using FORECAST. Forest Ecology and Management. *Forest Ecology and Management*, 94, 61-78.

Nambiar, E. K. S., and Sands, R. (1993). Competition for water and nutrients in forests. *Can. J. For. Res.*, 23, 1955-1968.

Nambiar, E. K. S., Squire, R., Cromer, R., Turner, J., and Boardman, R. (1990). Management of water and nutrient relations to increase forest growth. *Forest Ecology and Management*, 30,1-486.

Parminter, J. (1998). Natural disturbance ecology. In: J. Voller and S. Harrison (Eds), *Conservation Biology Principles for Forested Landscapes*. ISBN:-10: 077480629x. Vancouver: UBC Press.

Perera, A.H., and Buse, L.J. (2004). Emulating natural disturbance in forest management: an overview. In A.H. Perera, L.J. Buse and M.G. Weber (Eds), Emulating Natural Forest Landscape Disturbances: Concepts and Applications. ISBN 0-231-12915-5. New York, Columbia University Press

Poff, N. L., Allan, J. D., Bain, M. B., Karr, J. R., Prestegaard, K. L. Richter, D. D., Parks R. E., and Stromberg, J. C. (1997). The natural flow regime, a paradigm for river conservation and restoration. *Bioscience*, 47, 769-784.

Pojar, J. (1985). Ecological classification of Lodgepole pine in Canada. In: D. M. Baumgartner, R.G. Krebill, J. T. Arnott, and G.F. Weetman (Eds), Lodgepole Pine-The Species and Its Management (Symposium: May 8-10, 1984, Spokane, Washington, USA and repeated May 14-16, 1984, Vancouver, British Columbia, Canada), Washington State University, USA.

Powers, R. F., Mead, J.A., Burger, J.A. and Ritchie, M.W. (1994). Designing long-term site productivity experiments. In: W.J. Dyck, D.W. Cole, and N. B. Comerford (Eds), *Impacts of Forest Harvesting on Long-Term Site Productivity*. ISBN-10: 0412583909. ISBN-13: 978-0412583902. London: Chapman and Hall.

Raphael, M.G., and White, M. (1984). *Use of snags by cavity nesting birds in the Sierra-Nevada*, California. Wild. Monogr., 86, 1-66.

Reynolds, R. T., Graham, R. T., Reiser, M. H., et al. (1992). Management recommendations for the northern goshawk in the southwestern United States. Gen. Tech. Rep. RM-217. Fort Collins, CO: U.S. Department of Agriculture, Forest Service, *Rocky Mountain Forest and range Experiment Station*. 90 p.

Richter, B. D., Baumgarten, J.V., Wigington, R., and Braun, D. P. (1997). How much water does a river need? *Freshwater Biology*, 37, 231-249.

Roskoski, J. P. (1980). Nitrogen fixation in hardwood forests of the Northern United States. *Plant and Soil*, 54, 33-44.

Sachs, D. (1992). Calibration and initial testing of FORECAST for stands of lodgepole pine and Sitka Alder in the interior of British Columbia. *Ministry of Forests*, British Columbia, Victoria.

Seely, B., Kimmins, J. P., Welham, C., and Scoullar, K. (1999). Defining stand-level sustainability. Exploring stand-level stewardship. *J. For.*, 97, 4-10.

Seely, B., Welham, C. and Kimmins, H. (2002). Carbon sequestration in a boreal forest ecosystem: results from the ecosystem simulation model, FORECAST. *For. Ecol. Manage.*, 169, 123-135.

Sheng, W., and Xue, X. (1992). Comparison between pure stands of Chinese fir, Fukiencypress and mixed stands of these two species in growth, structure, biomass and ecological effects. *Sci. Silv. Sin.*, 5, 397-404 (in Chinese with English abstract).

Sollins, P. (1982). Input and decay of coarse woody debris in coniferous stands in western Oregon and Washington, *Can. J. For. Res.*, 12, 18-28.

Spano, S. D., Jurgensen, M. F., and Harvey, A. E. (1982). Nitrogen-fixing bacteria in Douglas-fir residue decayed by Fomitopsis pinicola. *Plant and Soil*, 68, 117-123.

Spies, T. A., Franklin, J. F., and Thomas, T. B. (1988). Coarse woody debris in Douglas-fir forests of western Oregon and Washington. *Ecology*, 69, 1689-1702.

Squire, R. O. (1983). Review of second rotation silviculture of Pinus Radiata plantations in southern Australia: establishment practice and expectations. In: R. Ballard and S. P. Gessel (Editors), IUFRO Symposium on Forest Site and Continuous Productivity, USDA For. Ser. Proc. N. W. For. Range. Expt. Stn., Portland, Ore. *Gen. Tech. Rept. PNW*-163. pp. 130-137.

Steen, O., and Demarchi, D. (1991). Sub-Boreal Pine Spruce Zone. In D. Meidinger and J. Pojar (Eds), Ecosystems of British Columbia. *Ministry of Forests. Special Report Series*: ISSN 0843-6452. Chapter 13.

Swift, M. J., Heal O. W., and Anderson, J. M. (1979). Decomposition in Terrestrial Ecosystems. *In: Studies in Ecology (Volume 5).* Berkeley and Los Angeles: University of California Press.

Wang, J.R., Comeau, P., and Kimmins, J.P. (1995). Simulation of mixedwood management of aspen and white spruce in Northeastern British Columbia. *Water, Air and Soil Pollution*, 82, 171-178.

Wei, X. (2003). Wildfire disturbance, large woody debris and aquatic habitat. *In: proceedings of the Canadian Water Resource Association Conference.* Vancouver, June, 2003, Vancouver.

Wei, X., Liu, W., Waterhouse, M., and Armleder, M. (2000). Simulation on impacts of different management strategies on long-term site productivity in lodgepole pine forests of the central interior of British Columbia. *Forest Ecology and Management*, 133, 217-229.

Wei, X., and Kimmins, J.P. (1998). Asymbiotic nitrogen fixation in disturbed lodgepole pine forests in the central interior of British Columbia. *Forest Ecology and Management*, 109, 343-353.

Wei, X., Kimmins, J.P., Peel, K., and Steen, O. (1997). Mass and nutrients in woody debris in harvested and wildfire-killed lodgepole pine forests in the central interior of British Columbia. *Can. J. For. Res.*, 27, 148-155.

Wei, X. and Kimmins, J.P. (1995). Simulations of the long-term impacts of alder-Douglas-fir mixture management on the sustainability of site productivity using the FORECAST ecosystem model. In: P. Comeau and K,D. Thomas (Eds), Silviculture of Temperate and Boreal Broad-Leaved-Conifer Mixtures, *Research Branch*, Ministry of Forests, British Columbia, Victoria.

Welham, C., Seely, B. and Kimmins, J. P. (2002). The utility of the two-pass harvesting system: an analysis using the ecosystem simulation model FORECAST. *Can. J. For. Res.*, 32, 1071-1079.

Wilkinson, L. (1990). *SYSTAT: The System for Statistics*. SYSTAT Inc., Evanston, I11.

Whyte, A.G. D., (1973). Productivity of first and second crops of Pinus radiata on the Moutere soils of Nelson. *N. Z. J. For.*, 18, 87-103.

Yu, X. T. (1988). Research on Chinese-fir in China. *Journal of Fujian College Forestry*, 8, 203-220. (in Chinese with English abstract).

6

Sustainable Forest Management in Rural Southern Brazil: Exploring Participatory Forest Management Planning

André Eduardo Biscaia de Lacerda et al.*
Embrapa Forestry, Paraná
Brazil

1. Introduction

Historically, agriculture and livestock farming have been the main drivers of land cover conversion replacing natural forests in tropical and sub-tropical Brazil. The consequences for the landscape are well known: habitat fragmentation, biodiversity loss, and reductions in the quality of environmental services. The intense exploitation of natural forest resources tends to generate immediate, but limited, short-term economic wealth, which is generally very poorly distributed. In the long-term, forest resources are depleted thus reducing the ability of small rural owners to move out of impoverished situations. Therefore, while conversion of forest to agriculture can in some cases improve rural incomes, all too often deforestation leads to impoverishment of both ecosystems and communities. In Brazil, forest displacement in favour of agriculture and livestock has occurred since early in its colonization; in the Southern region – the principal agricultural area – this process took place in the late XIX and XX centuries. In this part of the country, past forestry practices such as clear-cutting and predatory harvesting, combined with social and legal encouragement, produced scenarios in which forested lands are now mostly degraded, not fulfilling their ecologic, social or economic roles in our society. In spite of the challenges that forest management faces in sub-tropical Brazil, some promising experiences and experiments are helping to create an environment receptive to the reintroduction of sustainable forest management (SFM) as a means to enhance economic incomes for rural property. Herein, we explore the obstacles related to the adoption of SFM as an economic alternative and propose technical opportunities for both small and large rural properties by presenting two case studies.

* Maria Augusta Doetzer Rosot[1], Afonso Figueiredo Filho[2], Marilice Cordeiro Garrastazú[1], Evelyn Roberta Nimmo[3], Betina Kellermann[1], Maria Izabel Radomski[1], Thorsten Beimgraben[4], Patricia Povoa de Mattos[1] and Yeda Maria Malheiros de Oliveira[1]
[1]*Embrapa Forestry, Paraná, Brazil*
[2]*Midwest State University in Irati, Brazil*
[3]*University of Manitoba, Canada*
[4]*Rottenburg University of Applied Forest Sciences, Germany*

When the process of land conversion and the introduction of intensive land activities reached Southern Brazil, it found prosperous ground: timber harvested from the sub-tropical forests – especially the conifer *Araucaria angustifolia* – was Brazil's main export product during late XIX and early XX centuries. Sub-tropical forests were gradually reduced, plummeting to levels as low as 1-5% of primary forests. Today, the remaining forested areas are mostly secondary and in early or intermediary successional stages which are profit-limited in the short-term. However, the SFM potential of these forest areas should not be underestimated; they account for approximately 30% of the lands originally covered with forests in Southern Brazil. Environmental laws currently in place aim to protect forest cover through rigid control and bans on forest management, with a few exceptions at the small scale level, such as firewood collection for small farmers. Although environmental laws were mostly unsuccessful in avoiding deforestation, secondary forests have increased in the region.

The current set of state and federal legislation requires that at least 20% of the surface of most rural properties must be covered with forests and places severe restrictions on their use, while allowing for some agroforestry activities. Additionally, any waterway must have a forested buffer zone. Paradoxically, while aiming to prevent further deforestation, environmental legislation created an antagonism between forests and landowners to a point in which forest regeneration is avoided. In fact, rural properties that still have forested areas are drastically reduced in market value; the ultimate consequence is continued poverty in rural areas with an increasing economic disparity between urban and rural communities. This process contradicts the perception that forests should help in providing for basic needs of small landowners and forest communities, as well as the idea that the benefits and costs originating from maintaining forests should be shared by society as a whole and should not be a burden exclusively imposed on those remaining in rural areas. As a consequence, the restrictive legislation prevents forests from being used as a source of income while blocking any SFM initiative that, in the broad sense, includes recovery, conservation and long-term use. Although more recent experiences with payment for environmental services (PES) have helped to counter-balance the distortion in relation to sharing the costs of maintaining forests, such payments seem to be unfeasible even for family farms and forest communities.

Although the challenge for meeting people's needs in rural areas and managing forests is not a problem restricted to Southern Brazil, it is particularly relevant as the region is characterized by a severely threatened forest type in an area where more than half a million small rural properties (< 50ha) are subjected to near poverty conditions. The production of family farms accounts for about 10% of gross domestic product (GDP) and currently small farms account for 70% of food production. This figure demonstrates the economic importance of the sector.

In this paper we will explore some of the legal, social, economic and environmental issues related to the reduction of the forests in Southern Brazil and propose the implementation of a "locally adapted participatory sustainable forest management" (lapSFM) system focusing on reducing both rural poverty and deforestation. Finally, we discuss two case studies of participatory forest management in the south of Brazil. This paper aims to deliver scientific expertise translated into practical solutions related to land use and participatory SFM, considering a landscape approach for both small and large properties. The intent is to provide an evidence basis for changes in environmental policy to better reflect the enhancement of SFM in line with agroforestry and the use of tree genetic resources across the landscape, from forests to farms.

2. The *Araucaria* forest and the fragmentation process

According to the "Ecosystems of the World Classification", edited by Lieth and Werger in 1989, the Subtropical Evergreen Seasonal Conifer Forest or Mixed Ombrophilous Forest is typically dominated by the species *Araucaria angustifolia* (Bertoloni) O. Kuntze (paraná-pine) (also known as Araucaria Forest). Although the species is predominant, the forest type also supports complex, variable and regional ecosystems commonly composed of more than one hundred woody species, some of which are endemic to this forest type. Araucaria forest occurs naturally in an area of 216,100 square kilometers, encompassing a region of mountains and plateaus throughout Southern Brazil (Figure 1). The region is characterized by altitudes above 500 meters elevation and a subtropical highland climate (Cfb), where frosts might occur during the winter months or, less frequently, light snowfalls in the highest areas. Annual precipitation is high, ranging from 1,300 to 3,000 millimetres, without a dry season (Oliveira, 1999).

During the last century, Southern Brazil has experienced a rapid deterioration of its forest resources mainly due to land conversion, displacing forests for agriculture purposes as part of the colonization process and unsustainable selective logging of its commercial species. A central problem that is prevalent is determining how to manage the natural Araucaria forest fragments that remain. Some of these fragments are very small and are becoming poorer in terms of biodiversity because of intense human interference. The challenge is reconciling economic development and the conservation of biological resources and using the natural resources without destroying the possibilities for future generations. An important element in the efforts to save natural biodiversity is the establishment and maintenance of protected areas, as well as the sustainable management of the remaining areas.

Given the current situation, management strategies should be developed and applied to forest fragments in Southern Brazil in order to prevent the continuation of current processes of forest degradation and loss of biodiversity (Viana et al., 1992). Untended forests are more prone to disappear as they are gradually converted into other land uses that provide lower levels of ecosystem services and goods (Mc Evoy, 2004). In regions with intensive agriculture, protection against anthropogenic disturbance of these fragments is unlikely to be sufficient. A change from a top-down social relationship in which farmers are not sufficiently engaged in the process of developing environmental policies to a system that creates alternatives for natural resource use is likely one of the biggest challenges managers face in Brazil. There is an urgent need to reconcile local ecological knowledge (LEK) with environmental policies and natural resources protection with economic prosperity.

3. Legal issues concerning forest management in southern Brazil

Environmental law in Brazil is expressed mainly through the current Forest Code (Brasil, 1965) and subsequent regulations. The Forest Code considers that interventions in forested areas should be prescribed according to approved Management Plans (MP). However, for many years and in most cases, MPs have become synonymous with illegal logging practices. It was not until 1994 that the government defined SFM in practical terms through Decree 1.282 (Brasil, 1994). In establishing an official SFM policy, Brazil adopted the reduced-impact logging (RIL) concept as the basis for forest management (for the development of RIL see Putz & Pinard, 1993; The International Tropical Timber Organization [ITTO], 1990; and Food and Agricultural Organization of the United Nations [FAO], 1993). The definition of a policy for forest management enabled Brazil to sign the Tarapoto Proposal of Criteria and Indicators (C&I) in 1995, which forms the basis of sustainable management in Brazil's tropical forests. However, Brazil's legislation and signed international agreements focus on management of tropical forests; the need for regulating the use of forests in Southern Brazil (which are mostly subtropical) only began to be addressed more than a decade after the adoption of SFM in the Amazon.

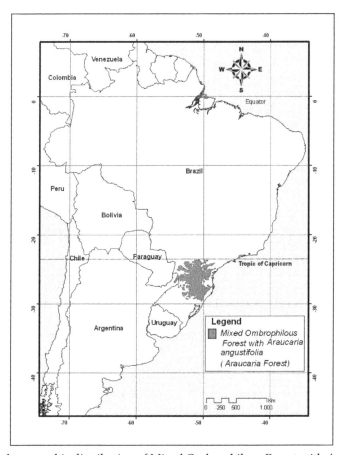

Fig. 1. Original geographic distribution of Mixed Ombrophilous Forest with *Araucaria angustifolia*

The first specific legislation concerning tropical and sub-tropical forests outside the Amazon biome was enacted in 2006 and is known as the Atlantic Forest Law (Brasil, 2006; 2008). This legislation, aiming to protect forest cover through rigid control, banned any land use that could potentially cause deforestation and restricted forest use to only non-commercial purposes based on the following regulations: a) management is permitted only when it does not produce tradable products or sub products, directly or indirectly; b) sustainable agroforestry management may be carried out in consortium with exotic species, in forestry or agricultural models (however commercial use of the wood from native tree species is forbidden); and c) forest management is forbidden unless the forest is composed of at least sixty percent of native pioneer tree species (and therefore restricted to very early successional stages). Furthermore, all forest management is subjected to environmental agency authorization. Additionally, *Araucaria angustifolia* and other important species are included in the list of "Endangered Species of the Brazilian flora" by the Ministry of Environment (MMA, 2008).

Given all the impediments to which Araucaria Forest is subjected, it is no wonder that many have given up on it. Landowners no longer consider the forest as a source of products and services; in many cases it is seen as an obstacle to other economic activities, especially agriculture. Although broad and specific legislation (especially the Atlantic Forest Law) on the management of Araucaria Forest resources have attempted to promote sustainable management through the diffusion of technologies, the management of forested areas is mostly forbidden. Therefore, it is a critical moment in a growing effort, championed by a group of researchers, to show that natural forests in Southern Brazil can be recovered and can be an important part of rural life.

4. The development of a system for engaging local communities in natural resources management

Before introducing the main components of our approach, it is important to clarify the terminology used. Initially, we understand that landscape level planning and management for natural resource governance (notably SFM) is the foundation of territorial zoning and follow the 12 principles, discussed and adopted by countries at the Convention of Biological Development (CBD). In applying the principles of the ecosystem approach, the following five points are proposed as operational guidance: a) focus on the relationships and processes within ecosystem; b) enhance benefit-sharing; c) use adaptive management practices; d) carry out management actions at the scale appropriate for the issue being addressed, with decentralization to lowest level, as appropriate; and e) ensure inter-sectoral cooperation.

Following FAO, SFM is defined as the stewardship and use of forests and forest lands in a way, and at a rate, that maintains their biodiversity, productivity, regeneration capacity, vitality and their potential to fulfill, now and in the future, relevant ecological, economic and social functions, at local, national, and global levels, and that does not cause damage to other ecosystems (FAO, 2005). This leads us to the idea behind "forest lands", which are defined by the US Forest Service as land at least 10% stocked with live trees, or land formerly having such a tree cover, and not currently developed for non-forest use. The minimum area of forest land recognized is 0.40 ha (Smith et al., 2009).

The development of a land management system that could integrate multiple uses of natural resources with a participatory approach in a social and politically complex context was one of the first and most difficult challenges to overcome. By using traditional forest management concepts, adapted to the current stage of scientific knowledge and societal comprehensiveness, we introduced a participatory approach as a means to engage local communities in order to build a management plan that could provide landowners with ways to plan the use of their properties by combining agroforestry, forest management and natural resource conservation. The decision-making process uses local ecological knowledge (LEK) as part of the input necessary for establishing the goals and objectives and is based on the demands and interests of landowners.

The conceptual foundation for the development of the method used in this paper is called *Regeneration by Stands*, a method focused on the management of forest resources. Forest management, for the purpose of this chapter, can be defined as the set of actions related to forested areas with a focus on its silviculture that aims to optimize the production of goods and services in a sustainable manner over time (Rosot, 2007); such optimization relies on identifying potential land uses which extends beyond currently forested areas. When planning the use of an area we must consider that multiple uses and functions cannot coexist simultaneously in the same place at the same time, consequently requiring the prioritization of tasks, the identification of preferred uses and the analysis of their compatibility and zoning (Gonzales et al., 2006). Furthermore, while designing an integrated plan we must consider all the resources and limitations that a property (or an area) may have. This process establishes a baseline considering multiple alternatives for managing land resources while decision-making takes advantage of the information available to determine specific actions for different areas considering a landscape approach (Rosot et al., 2006). By deciding the areas in which forests will play an economic and ecological role, a sustainable forest management plan can be defined as a means to organize the use of forest resources, which is intended primarily to ensure its perpetuity (Gonzales et al., 2006).

Here, we adapt Regeneration by Stands, one of the most successful forest management methods, as a means to integrate a participatory approach to the decision-making process with sustainable natural resources use. Regeneration by Stands has been widely implemented throughout Central Europe and has its historical origins in the work of Friedrich Judeich, in late nineteenth-century Germany. The method can be considered one of the most advanced planning systems currently available as it allows forest areas to respond effectively to the challenge of multifunctional management and conservation of forest resources (Gonzales et al., 2006). The system is based on maintaining a balance of age classes and the ability to generate goods and services from forests rather than turning it into a predetermined rigid pattern. This method differs from traditional methods of forest management mainly in relation to its short planning period, typically 10 years. As a consequence, management is based on continuous re-assessments in which the stands are evaluated in terms of their resources and respective use (i.e. objective and management applied); the system is therefore more flexible to changes such as forest fires, market demand, and land-owner interests, among others.

Hernando et al. (2010) applied the method of Regeneration by Stands for managing Natura 2000 forest sites in Spain, considering two different phases. In the first phase, the study area was divided in "stands", considered as any homogeneous patch of vegetation using Geographic Information System and Remote Sensing technologies and detailed fieldwork. The second phase evaluated the conservation status of each stand; the conservation status of the habitat was then obtained by integrating these values. Finally, forestry management measures were recommended for maintaining the favorable conservation status of the study area. These measures included consumption of the forest resources in such a way as to satisfy the objectives of both landowners and society (Brunson & Huntsinger, 2008; Davis & Johnson, 1987; Irvine et al., 2009). In Latin America the Regeneration by Stands method has not been implemented, with the exception of some studies in temperate forests in Chile (Rivera et al., 2002; Cruz et al., 2005).

Although some improvements in developing a sustainable forest system have been achieved, harmonizing different productive (economic) and conservation goals is still difficult to obtain. As stated above, in an ideal scenario a land management system should integrate multiple uses of natural resources which require knowledge from different areas of expertise. As a consequence, techniques applied to one activity should be balanced considering the outcomes in other areas. For example, conservation techniques used in agriculture should consider its effects on downstream water quality (e.g. no-till farming, green manure, among others) or the effects of agroforestry systems on increased crop pollination (and overall biodiversity). The multidisciplinary approach necessary for achieving integrated management might require specific solutions that can be designed using different expertise in a participatory system. However, in this paper we do not aim to detail agricultural alternatives, but rather consider some general principles that can be used to integrate agriculture with other land uses.

4.1 Locally adapted participatory sustainable forest management system – lapSFM

The definition and implementation of the "**locally adapted participatory sustainable forest management - lapSFM**" system followed two main steps that are common to both scenarios, encompassing landscapes with both small and large rural properties. The system aims to deliver a Management Plan, as a part of a Roadmap (Figure 2), composed of different stages. The first phase – the *Ecosystem Analysis* – is related to the landscape as a whole, which can be a property, a set of properties, a municipality or a watershed, for example. The main purpose of this phase is to design territory zoning, based on spatially organized available information. For the purpose of this paper the second phase – the *Management Plan* – is related to the rural property and will focus not only on forested areas (forest) but also on forest lands, in which the forest component includes agroforestry, forest plantations and management of native fragments. Other designated zones, such as agriculture, are noted in the lapSFM but not discussed in this paper. The second phase also encompasses the monitoring activities of the implemented silvicultural treatments.

Phase 1: The initial step in the participatory forest management system refers to a broad analysis of the current environmental state of the area under consideration. It can be defined as a territory zoning which is based on the compilation of environmental data obtained from primary or secondary sources and requires initial land-use/land-cover (LULC) mapping. When possible, a landscape (or an ecosystem) approach is always the best way for dealing with large areas because of the managerial possibilities that planning at this level has for natural resource governance (notably SFM). Such a map should locate different land uses and different forest types and can be obtained either from satellite imagery (including Google Earth) or based on available ancillary information. In both cases a field verification of the classification is recommended as a means to update and check the gathered information. Additional cartographic, soil and hydrography layers can also be used and integrated into a GIS platform. The LULC mapping with the goal of creating territory zoning is a participatory process. Different stakeholders, including landowners, local government representatives, environmental agencies and academic institutions, for example, help in defining land use priorities and identifying the consequences of different decisions on the landscape configuration. LULC classes that are defined by environmental law as "restricted use" must also be mapped. For the forested areas or for forest lands, a procedure performed in a GIS platform can define the "forest stands", based on cross referencing the "territory zoning" layer and the "forest sub-typology" layer.

Phase 2: Following the territory zoning and the definition of the areas in which agroforestry (for forest land) and forest management (for forest) will take place either directly or indirectly, we define the techniques to be applied in forest management or other forest-based activities. This step involves the characterisation of forest stands (including areas to be planted or restored) and the definition of specific silvicultural treatments and/or other long-term actions to be put into practice for each stand. The characterization of forest stands is an initial step that aims at creating a rule of thumb for correlating the types of forests found in Southern Brazil and potential uses and management options. In this system, the vegetation present in an area is characterised by its structure, species composition, successional stage, threats and levels of degradation; this information is then correlated with the landowners' views in relation to the area and includes economic expectations and potential management practices. As a result, we can classify the current stage of conservation of most Araucaria Forest into five different vegetation types (discussed further below) which we denominate as Management Units – MUs.

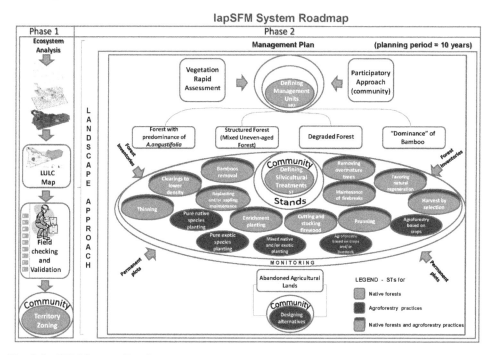

Fig. 2. lapSFM System Roadmap

The broad characterisation of each MU allows for the definition of a set of specific management actions aimed at achieving their pre-defined goals. Each MU is defined by the available information and by in-site checking which confirms the groups and general management actions; in each area (or stand) it is recommended that a rapid forest inventory take place to complement the information on the available resources which will support the actions to be applied. Based on the goals of each MU, general and specific silviculture approaches are proposed. Depending on the MU, different actions are expected to take place, such as thinning (and definition of which plants to be favoured), planting, pruning, removal/introduction of trees or species or no treatment at all. Specific decisions such as plants to be removed or planted are made in the field during designated field visits established in the lapSFM.

As the classification of land cover into MUs is a participatory process which allows for the inclusion of historical, cultural and economic inputs, its application is not only restricted to a vegetation classification but instead aims at integrating social, environmental and economic variables.

The five MUs are:

Forest with Araucaria predominance MU1 – this MU is a general designation for all places whose canopy is dominated by *Araucaria angustifolia* (paraná-pine) regardless of the forest structure, composition, and current use. However, under this land use classification it is possible to identity different levels of canopy cover which are related to forest dynamic processes, especially for natural regeneration. In situations where the canopy has a higher density (> 70%) of predominantly overmature (remnant) paraná-pine trees, a near absent natural regeneration is observed. This process seems to be related to paraná-pine life cycle (~ 400 years) which creates light conditions on the forest floor below the limits required for seedlings to thrive; this situation is sometimes exacerbated by the presence of *Dicksonia sellowiana* and other species of fern. On the other hand, when a more open canopy is present, the ability for a regular natural regeneration to occur depends on whether bamboo species *Merostachys skvortzovii* and others are present or not as they take advantage of favorable light conditions and form a homogenous stratum that prevents any regeneration. Although both situations might require natural resources management through direct intervention, legal and political factors constrain most initiatives, leading to land abandonment. Such a situation tends to cause a gradual degradation of forest structure and floristic composition as a result of trees senescence and lack of regeneration accompanied with further bamboo spreading. Although current legislation was designed to halt deforestation it also resulted in farmers no longer holding decision-making power over land use. Their reaction was to convert forested areas into other uses (also by illegal means); this counter-productive situation if not tackled will result in more natural resources degradation while contributing to continued legal, economic and ownership insecurity.

Structured Forest (Mixed Uneven-aged Forest) MU2 - Structured Mixed Uneven-aged Forests correspond to secondary forests – regardless of whether or not they contain remnant trees – that have a diversified vertical structure, high tree species richness and in some cases, accumulation of forest biomass and volume. This forest is multi-strata structured with a canopy dominated by shade-tolerant and remnants of intolerant species with one or more additional tree strata. Natural regeneration is abundant with high species diversity, although the development of saplings is severely reduced when the understory is dominated by bamboo. For its structural complexity and the available forest biomass, classical methods of forestry (e.g. group or single-tree selection) may be applied to improve its composition and optimize biomass production with reduced impacts.

The MU2 is considered an ideal type for forest management as it has a developed horizontal and vertical structure. Although it is a secondary forest, its management could include the removal of trees for lumber and firewood as a result of the thinning of trees with undesirable characteristics (multiple stems, rotten, broken) as well as removing senescent pioneer trees (e.g. *Ocotea puberula*). The main goal of the management is to provide more light for regeneration and the development of commercial or ecologically important trees. In addition to thinning, controlling bamboo might be necessary as some species commonly find ideal conditions and tend to become invasive. In the early stages of bamboo development, mostly during seed germination following entire population die-offs, it is possible to control the re-population by a manual sapling removal. However, in most conditions repeated cutting using brushcutters with or without additional procedures such as chemical control might be necessary.

Degraded Forest MU3 - in the context of this project Degraded Forests are considered as forests that have suffered intensive logging and may have been affected by forest fires which caused a significant reduction in biomass and volume, as well as substantial changes to their structure and floristic composition (see Lund, 2009 for degradation definition). Remnant trees are of low commercial value (hollow, broken, burned) and possibly rejected during earlier harvestings; such trees are mostly old-growth usually in the senescence process and may occur in densities ranging from 5 to 15 trees per hectare. Under an open canopy there is a variety of possible situations: a) gaps are invaded by short grass under which scarce regeneration is observed (estimated between 500 to 1,500 plants/ha); b) pioneer trees, such as *Mimosa scabrella, Vernonia discolor* and others, dominate and might advance in succession or might be invaded by bamboo species; c) a mix of vines, pioneer shrubs and grasses dominate the gaps in which few trees are observed. The relatively open canopy of this type of forest is the main factor that allows bamboo to develop into near-exclusive populations. Such invasive behavior has been observed throughout Southern Brazil. In this kind of forest, natural regeneration is in most cases absent and therefore requires intense understory and bamboo management in order to allow for any commercial or ecological purpose. However, considerable initial income might be possible by removing fallen wood and any wood resulting from the thinning of undesirable trees.

Dominance of bamboo MU4 – this management unit is characterized by the dominance of bamboo species (especially *Merostachys skvortzovii* Send.) that show an invasive behavior (although it is a native species) forming highly populated, near-exclusive communities under a very open canopy. When trees are present, it might be related to previous logging or pioneers that took advantage of the last bamboo die-off event; arguably this type of vegetation might not be considered a forest. Large volumes of wood remnants from harvesting and wood debris might be found which can be used for firewood and sometimes even for sawmill processing. This sub-type requires intensive bamboo management. The specific procedures will depend on the degree of bamboo development, but in general it is expected that controlling bamboo populations will increase natural regeneration. Just as in MU3, the control of bamboo can be done manually and with a brushcutter or in more extreme cases, through chemical control or even by using bulldozer blades to remove the plant and/or root systems. After bamboo removal, the development of pioneer species, especially *Mimosa scabrella* is intense depending on seed availability in the seed bank or nearby dispersion. The management of pioneer species, including practices such as thinning, is possible as early as the third year; pioneer communities can be further managed for agroforest or forest systems.

Abandoned agriculture land MU5 – this type of land cover is characterized by abandoned agricultural areas and is observed especially in small farms. The vegetation is variable and covered by herbaceous (including crops and invasive species regeneration), shrubs and pioneer trees which characterizes early stages of ecological succession. Generally these areas are no longer suitable for annual crops due to depletion in soil fertility following successive crop rotations, although strict environmental legislation and rural out-migration might also be factors in abandonment. Thus, depending on the situation, the state of fallow might last until soil fertility is restored, allowing for new crops, or abandonment continues and forest succession advances (note that legislation severely restricts any land use if succession reaches a stage dominated by pioneer trees). As an alternative to traditional agriculture, agroforestry can be adopted following a wide variety of practices characterized by the establishment of pure or mixed plantations that might include traditional crops, medicinal plants, trees (both native and exotic) and livestock, or simply allowing forest succession with or without management. In essence, agroforestry systems are here considered as practices that combine the spatial and temporal management of tree species with horticulture and animal husbandry. The main goals of agroforestry is to allow for the diversification of production and income and to use the different ecological interactions that a multi-species system can provide which also tends to promote more environmentally sustainable practices.

Along with the management of natural resources using one of the Management Units, it is also essential to monitor the development of each system as specific ecological dynamics might lead to unexpected results. Any silvicultural prescription will necessarily affect the forest in many ways with varying levels of intensity. Therefore, monitoring is a means of assessing the forest response to management practices over time and space. Traditional monitoring systems include the adoption of permanent plots to monitor changes in vegetation structure and composition, soil fertility, water quality and other components that combined determine if the pre-defined goals will be achieved, both economically and environmentally. In forested areas, the implementation of forest inventories can be used as the basis for the monitoring and sometimes is restricted to one or few forest components (trees, regeneration, epiphytes, etc.). Ideally, forest inventories are conducted on a regular basis through permanent or temporary plots. Other methods such as rapid ecological evaluations are also useful tools for identifying general trends in the forest dynamics. Whichever the method chosen, monitoring is very desirable as it allows managers to determine if expected objectives will be achieved. It is worth pointing out that monitoring should not be a task restricted to academic interests but should be used as a tool for decision-making. As a consequence, local communities should be involved.

The steps and MUs described are part of a lapSFM that can be applied to landscapes characterised by both small and large properties. However, as land tenure is variable together with cultural and economic factors, adaptable approaches are necessary and should reflect the context of local communities. In the case of large properties, often with a single or reduced number of owners, the management based on the proposed method is also feasible. However, it is of extreme importance that nearby properties are also engaged in discussing potential involvement and natural resources management (e.g. allowing local communities to manage non-wood forest products – NWFP). Such an approach also has the advantage of reducing tensions often found between large and small landowners. On the other hand, regions characterized by small properties in which the management of natural resources lack the scale to reach intended markets, along with the difficulty to produce and implement management plans, local communities are encouraged to act in co-operative systems and manage their forests following a common lapSFM.

4.2 Case study 1 – Landscape characterised by small rural properties

The implementation of the lapSFM in a landscape characterised by a mosaic of small properties took place in the southwest of Paraná State, (Southern Brazil) in an area of approximately 1,200 ha within the Imbituvão River Basin, municipality of Fernandes Pinheiro (25°32'29.64"S, 50°33'44.58"W; Figure 3, A). The project was developed through an international cooperation initiative between the Rottenburg University of Applied Forest Sciences (Germany) and the Midwest State University in Irati (Brazil). The project was designed to test several alternative techniques for forest recovery and sustainable management, with the goal of producing timber and non-wood forest products (NWFP). The 40 properties that participated in the project are all considered small properties, based on State classification; however, a significant variation in size was observed as the properties range from 2.4 to 50.8 ha (average of 15.5 ha).

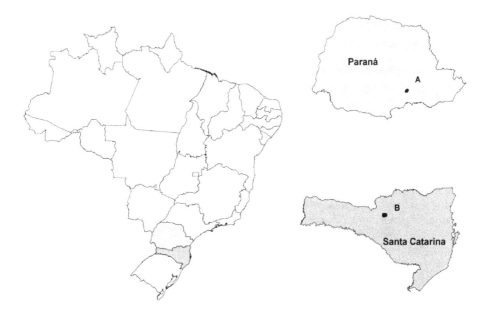

Fig. 3. Location of the study areas in Southern Brazil. A – Imbituvão River Basin. B – Caçador Forest Reserve.

The region was chosen as it represents typical conditions in rural Southern Brazil in terms of its economy (agriculture based), social (low Human Development Index- HDI) and land tenure (properties <50 ha) aspects. Additionally, the regional landscape is characteristic of Southern Brazil including a mosaic of forest and non-forest zones (Figure 4). Initially, properties were visited to introduce landowners to the project and, when possible, forest fragments and other land use areas were visited and described in order to conduct the Ecosystem Analysis (described above) and develop the land use and land cover (LULC) map. The process used tools such as a navigation GPS, questionnaires, cartographical maps and satellite images from Google Earth. An initial land use map was also prepared based on imagery and secondary data. The response from the local community was generally positive, with 40 small landowners engaged in the project. Interviews were carried out using structured questionnaires aimed at gathering data on property characteristics, current activities, social and economic factors, and views on natural resources issues (including expectations, difficulties, benefits, etc.).

In this project, permanent plots and on occasion forest inventories were implemented as part of the monitoring of the lapSFM and to gather information for the development of silvicultural treatments (STs). We engaged local communities in decision-making process by providing landowners with a tentative list of wood and non-wood species as a basis for defining species of interest and STs to be used. The STs were also presented as suggested options to be discussed and adapted if necessary. The information obtained from landowner's expectations and the Ecosystem Analysis was used to define a LULC map. As forest inventories and permanent plots were carried out by the supporting academic institutions, it is expected such information will be incorporated in the Management Plan revision as results become available.

Fig. 4. Photo of the Imbituvão River Basin landscape which is characteristic of Southern Brazil.

The project also included a conservation component which was based on a landscape ecology approach aiming at conserving water resources and promoting biological conservation. A landscape analysis supported by the LULC map provided the basis for discussing with landowners the best areas and practices for enhancing connectivity between forested areas in order to create ecological corridors and to protect riparian ecosystems. The process of engaging landowners in those specific goals occurs through a process of negotiation and should not compromise families' income and decision-making power over the land.

The project proposed five silvicultural treatments (STs) following the logic presented in the lapSFM system Roadmap (Figure 2; see Table 1 for a summary of proposed STs and their correlation to original LULC). Each ST is briefly described as follows:

- Exotic-native species system – this ST involves reforesting areas based on combining a eucalypt species (*E. dunnii* or *E. benthamii*) with approximately five native species. The main driver for this system was the need for landowners to comply with current legislation. As such, properties are required to have a minimum of 20% of their area covered with forests which also have very restrictive regulations regarding management; properties that do not comply are compelled to reforest in order to reach the required area. The need to put aside areas for forest protection usually causes dissatisfaction among landowners and reforestation systems that provide economical return are preferred. Due to the adaptation to regional climate and fast growth rates, eucalypt species have been planted in large scale commercial plantation programs in Southern Brazil and are also preferred by many landowners. Legislation permits use of exotic species if accompanied with native and different wood and non-wood species in the system.

- Pioneer species system – the system is based on the combination of bracatinga (*Mimosa scabrella* Benth.) – a fast-growth pioneer species – with other native species (minimum 5). This species has been traditionally managed through 30-year cycles after which the trees are logged and the land is burned to promote species regeneration. The species can be introduced by sowing, planting or simply managing the natural regeneration; other species included in the system are usually planted but managing natural regeneration is also possible. It is worth mentioning that *M. scabrella* is a leguminous species that allows nitrogen fixation in the soil and accelerates soil restoration.

- Mix of native species – this treatment combines different native species that are planted in mosaics based on their ecological requirements in order to hasten succession. This model is used mainly for restoration purposes as management might be restricted by current legislation.

- Enrichment planting – the system is based on planting native species in already forested areas in which commercial species can be planted. Usually the model is used in degraded forests and management is restricted by current legislation.

- Non-wood forest management – this is an agroforestry system that integrates management of forest species with non-wood products. There are different alternatives to be employed such as apiculture (native bees and *Apis melifera*), fruit production, *Ilex paraguariensis* (tea, *mate* and other products) and medicinal, aromatic and seasoning plants.

The challenges faced by the project include different social, economic and legal factors. As noted above, natural resources management is subject to restrictive legislation that constrains most technical alternatives. Therefore, landowners tend to avoid adopting any alternative that involves planting native trees due to the insecurity regarding future use. They may even feel compelled to reduce forested areas on their properties (which is in direct conflict with legislation). Additionally, the lack of a tradition of forest management tends to drive landowners to livestock or traditional agriculture which is perceived as unrestrained by the environmental legislation applied to forests. Finally, the small-scale production related to small properties usually does not find a market unless the owners are organised into co-operatives. Traditionally, rural communities are not well organized in Southern Brazil and several past experiences in the region show that in the long-term co-operatives rely on external assistance to maintain the organization.

Current land use and/or land cover	Silvicultural treatments (ST)
Degraded (disturbed) forests, without legal use restrictions	Planting pure exotic/native or a combination; thinning trees with unwanted characteristics and planting NWFP; manage pioneer species for timber and encourage succession advancement.
Degraded (disturbed) forests with legal use restrictions (riparian zones)	Recovering using native species and an indirect use of the area (Example: honey production or decorative plant production; removal and use of invasive tree species for lumber, such as *Hovenia dulcis* (oriental raisin tree))
Well structured forest fragments	Thinning trees with unwanted characteristics to improve timber stock; manage NWFP.
Non-forest or early stage of native vegetation regeneration	Enrichment planting (Example: *Mate* (tea), fruit trees, Eucalyptus, Pine); mixed or pure stands
Non-forest (land use based on agriculture and pasture)	Agroforestry (Example: trees in contours, shelterbelts, windbreaks or hedgerows; trees along internal roads)

Table 1. Summary of the land use and land cover (LULC) and respective silvicultural treatments (STs) proposed for the Imbituvão River Basin case study

4.3 Case study 2 – A large rural property in a small farmers mosaic

In this case study, the "core" area chosen for implementing a IapSFM is the Caçador Forest Reserve, a 1,000-hectare Forest Reserve (not a conservation unit) which belongs to the Brazilian Corporation of Agricultural Research (Embrapa) and is located in the Araucaria Forest region (25°32'29.64"S 50°33'44.58"W; Figure 3). Although the area was selectively exploited in the past, the Reserve has received no silvicultural interventions for the last 20 years as it was treated as a protected area (Kurasz, 2005). It is recognized as one of most well conserved forest remnants of Araucaria Forest and two of the most valuable timber species in the south can be found in the Reserve: *Araucaria angustifolia* (Araucariaceae), and *Ocotea porosa* (Lauraceae). Land tenure in the surrounding region consists almost exclusively of small farms and forest companies (Figure 5). Successive three-year research projects focusing on monitoring the Forest Reserve have been in-place since 2002 by Embrapa and other research and academic institutions. A significant amount of data has been collected over the years.

Following the Roadmap outlined above, the goal of this study is to develop a management plan, following Rosot et al. (2006). The first phase of the Roadmap, which includes land use planning/zoning in collaboration with local stakeholders has been completed. Currently we are in the midst of rolling out phase 2 by developing the forest management plan and associated monitoring activities. All MUs (described above) are represented in the Forest Reserve and therefore provides an excellent case study to examine the diverse forest types currently seen in Southern Brazil. The availability of data was a determining factor for the selection of the study area, as various studies have been conducted in the property and in its surrounding regions in the last ten years. A Geographic Information System (GIS) served as a basis for the territorial planning of the Reserve (Kurasz, 2005), which used mainly legal and environmental criteria – along with the expectations of the community – to define specific zones. These criteria led us to define Areas of Permanent Preservation (riparian vegetation), Areas of Restricted Use (like the Legal Forest Reserve) and unrestricted zones (Figure 6).

Fig. 5. Typical small farm property surrounding the Caçador Forest Reserve. Note the combination of different crops (tomatoes, cabbage) with hay for livestock and Degraded Forest MU3 with use limited to NWFP in the background.

The proposed zoning cannot be considered definitive, but it addresses legal and environmental constraints that must be considered when planning management activities and assigning uses to the area. Once the zones were mapped, landowners from the vicinity were invited to take part in the field checking and validation process.

The next phase involves the development of the forest management plan for the whole area, according to the previously defined zones. As a first step the area was divided into homogeneous management units (forest sub-typologies) by means of on-screen photo interpretation of Ikonos imagery and incorporating the information provided by the vegetation layer and the zoning layer available in GIS. The management units (MUs) represent groups of stands based on forest physiognomy for which the same type of silviculture could be applied (see above for detailed descriptions of MUs) (Figure 6). The initial and general objective is the recovery and improvement of the forest in terms of species composition and structure and to ensure the maintenance or rehabilitation of natural processes of plant succession.

The next step was to split the Forest Reserve into stands based on the intersection of MUs and the zones defined on the territory planning. Neighboring polygons belonging to different MUs or non-contiguous polygons belonging to the same MU constituted different stands. When a stand was crossed by rivers, roads or other physical obstacles, it was subdivided, thus generating two or more different stands (Figure 7).

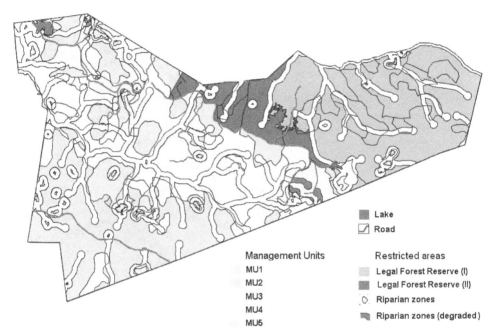

Management Units	Restricted areas
MU1	Lake
MU2	Road
MU3	Legal Forest Reserve (I)
MU4	Legal Forest Reserve (II)
MU5	Riparian zones
	Riparian zones (degraded)

Fig. 6. Management Units and Zones, including Areas of Permanent Preservation (riparian zones), Areas of Restricted Use (like the Legal Forest Reserve), at Caçador Forest Reserve.

Silvicultural Treatments to be applied in each stand are composed of general silvicultural regimes for the respective type of vegetation, plus the special features that require the management of each stand based on its current situation. Site-specific silviculture is defined by the needs identified during the definition of management units and the objectives and constraints of management.

Although the Roadmap suggests that forest inventories should be carried out after MUs are defined, an inventory for Caçador Forest Reserve was already available due to previous research in the region. The existing survey aimed to assess the species composition and structure of the forest. During the survey a stratified random sampling was applied, considering 13 different strata which combined different slope and aspect classes. In addition, seventy-two temporary sample plots of 500 m² (64 plots) and 250 m² (8 plots) were distributed proportionally to the strata areas. In order to evaluate the composition, the horizontal and vertical structure of the forest, plant samples were collected for the species identification.

One of the preliminary outcomes of the integration of MUs with the inventory was to further define stands within the forest using a dynamic correlation in the GIS. One aspect of the analysis was a verification of the defined stands and an analysis of existing stock within each stand. As as a result, average timber stock and increment rates in terms of diameter, volume and basal area (Figure 8) are now available for each stand and together with other relevant information will further support management planning in the area.

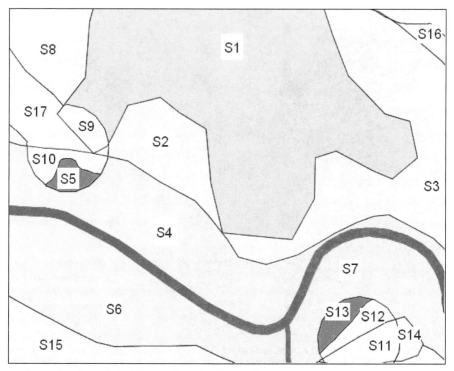

Fig. 7. Example of stand division at Caçador Forest Reserve.

Other preliminary results include:

- Local farmers showed interest in adopting forest management concepts in their properties, mainly through the implementation of agroforestry. However, there is still a general reluctance in engaging in forest management because of environmental law restrictions.
- There is widespread occurrence of native bamboo species that are impeding the development of forest species and causing the degradation of forest communities. This pattern was observed in all different MUs and in the most extreme cases, it was found that the phenomenon is restricting any forest regeneration. Intense human effort is necessary for controlling these species with promising initial results using a brushcutter. On the other hand, economic use can be considered for the removals.
- There is an important stock of firewood previously underestimated and not being considered as a source of income: more than 47m³/ha of firewood and 2.5 m³/ha of lumber were found in the inventory.
- A comprehensive monitoring program is being carried out aiming to assess the response of forest stands to different silvicultural prescriptions through the observation and/or measurement of permanent plots.

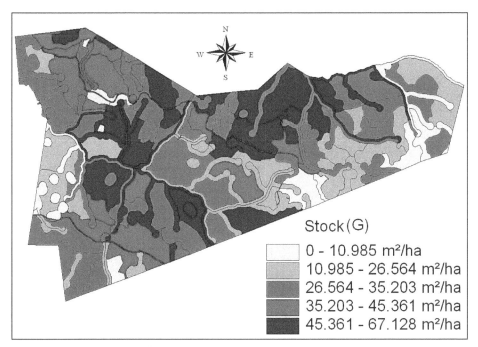

Fig. 8. Basal area (G) classes for the Caçador Forest Reserve

5. Conclusion

In this paper we explored some of the legal, social and environmental issues related to the reduction of the forests in Southern Brazil and propose the implementation of a "locally adapted participatory sustainable forest management" system focusing on reducing both rural poverty and deforestation. Thus, two of the three main components of sustainability, the environmental and social aspects are being taking into account in the lapSFM roadmap. The economical aspect – the third component of sustainability – still needs to be fully developed in order to establish a complete framework for integrated natural resources management. However, in this chapter we discussed some of the problems that influence the economic environment that characterise rural properties (especially small ones) while introducing technical solutions for the management of natural resources. Finally, we aimed at shedding light on the discussion related to the current environmental legislation that we believe should be improved in order to achieve a more practical and homogenous accountability for the protection of the natural resources.

We discussed two case studies of participatory forest management in Southern Brazil through a Roadmap, built to customize practical solutions related to land use and participatory SFM, considering a landscape approach for both small and large properties. The intent was to provide a basis for changes in environmental policy to better reflect the enhancement of SFM in line with agroforestry, forest and non-wood forest resources use found throughout the landscape.

We addressed initially the lack of technical foundation for an integrated natural resources management (especially regarding forested areas) by introducing the lapSFM system. This system promotes a participatory approach through the engagement of landowners into the decision-making in order to combine land use interests with best practices. Furthermore, the system proposed integrates Local Ecological Knowledge – LEK – into the technical framework; as such, the result is a locally adapted management of natural resources. However, new approaches such as that outlined here face important challenges. Initially, it requires knowledge from different areas of expertise that is not always available, mainly in more isolated communities. Secondly, changing the top-down approach usually employed by technicians demands further training and willingness to share authority over decision-making. Thirdly, a turnover in mentality towards a group commitment for achieving common objectives might be a slow process; however, the introduction of co-operatives supported by adequate public financing and technical institutions is likely one of the best solutions for overcoming problems such as scale production. Finally, the most important contribution of this project is proposing solutions focused on enhancing economic prosperity tied to conservation.

By introducing production diversification, landowners can reduce their dependency on the price of globalized commodities that are mostly driven by policies at the national level focused on large-scale productions for international competition. In such situations, small-scale, unsubsidized agriculture (as subsidies and financing are generally designed for large-scale business) has little chance of success. Managing the land in integrated systems of production that involve forests and agroforestry and allow for forest and non-forest wood products to be produced together with crops and livestock has the potential to lead to prosperity while protecting the natural resources.

The lapSFM system also re-introduces the forest component as an economic alternative for landowners. Current legislation that restricts land use together with antiquated ideas related to conservation and responsibility regarding natural resources has generated antagonism between farmers and forests. Although most forest fragments in Southern Brazil are found in small properties, the current legislation does not provide incentives for landowners to protect natural resources but rather only restricts land use and increases economic insecurity. As a consequence, while landowners are key elements in forest conservation, they solely carry the burden for maintaining land under protection. Altering this situation involves various strategies and includes changes in the legislation, public education and pro-active government policies. In the last few years, new public policies in Brazil have introduced programs for financing small-scale agriculture with clear beneficial consequences. Other policies such as the payment for environmental services (PES) have only recently started to be regulated and require further development and study. While PES has been used to promote reforestation and agroforestry, it is mostly used in the context of water protection. Many initiatives are now being implemented by states and municipalities and a National Policy of Environmental Services is being discussed in Congress. Ultimately such initiatives are helping to create a common ground on which relationships between landowners and forests can develop.

Other alternatives derived from international agreements such as REDD (Reducing Emissions from Deforestation and Forest Degradation) are important international policies and positive incentives relating to reducing emissions from deforestation and forest degradation in developing countries. In the second part of United Nations Framework Convention on Climate Change [UNFCCC] Conference in Copenhagen (2009), the importance of including the role of conservation, sustainable management of forests and enhancement of forest carbon stocks in developing countries was recognized in its initial rationale (called REDD+). Currently, REDD++ is being conceptualized and includes agroforestry, following Word Agroforestry Centre [ICRAF] reasoning (Akinnifesi, 2010).

The PES, REDD and ecosystem-based mitigation of greenhouse gases are all instruments of finance transfer between industrialized and developing countries in exchange for emission reductions associated with improvements in forest protection and management. However, commonly these international agreements get 'lost in translation' from the international and national level to forest landowners and communities. LapSFM is being built as a tool to help landowners in Southern Brazil, who are stewards of natural forest patches, be part of this new era, acknowledging that coherent public policies and legislation are necessary to link different levels of decision making – the international to the local.

6. Acknowledgements

This project was made possible through the cooperation of a number of agencies and institutions including: the Agricultural Research and Rural Extension Corporation of Santa Catarina (EPAGRI), Federal Technological University of Paraná (UTFPR), Rural Extension Institute of Paraná (EMATER/PR), and Environmental Institute of Paraná (IAP). We would like to thank the following institutions for their kind financial support: Brazilian Agricultural Research Corporation, Rottenburg University of Applied Forest Sciences, the Secretary of Science, Technology and Post-Secondary Education (SETI) of Paraná State, and the National Council for Scientific and Technological Development (CNPq). Finally, we would like to thank our colleagues for their assistance in developing and carrying out the research described herein, including: Ana Hernando, Arnaldo de Oliveira Soares, Carlos Henrique Nauiack, Carlos Roberto Úrio, Evaldo Muñoz-Braz, Fernando Luis Dlugosz, Flavia Colla, Gabriel Berenhauser Leite, Lisâneas Albergoni, Luziane Francicson, Nelson Carlos Rosot, Osni Ruppel, Pablo Cruz, and Pilar Gallo.

7. References

Akinnifesi, F. (2010). Reducing emissions through agroforestry Research summary. Jotoafrika adapting to climate change in Africa. *World Agroforestry Centre* (ICRAF), No.4. pp. 5, ISSN 2075-556

Brasil. (1965). Lei n° 4771, de 16 de setembro de 1965. Institui o novo Código Florestal. Diário Oficial [da República Federativa do Brasil], Brasília. pp. 9529

Brasil. (1994). Regulamenta os arts. 15, 19, 20 e 21 da lei 4.771, de 15/09/1965, que institui o novo código florestal, e da outras providências. Diário Oficial [da República Federativa do Brasil], Brasília

Brasil. (2006). Decreto n° 11428, 22 de dezembro de 2006. Dispõe sobre a utilização e proteção da vegetação native do Bioma Mata Atlântica, e dá outras providências. Diário Oficial [da República Federativa do Brasil], Brasília, pp. 1

Brasil. (2008). Decreto n° 6514, de 22 de junho de 2008. Dispõe sobre as infrações e sanções administrativas ao meio ambiente, estabelece o processo administrativo federal para apuração destas infrações, e dá outras providências. Diário Oficial [da República Federativa do Brasil], Brasília. 23/07/2008, pp. 1

Brunson, M. W., & Huntsinger, L. (2008). Ranching as a conservation strategy: can old ranchers save the new west? *Rangeland Ecol Manag*, Vol.61, No.2, (March 2008), pp. 137–147, ISSN 1550-7424

Cruz, P., Honeyman, P., & Caballero, C. (2005). Propuesta metodológica de ordenación forestal, aplicación a bosques de lenga en la XI Región. *Bosque*, Vol.26, No.2, (August 2005), pp. 57-70, ISSN 0717-9200

Davis, L. S., & Johnson, K. N. (1987). *Forest management*. McGraw-Hill, ISBN 634-928-000-000-000-0, New York

FAO. (1993). Forest resources assessment 1990: Tropical countries. FAO *Forestry Paper*, No. 112, ISBN 92-5-103390-0

FAO. (2005). *State of the world's forests 2005*, Food and Agriculture Organization of the United Nations, ISBN 97-892-510518-70, Rome, Italy

Gonzáles, J. M., Piqué, M., & Vericat, P. (2006). *Manual de ordenación por rodales*. Gestión multifuncional de los espacios forestales. Norprint, ISBN 84-690-3133-3, Barcelona

Hernando, A., Tejera, R., Velázques, J., & Ñuñez, M.V. (2010). Quantitatively defining the conservation status of Natura 2000 forest habitats and improving management options for enhancing biodiversity. *Biodiversity and Conservation*, Vol.19, No. 8, pp. 2221-2233, ISSN 0960-3115

Irvine, R.J., Fiorini, S., Yearley, S., McLeod, J. E., Turner, A., Armstrong, H., White, P. C. L., & Van Der Wal, R. (2009). Can managers inform models? Integrating local knowledge into models of red deer habitat use. *J. of Appl. Ecol.*, 46, 2, pp. 344–352, ISSN 0021-8901

ITTO. (1990). ITTO guidelines for the sustainable management of natural tropical forests. *Technical Series 5*. International Tropical Timber Organization, Yokohama, Japan.

Lund, H. Gyde. (2009). *What is a degraded forest?* White paper prepared for FAO. Retrieved from http://home.comcast.net/~gyde/2009forest_degrade.doc

Mc Evoy, T. J. (2004). *Positive impact forestry: a sustainable approach to managing woodlands*. Island Press. pp 268, ISBN 1-55963-788-9 Washington, D.C.

MMA. (2008). Instrução Normativa n° 6, de 23 de setembro de 2008. Reconhece espécies da flora brasileira ameaçadas de extinção e revoga a Portaria Normativa Ibama no 37-N, de 3 de abril de 1992. Diário Oficial [da República Federativa do Brasil], Brasília de 24/09/2008, n° 185, seção 1, pp. 75

Oliveira, Y. M. M. de. (2000). Investigation of remote sensing for assessing and monitoring the Araucaria Forest region of Brazil. Thesis (Doctor of Philosophy). Department of Plant Sciences, University of Oxford. Oxford.

Putz, F. E., & Pinard, M. A. (1993). Reduced-impact logging as a carbon-offset method. *Cons. Biol.* 7, pp. 755-757, ISSN 0888-8892

Rivera, H. (2007). Ordenamento territorial de áreas florestais utilizando avaliação multicritério apoiada por geoprocessamento, fitossociologia e análise multivariada. Curitiba. Thesis (MSc). Department of Forest Engeneering. Universidade Federal do Paraná, Curitiba

Rivera, H., Rudloff, A. & Cruz, P. (2002). *Plan de Ordenación de la Reserva Nacional Valdivia*. CONAF. Santiago

Rosot, M. A. D. (2007). Manejo florestal de uso múltiplo: uma alternativa contra a extinção da floresta com Araucária? *Pesquisa Florestal Brasileira*, Colombo, n. 55, pp. 75-85, ISSN 1809-3647

Rosot, M. A. D., Oliveira, Y. M. M. de, Rivera, H., Cruz, P. & Mattos, P.P. (2006). Desarrollo de un modelo de plan de manejo para áreas protegidas en bosques con araucaria en el sur de Brasil, *Proceedings of IUFRO Second LatinAmerican Congress*. La Serena, Chile, October, 2006

Smith, W. B., Miles, P. D., Perry, C. H. & Pugh, S. A. (2009). *Forest resources of the United States, 2007*. WO GTR-78. USDA Forest Service, Washington Office, Washington, D.C., USA. Retrieved from http://www.fs.fed.us/nrs/pubs/gtr/gtr_wo78.pdf

Part 3

Asia

7

Sustainability of an Urban Forest:
Bukit Timah Nature Reserve, Singapore

Kalyani Chatterjea
National Institute of Education
Nanyang Technological University
Singapore

1. Introduction

Singapore, an island republic, is situated south of the Malay Peninsula, between 1°09′N and 1°29′N and longitudes 103°38′E and 104°06′E. The main island and 60 small islets cover an area of about 710.2 sq km and support a humid tropical type of vegetation. At the time of the founding of modern Singapore in 1819, practically the whole of the main island was forest covered. Land clearance for development was done in massive scale during the colonial times. After the first forest reserves were set up in 1883, efforts to conserve parts of the forested areas have evolved. In 1951 legal protection was given to Bukit Timah, Pandan, Labrador and the water catchment areas. When Singapore became an independent state in1965, there were five nature reserves in all (Wee & Corlett, 1986). Since its independence in 1965, in an effort to develop its economy and infrastructure, Singapore has continued to clear forests to provide land for industries, residential use, military purposes, and infrastructure. With one of the highest population densities in the world, pressure on land is the driving force that has influenced the extents of the forests. But Singapore has managed to provide legal protection to retain some land as reserve forests. Till the 90's nature conservation was a mere governmental task to maintain the forested areas of the island. About 4.5% of the total land area is given to forests and there are a total of four protected nature reserves in Singapore. Of these, Bukit Timah Nature Reserve and Central Water Catchment Reserve are the inland tropical rainforests, with some interior areas of primary rainforest. Protected under the Parks and Trees Act of 2005 for the protection of the native biodiversity, a total area of 3,043 ha is given to these two forests which were contiguous till 1995. Since then a six-lane highway cut through the heart of the forest, segregating Bukit Timah from its much bigger counterpart, into a 164 ha of some secondary and some primary forest. The actual closed forest covers only 75 ha of this. Though small, it is recorded as having 1000 species of flowering plants, 10,000 species of beetles, and many other organisms and does retain an authentic 'feel' of a primeval rainforest in the interiors. The forest is a mixture of lowland and coastal hill dipterocarp forest and some secondary forest, lying on the flanks of the highest (163.6m), mostly granitic hill in Singapore. It is only 12 kms away from the city centre and is surrounded by a fast-growing condominium belt of Singapore. Tagged as the country's flagship nature reserve, Bukit Timah represents the constant struggle and compromise between increasing pressures of urbanization and the commitment towards nature conservation currently faced by all countries.

Expanding populations, increased income and leisure time, altered consumer demands, increasing media and commercial propaganda of nature-based attractions, higher awareness of Nature and natural landscapes, increasing interests in non-urban environments and off-the-beaten track places, as well as proliferation of urban lifestyles (Dotzenko et al., 1967; Poon, 1990; Sutherland et al., 2001) in the past decade have seen a rapid rise in the interest in Nature areas. Whether for a piece of Nature or to satisfy the adventurous, nature areas have become common destinations for people during their leisure times. This evolution of tastes has produced growing pressures on natural landscapes, adversely affecting the natural habitat conditions and causing degradation. Visits to remote forests have seen an unprecedented rise as conducted and organized tours to remote areas, sometimes even with comfortable facilities, have allowed otherwise sedate travelers to choose these Nature areas over other choices as their preferred destinations. Large-scale outdoor recreation leads to greater and more widespread ecological impacts on natural ecosystems (Lynn & Brown, 2003) and forests near urban centres have been the worst affected, as these are seen as places of relaxation and physical exercise by increasing number of urban dwellers.

Being close to home, these urban forests also are visited on a regular basis and hence, the impact to such natural sites is more protracted. This paper discusses the case of Bukit Timah Nature Reserve, the 75 ha of partly primary and partly secondary low land Dipterocarp forest, located alongside the popular condominium belt in central Singapore and which is frequented by an unprecedented number of people from the nearby residential areas. The forest has been accorded protection by law as a gazetted Nature Reserve. As for the interiors, the forest in question is crisscrossed by a network of walking trails and the increased popularity of this forest has led to severe degradation of some of the trails. At the outer boundaries, the forest is getting encircled by the ever-encroaching urban residential and infrastructural developments, altering the peripheral environment and steepening the environmental gradients from forest boundaries to the interiors. As for its patrons, the visitor numbers have gone staggeringly high over the decades when the forest came to be more popular. The service demands of this clientele have also seen a distinct transformation and forest management, so far, has focused on keeping up with popular demands by providing various people-friendly facilities and amenities. Bukit Timah Nature Reserve, therefore, exemplifies the aspirations of nature conservation, as well as nature usage and exploitation that many other forested areas near urban developments may go through. Sustainability of such areas depend greatly on the analysis of the issues involved and assessment of the extent of problems brought about by such invasion from urban development.

2. Methodology for the study

Sustainability of BTNR is impacted upon by many factors, such as the increasing numbers of urban development around the forest and its resultant edge effects, the rising numbers of forest visitors and their impact on the forest interiors, the perceptions of the urban visitors who shape the forests' future through public participation. This study looks at these issues and analyses (i) the encroachment from non-forested urban development that impacts the forest peripheral environment (ii) the physical impacts of the heavy usage in the forest interiors, and (iii) the perceptions of stakeholders towards this remnant forest that may well influence the future sustainability of the forest.Forest peripheral environment was assessed by measuring the ambient temperature, and relative humidity along the forest boundaries, skirting the urban landuse such as the residential developments and the roads around the forest. Data was then mapped using GIS to ascertain the environmental gradient resulting from edge effects of the development.

To quantify the impact of hiking, jogging on soil properties, a post-impact sampling framework was employed, covering forest, trail, and trail-side segments along the forest paths. Following park-designated hiking trails, measurements were taken along transects through forest, trail and trail-sides, mostly at regular 100m intervals. The results were then compared with adjoining undisturbed forested slopes (used as a control), to ascertain the degree of compaction and other changes trails and trail-side sites have gone through.

Visitors to the forest are an increasingly important factor in determining the future sustainability of the forest. Hence, apart from getting data on visitors visiting the forest, their profiles, usage preferences, surveys were also conducted to assess the perceptions these people have about the forest, its value as a nature reserve, and its services as a nature reserve. It is thought that such perceptions among the public users of the forest may well influence the sustainability of the forest.

3. Background and site details

Bukit Timah Nature Reserve (BTNR) is the only primary rainforest in Singapore, housing innumerable species of tropical trees and animals, of which a number are in the endangered list. Originally part of the much larger Central Water Catchment area, this forest was truncated from the bigger part of the forest in 1985 when the six-lane Bukit Timah Expressway (BKE) was constructed to run right through the heart of the forest. In addition to the expressway, currently there are more roads around the forest. As a result BTNR is physically fragmented, permanently severed from the bigger counterpart.

Presently the Nature Reserve has a designated size of some 164 ha of land area although the real forest is only about 75 hectares. In spite of the non-urban landscape around the forest, parts of the inner forest still maintain its primeval characteristics (Fig. 1) and offer the urban population much respite from the stresses of city life.

A metalled road (not open to vehicular traffic) runs up to the summit, while 10.6 kms of dirt trails, often paved with rocks, concrete steps, or even wooden boardwalks criss-cross the forest, over steep slopes, and rocky surfaces. In addition to these, there are mountain biking trails (6kms) that offer varying degrees of challenge to bikers across steep and rough slopes. The forest interior has small streams and caves that are in the more remote areas and some parts are kept off from visitors by not having trails running through them. These interiors still show characteristics of undisturbed primary rainforest. The forest itself is quite dense with close canopy cover, complete with tall trees, heavy growth of lianas and epiphytes, middle and lower shrub layers with abundant supply of fresh ground litter. Studies on geomorphological processes in the past (Chatterjea, 1989a, b) recorded slow movement of surface sediments and a well-balance sediment budget on the well-drained forest floors of Bukit Timah, providing adequate nutrients to the dense vegetation and support to the forest slopes. Till the 1980s and early 1990s the forest was an unknown natural landscape, visited only by researching scientists, and occasional visitors who came there either for intellectual pursuits or to enjoy the aesthetic appeal of an undisturbed forested environment.

Fig. 1. The forest interior, with a walking trail

This close-to-nature status, however, attracted the attention of urban developers who targeted this location to cater to the home choices of an increasingly prosperous and more discerning local population who preferred to be close to nature as a respite from their hectic urban lifestyle. The urban land use plan of Singapore demarcated areas around BTNR for private residential development and soon condominiums filled up the area. In land-starved Singapore this newly-available proximity to Nature created opportunities for private land developers to develop residential properties very close to the Reserve, just outside the officially designated boundary of the forest. Several high-rise and low-rise condominiums have been developed even within 200 meter periphery of the forest (Fig. 2).

Fig. 2. One of the condominiums just outside the forest

Figure 3 shows the land use of the areas around the forest in 1951, with open forest and fruit trees scattered with some low-impact residential development. At this time the only infringement was the railway line and one major road on one side of the forest. Figure 4, based on land use of 2009 however, shows how the urban development has encroached upon the outer boundaries and how transport infrastructure has kept pace with that.

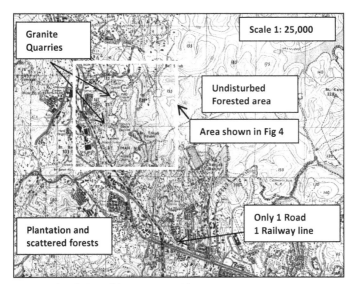

(Adapted from Topographical Map of Singapore, 1951)

Fig. 3. Land use around BTNR in 1951

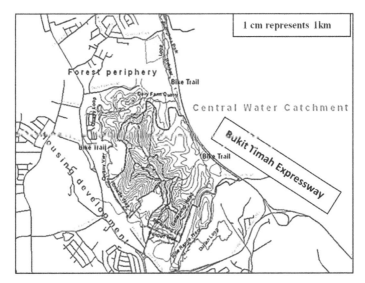

Fig. 4. Land use in 2009 (Based on author's survey): Area covered marked in Fig. 3

The 1990s saw the rapid development of the residential areas and Fig. 5 shows the numbers of such development, all within a 2km radius, which may be considered a walking distance from the forest. Residential development in such close proximity invariably lures the city dwellers and the natural environment of the forest is always a point of attraction, much advertised by the developers. From one condominium in 1984 with 157 residents near this forest, 2009 saw 24 condominiums and more than 21,000 residents, all within a short walking distance away from the forest boundaries (Fig.5) and the number is increasing with two very big developments launched in 2011 barely 300 meters away from the forest.

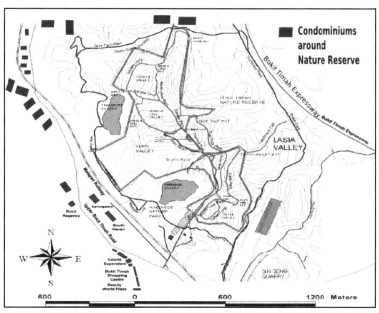

Fig. 5. Residential development within 2km of BTNR

The forest boundaries also have urban infrastructural set-ups such as major roads, a railway line, a rifle range and even a golf course around it. 1985 saw the opening of a six-lane highway connecting the northern parts of the island with the rest, running through the dense forested areas of the Central Water Catchment, literally severing all forested connections between the bigger forests of the catchment with the now truncated BTNR (Fig. 4). So while the urban infrastructural growth connected the island's north and south, the island's most significant forest reserve got fragmented. BTNR now resides as a truncated urban forest, full of original wilderness properties that attract large numbers of newly-aware visitors to the interiors but is ill-placed to handle such large degree of impact.

4. Effects of high density residential development

4.1 Impact on the forest peripheries and forest interiors
Figure 6 shows the temperature differences between the forest peripheries and the forest interiors, while Fig. 7 shows the differences in relative humidity in the same locations.

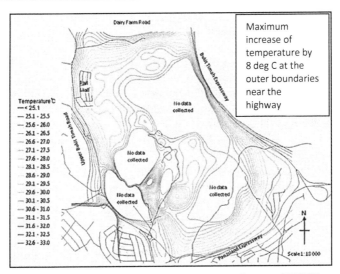

Fig. 6. Isotherms showing distribution of temperature in and around BTNR

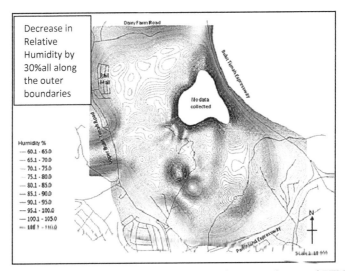

Fig. 7. Isohumes showing distribution of Relative Humidity in and around BTNR

From the data obtained from 162 locations reveal that there is a sharp increase in temperature by 8 deg C in the outer boundaries and that the gradients are very steep in some locations, such as the areas skirting the Bukit Timah Expressway (BKE) as well as along the Dairy Farm Road, another dual-carriageway. Gradient is particularly steep all along the BKE –BTNR interface, with temperature increase of 8 deg C within 50m at the edge of the forest from 25 deg C to 33 deg C. Similar effects are felt for relative humidity. Along the boundaries there has been a recorded drop of 40% in the humidity levels along the BKE-BTNR interface within a horizontal distance of 70m. Along the Dairy Farm Road average relative humidity drop was recorded to be about 30% all along the outside boundary. The slightly lower value recorded along the Dairy Farm Road is thought to be because of the presence of more trees along the outer boundaries of the forest and also because of some planted trees along the roadside. This is not the case along BKE, where the forest ends abruptly at the expressway. All data show clearly that the boundary environment is very different from the inner forest and the outer peripheries of this small forest are subjected to much harsher environment than the forest interior, creating a steep edge: interior difference. This is obviously detrimental to the well-being of the forest vegetation in particular and the entire forest ecosystem in general.

4.2 Impact on forest visitor arrivals

Apart from the other impacts from such urban landscape, residential developments close to the forest boundaries do provide a heavy visitor base to the forest. Figure 8, based on data released by the forest management shows the unprecedented rise in the number of visitors to the forest in recent years and numbers are increasing.

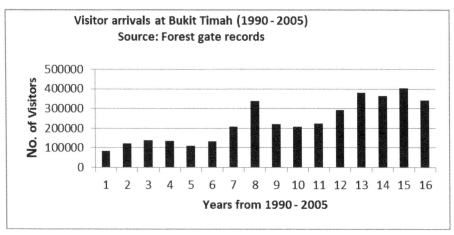

Fig. 8. Increase in number of visitors to BTNR (1990-2007): Overall percentage increase in annual number of visitors from 1990 to 2005 > 400%

The figures, however, are only from records maintained on the visitor arrivals at the main gate. It needs to be mentioned here that BTNR has six other entrances where no records can be maintained because of the open access to those places. Therefore, it can be safely assumed that the actual visitor numbers are much higher than is seen in the official records. It is clear that the forest is a popular recreation getaway for the nature-loving urban dwellers, most of whom are local residents. Visitor surveys were done with 850 persons in total in 2004 and again in 2006 (Table 1).

The 2004 data showed that 68% preferred the trails, while 8% had no clear preference, though they often frequented the trails. This could well mean that these visitors also visited the trails, making the figure a staggering 76%. The 2006 survey revealed similar but increasing trend of people visiting the trails, a possible 83 - 92% of the total visitors. With annual figures going as high as 402621 persons (2004), there are more than 1103 visitors per day and with 83% going to the trails, the 10.6km walking trails alone would see a staggering figure of > 915 visitors per day. This amounts to >86 persons trampling for around 2 hours per km of the trails, which is alarmingly high. While in 2004, 46% percent had visited most of the trails, the 2006 survey revealed that at least 92% had ventured into at least one or more of the walking trails; 78% said they stay for two or more hours at each visit in 2004 while in 2006 surveys revealed that 83% of the visitors spend 1-2 hours in the forest..

Type of observation	Categories surveyed	2004	2006	Comments
Frequency of visit in a year	>10 times	31%	64%	There has been a drastic increase in the frequency of visits
	3 – 10 times	39%	25%	
	1 – 2 times	30%	11%	
Length of each visit	>2 hours	17%	16%	Many people are staying in the forest for a shorter time
	2 hours	61%	42%	
	1 hour	22%	42%	
Destinations within the forest	Main Road	24%	17%	Percentage of people visiting trails might have gone up
	Trails	68%	60%	
	No Preference	8%	23%	
Favourite activities while in the forest	Nature Watch (bird watching)	19%	11%	Solitary activities have seen a decline
	Jogging, exercise, walking	81%	89%	Exercise activities on the rise

Table 1. Results of surveys done for visitors to BTNR (n=850)

The more revealing information relating to altered usage of the forest comes from the 2006 data which show that 89% of the visitors visit the forest for exercise, trail walk, jogging, and biking, with only 11% of the visitors having nature and bird watching in their mind. It is quite obvious that over the years BTNR is seen more as an exercise spot than as an area for quite appreciation of Nature.

4.2.1 Impact of high visitor arrivals

In response to this rising demand for nearness to Nature, Forest managers usually respond with more inroads into the forest and more alluring facilities, which unfortunately, not only take more number of visitors to the interiors, but also allow them to stay longer in the forest. Research shows that the fastest growing recreational activities are associated with forest trail use (Lynn & Brown, 2003) as visitors use these inroads to venture into the remote wild, which may otherwise be non-passable. In BTNR, 84% of the visitors stay for 1-2 hours and all that time are spent on the trails. Trail degradation is established as one of the most evident consequences of expanding visitor numbers in nature parks. While trails are intended to confine visitors to predetermined paths and, thereby, minimize the negative impacts of human visits, additions to the trail network are an inevitable response to the increased movement of people within the forest.

Trampling on trails by increasing number of visitors alters the soil surface characteristics, mechanical properties and hydrophysical behavior. Severe compaction of the top surface leads to greater trail-surface erosion and changes in the forest floor environment, with top soil washed out, tree roots exposed, litter protection removed, and very importantly, compacting the soil beyond root penetration. Over time, the concentration of such activities on pre-defined trails result in extensive surface degradation and wildlife habitat disruption. This reduces the quality of wilderness experience of the visitors, in addition to degradation of the forest environment. As the recreational value of nature trails grows with public engagement in outdoor activities (Bhuju & Ohsawa, 1998), park managers are faced with the dilemma of satisfying demands for opportunities of experiencing Nature from the growing number of visitors and the progressively deteriorating trails. Extensions of trail lengths have been the usual response to reduce impact per length of the trail. However, with the increasing number of visitors, this does not seem to ameliorate the problem but further impact the forest by taking visitors more and more into the forest interiors. Research has also established that trails are not only heavily used, but also widened through overuse, thus degrading the immediate peripheries of the trails as well (Chatterjea, 2007). With mainly recreational use of the forest, as Sutherland et al. (2001) suggest, it is important to examine the impact of recreational hiking on soil hydro-physical characteristics as well as to identify indicators of trail degradation so that the data can be used by land managers to assess the degree of stress imposed by recreational impacts.

Measured values of penetration resistance and shear strength, infiltration rate (Gardner and Chong, 1990) of the top surface soil can be analysed to show how increase in trail use leads to increased surface soil compaction and thus alters the original forest surface hydrological characteristics. Chatterjea (2007) gives details of the trail degradation and provides spatial information on hazardous trail segments which can be used by the forest management to target their repair works to affected sites as well as to plan future repairs, depending on requirements. Data regarding trail degradation along the well-used forest paths in BTNR are compared with the original undisturbed forest surfaces (Table 2). Impact of the heavy use is most obvious in the altered surface compaction. Readings of surface penetration resistance at 406 locations on all trails show alarming rise, as seen in Table 2 below.

Surface Resistance	Forest	Trail	Trail-side
Maximum Resistance (kPa)	965 .3	2068.4	2137.4
Minimum Resistance (kPa)	68.9	137.9	68.9
Mean Resistance (kPa)	419.2	1238.8	912.0

Table 2. Average Surface Penetration Resistance values, indicating surface compaction on trails in BTNR (From 406 locations)

Some trails such as Seraya Loop, being closer to one of the accessible boundary paths, has been widened beyond 3.5m and illustrates how the trails develop severely compacted surfaces, give rise to ready surface wash during frequent rainstorms, wash off the top soil on the trail as well as from the surrounding slopes, become extremely slippery and hence highly hazardous. Surface compactions from two severely degraded trail segments are shown in Fig. 9 and Table 3. They show the differences in surface compaction on trails and undisturbed forest slopes, to illustrate the changes brought about by excessive use of the trails by visitors.

Location	Forest	Trail	Root Steps	Trail-side	Comments
Seraya Loop: Site1	689.4	1323.7	1613.2	861.75	• In this section, there is room for straying on the side of the trail
Seraya Loop: Site2	103.4	1116.8	1654.6	689.4	
Seraya Loop: Site3	386.1	2171.6	2668.0	930.7	• The trail has many bike tread marks and deep grooves
Seraya Loop: Site4	344.7	1034.1	1964.8	758.3	
Seraya Loop: Site5	344.7	1447.7	1778.7	896.2	
Seraya Loop: Site6	137.8	792.8	1916.5	654.9	• The mean width of this section of the trail is 3.4m
Seraya Loop: Site7	193.0	2344.0	1379.0	978.9	
Dairy Farm Loop: Site1	322.0	1623.5	1814.2	761.8	• This section of the trail is very steep – in places > 35 deg
Dairy Farm Loop: Site2	104.4	2116.8	2655.0	859.4	• The trail has many exposed rocks, tree roots, and pot holes
Dairy Farm Loop: Site3	295.1	1971.6	2668.0	980.0	
Dairy Farm Loop: Site4	231.7	1641.1	2660.3	758.8	• The trail has very limited room on the side for sideways extension
Dairy Farm Loop: Site5	244.4	2400.7	2195.7	996.0	
Dairy Farm Loop: Site6	136.8	1752.8	2016.5	723.9	
Dairy Farm Loop: Site7	153.0	2405.0	1878.0	1578.9	

Table 3. Mean Penetration resistance values (kPa) of surface soils on forest, trail, trail-side and root steps in Bukit Timah Nature Reserve

Due to excessive compaction, as shown in the surface penetration data, such trails go beyond repair, disallowing any root penetration, even after being left unused for some length of time. What is alarming is not just the degradation of the demarcated trails, but the increasing penetration data obtained from the trail-sides. These tracts running along the designated trails are initially covered often with indigenous ground vegetation, young saplings and fresh leaf litter, a major source of nutrients to the forest vegetation. With very high human traffic on the trails, visitors tend to go beyond the designated trails and this problem is exaggerated during heavy rains when the trails, with their compacted surfaces

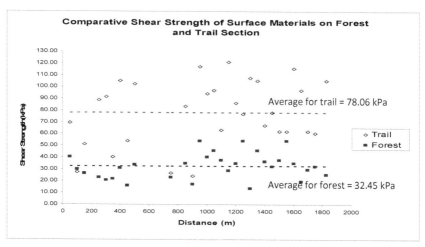

Fig. 9. Comparative Shear Strength of surface materials on different surfaces, indicating compaction: higher shear strength indicates greater compaction

are sites of heavy surfacewash. Top surfaces get muddy and subsequently the slippery surfaces are avoided by the joggers. This leads to widening of the existing trails. Fig.10 shows the width of one of the trails in BTNR, as measured and there is a marked increase in the width in all of the trails from the original layout of 0.5m.

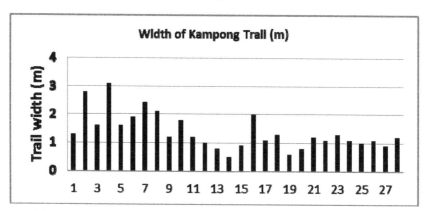

Fig. 10. Width of trail along Kampong Trail, a newly laid out trail at BTNR: x6 increase in some sections

With prolonged impact on the trails, many of the trail segments are ridden with rills, pot holes, exposed tree roots, exposed and lose rock fragments, and smoothened soil surfaces (Fig. 11). Trails with such severe surface degradation and high surface compaction lead to permanent damage to the forest floors, inhibiting further root generation and extension. Plant saplings along the trail-sides often get smothered as trails are extended beyond the original layout. This impact, therefore, is extended to the immediate forest interiors.

Many trails get entrenched with bike tire treads and thus assist in channelizing the developed surface wash (Fig.11). Such damages cannot be easily repaired as trails are heavily used every day, leaving little time for natural regeneration or management intervention. Figure above shows the spatial extent of one of the damaged trail sections in BTNR and even newer trails such as Kampong Trail are not spared with rapid degradation subsequent to the opening up for public use. As these trails run through much of the forest interiors, trail degradation can be seen as a cause for concern as far as the health and sustainability of the forest is concerned.

Fig. 11. Surface degradation along trails

5. BTNR in the eyes of the visitors: Perceptions and views of stakeholders

As the forest became increasingly popular to the local visitors, it became increasingly clear that the views, perspectives of the visitors drive the way this forest is managed. With initiatives from the National Parks Board which manages the forest, the forest has gone through some major facelift in terms of facilities aimed at attracting visitors and it sees large numbers of visitors, mostly locals from the neighbourhood and organized groups from schools and other organizations. It also provides opportunities for local schools, universities and other research bodies to conduct forest ecology related research. The forest has therefore moved from an unknown forest to one that provides services such as recreation, education, and research opportunities.

To understand how this new involvement of the people and the continuous efforts to provide various services affects people's perception of the forest, surveys were conducted with 284 visitors in 2010. Table 4 gives a summary of the views received from the visitors about several aspects related to the forest.

Views and perceptions	Percentage of people interviewed (n=284)
Unfazed by development around the forest	43
Thought urban development too close to the forest was inevitable, given the shortage of land	51
Against development close to the forest	6
Supported idea of restricting visitors	0
Supported idea of paid entrance	0
Wanted more car parks	100
Those who said trails should be repaired	40
Those who said trails should be left as it is	24
Those who didn't notice the degradation of trails	36
Those who said the forest was too crowded but was inevitable	53
Those who said the forest was not too crowded	47

Table 4. Views and perceptions of visitors to BTNR (n=284)

The survey reveals the opinions of an urban population, whose outdoor pursuits do not allow any room for consideration for Nature conservation. Although a fair number of people mentioned that they value coming to the forest as it brings them closer to Nature, they do not see the urgent need to ensure that this piece of forest is conserved for what it is. Most people are comfortable to use it as a Nature Park that gives them a respite from urban stresses and yet they do value it as a piece of Nature.

6. Conclusion and comments

While BTNR falls under IUCN (World Conservation Union) Category IV and is classed as a Habitat/Species Management Area (Protected Area managed mainly for conservation through management intervention) (Australian Government Department of the Environment and Heritage, 2006; IUCN, 2006), its activities also include public participation whether for recreation or education. This dual role thrust on the forest, as a nature area as well as one for recreational and educational use has provided a greater responsibility on the park management, and the issue of survival of the forest environment in the face of rocketing interests in the forest and its natural environment has become a big challenge for environmental conservation in this urban forest remnant. Efforts of the forest management so far has been to offer varieties of activities for different age groups and inclinations and many of the trails have been covered with board walks and efforts have been taken to blend the additional features to the forest environment. In response to demands from educational institutions, the forest now houses an Education Centre at the western boundary that allows schools to hold field classes (Fig. 12). Hindhede Park with its rest houses, viewing platform overlooking the scenic quarry lake is another example of the forest management to divert the pressure of visitors from the real forest interiors. Currently there are strict rules regarding groups visiting the forest in large numbers and these are subject to prior approval but individual or small groups are exempt from such restrictions. Forest management also does trail management and trail maintenance, as and when necessary, using indigenous forest resources to blend with the environment. But such efforts are mostly ad-hoc and not done in a regular planned manner.

Fig. 12. Initiatives taken by the forest management in response to changing usage of the forest: (a) Rest Huts, (b) Education Centre, (c), (d), and (e) Trail management

However, as surveys suggest, rugged trails still seem to be the best attraction of this forest and 89% of the visitors go directly to the trails. New trails of all grades and gradients have been laid out through the forest to cater to the growing demand. One such example is the Kampong Trail which actually links BTNR to the next forest in the Central Water Catchment, the MacRitchie forest. Currently there are various networks of trails through varying gradients of the forest, running through small streams, caves, rock outcrops, dense vegetated slopes and deserted old hutments, giving the much desired rustic and wild environment that urban dwellers crave for during their leisure times.

With annual figures going as high as 402621 persons, the 10.6km walking trails alone would see a staggering 1103 visitors per day. This figure is not final as entrees from the six unguarded entrances are not included in this data. If the IUCN guidelines of 10 persons per hectare per day are followed, the 164 ha area of BTNR should be able to support a visitor figure of 1640 per day (World Tourism Organization, 1999). But the 75 ha closed forested area of BTNR that is actually visited by people can only support an optimum visitor number of only 750 per day. If the present visitor preferences for the trail (92%) are considered, the trails should only get a daily maximum of 690 visitors. The present figures (an estimated minimum of 1103 persons, not including the ones who come to the forest through the unmanned entrances) are markedly higher than this optimum value. As mentioned earlier, the trails do show the impact of such intensive usage.

When forest patches are under stress from overuse, the forest managers have basically two choices: (1) maintenance of trails with the help of engineering structures such as board walks, surface layering with forest/ artificial materials to ensure that the trails are in usable condition even under severe use, (2) controlling the number of visitors to the area. However the strategy of controlling the visitors is never seen as an acceptable alternative for two reasons. The forests are considered as nature areas for the recreation and enjoyment of the public. Any restrictions on their use may be seen as infringement on public rights. Restrictions also keep the public away from the problem, creating a forced avoidance and a possible apathy, which goes against the grain of conservation of public forest areas.

Surveys conducted for this research revealed that visitors to Bukit Timah Nature Reserve do not advocate regulating the number of visitors, although some do mention having restrictions on times to reduce the total number of visitors (Table 4). But, quite contrary to what John and Pang (2004) found out in Hong Kong, visitors to BTNR recorded their concerns for the forest and actually mentioned that the natural environment must be conserved, although they were not sure how. So the public sentiment towards conservation of this forest reserve is positive. Surveys show visitors opting for a combination of all types of management - but are reluctant to accept closing down of trails. Judging from the visitor responses collated for this study, the more acceptable way towards a managed forest is through close monitoring and response through management of the conditions, rather than restricting the visitorship. Therefore, choices left to the park managers have to focus on managing the existing forest paths with the given resources.

Although this forest trail management is generally left to the forest/ park managers, little institutional effort is made towards actually ensuring that it is done in an organized manner. Trail maintenance in many forests is left to the users and other voluntary organizations. While there are many guidelines on trail maintenance, trail layouts (U.S.D.A. Forest Service, 1991; Florida Trail Association, 1991), an effective management, especially in small forest reserves requires an organized schedule and planning based on site specific surveys and location-specific decisions. In the absence of organized scientific support and database, most of the maintenance is dependent on visual surveys by the rangers and is at best an ad hoc attempt to maintain an area under stress from overuse. While vast wildernesses have the resilience to bounce back from stressful conditions, if left to do so, smaller patches of nature around urban areas, such as BTNR, are the ones that are more vulnerable and are in danger of being reduced to degraded lands, especially if proactive measures are not emplaced to monitor the changes in the conditions.

With urban centres on the rise and populations and their lifestyles going through changes with the times, such dwindling nature pockets need to be brought to focus in order to sustain them even in the face of rapid intrusions from urban development. One of the recent notable developments is this growing acceptance for the need to assess and possibly reduce the impacts on nature areas for conserving these pockets of nature in and around urban areas. Assessment of the impacts, demarcation of the stress indicators, identification of the options to improve, prioritizing the amelioration, reducing or avoidance of the causes of threats and hazards are some of the issues that need to be stressed upon, if nature areas are to be conserved, either in big wilderness areas or in places near urban developments. As

Barrow (2006) mentions, environmental risk management incorporates estimation of risk, evaluation of risk and also includes response to risk. Hence, it is imperative that the lay-out of forest trails, regular monitoring of their condition, and their maintenance are carried out in a planned manner by the park managers to benefit the visitor and at the same time reduce the impacts of trail use.

While forests of all dimensions are subject to degradation due to increasing influx of visitors, it is the urban forest patches in close proximity of populated urban centres that are more at risk of such intensive usage and consequent degradation. Developing some long-term monitoring and management protocols, regular monitoring and updating of trail conditions, and constant alignment of usage and amelioration of its impacts are essential for long term sustainable use of the forest located close to urban areas.

At Bukit Timah, while urban development is already at stone's throw from the forest boundaries and this cannot be reverted, it is imperative that any further urban encroachment is kept at bay, through development of a conscientious public view and a committed governmental effort targeted towards long-term conservation of this forest patch. One workable plan could be achieved through managed and well-placed diversion of visitor traffic from the core of the forest by building forest corridors for people yearning leisure activities in green spaces. Several sites may be possible, such as connecting the nearby Bukit Batok Nature Park with the existing Bukit Timah Nature Reserve through building of a walking green corridor across one of the busy roads (Upper Bukit Timah Road). Similarly, the recently vacated Malayan Railway line that skirts BTNR can easily be used as not just a green corridor but also an effective diversion for visitor traffic directly from the main gate of BTNR, thus relieving BTNR from the strangling pressure of visitors. The popularity of the recently vacated railway line proves that it may prove successful. Public education regarding sustainability of BTNR is also essential to maintain the status of the forest as a nature reserve. The current thrusts of the forest management through some public talks, in-forest posters need to reinforced and involvement of NGOs such a Nature Society of Singapore needs to be more targeted if this effort is to be sustained.

Bukit Timah Nature Reserve, for its size, its rich diversity, its services to the local population, its vulnerability in the face of growing urbanization is a forest that is a case for concern. Issue of sustainability of forests has never been as acute as it is in the case of Bukit Timah Nature Reserve.

7. Acknowledgements

Part of this research on the forest environment was funded by a project on Trail Degradation at Bukit Timah Nature Reserve, NIE RP 19/02 KC. Many of the ground data could not have been collected without the help of students from various courses at National Institute of Education, Nanyang Technological University, Singapore. Support from National Parks Board, Singapore and help offered by the officers at Bukit Timah Nature Reserve for conducting this research are acknowledged. Visitors to the forest who helped by giving their views are also acknowledged.

8. References

Australian Government Department of the Environment and Heritage. (2006). Six IUCN Protected Area Categories, In: *IUCN category of reserve*, 02-03-2006, Available from: http://www.deh.gov.au/parks/iucn.html#IV

Barrow, C. (2006). *Environmental Management for Sustainable Development: Principles and Practice* (2nd Edition), Routledge, ISBN-10: 041536535X, London.

Bhuju, D.R. & Ohsawa, M. (1998). Effects of Nature Trails on Ground Vegetation & Understorey Colonization of a Patchy Remnant Forest in an Urban Domain. *Biological Conservation*, Vol. 85, No. 1-2, (July-August 1998), pp. (123-135), ISSN: 0006-3207.

Chatterjea, K. (1989). Surface Wash: The Dominant Geomorphic Process in the Surviving Rainforest of Singapore. *Singapore Journal of Tropical Geography*, Vol. 10, No. 2, (December 1990), pp. (95-109), ISSN: 0129-7619.

Chatterjea, K. (1994). Dynamics of Fluvial and Slope Processes in the Changing Geomorphic Environment of Singapore. *Earth Surface Processes and Landforms*, Vol. 19, No. 7, (November 1994), pp. (585-607), ISSN: 0197-9337.

Chatterjea, K. (2006). Public Use of Urban Forest, its Impact, and related Conservation Issues: The Singapore Experience, In: *Challenging Sustainability: Urban Development and Change in Southeast Asia*, Wong T.C. et al, pp. (53-79), Marshall Cavendish Academic, ISBN:13:978-981-210-307-9, Singapore.

Chatterjea, K. (2007). Assessment and Demarcation of Trail Degradation in a Nature Reserve, Using GIS: Case of Bukit Timah Nature Reserve. *Land Degradation and Development*, Vol. 18, No. 5, (September-October 2007) pp. (500-518), ISSN: 1085-3278.

Dotzenko A. D.; Papamichos N. T. & Romine D. S. (1967). Effects of Recreational Use on Soil & Moisture Conditions in Rocky Mountain National Park. *Journal of Soil & Water Conservation*, Vol. 22, No. 5, (September-October 1967), pp. (196-197), ISSN: 0022-4561.

Florida Trail Association, Inc. (1991). *Trail Manual of Florida Trail*, Florida Trail Association, Gainsville, Florida, USA.

Forest Services, U.S. Department of Agriculture. (1991). *Trails Management Handbook* (amended: FSH 2309.18), Government Printing Office, Washington, D. C. (FSH 2309.18), Washington, D.C." Government Printing Office.

Gardner, B.D. & Chong, S.K. (1990). Hydrological Responses of Compacted Forest Soils. *Journal of Hydrology*, Vol. 112, No. 3-4, (January 1990), pp. (327-334), ISSN: 0022-1694.

IUCN (World Conservation Union). (2006). Protection of Bukit Timah as a reserve, In: *IUCN Management Category*, 02-03-2006, Available from:
http://sea.unep-wcmc.org/sites/pa/0538v.htm

Lynn, N. A. & Brown, R. D. (2003). Effects of Recreational Use Impacts on Hiking Experiences in Natural Areas. *Landscape & Urban Planning*, Vol. 64, No. 1-2, (June 2003), pp. (77-87), ISSN: 0169-2046.

Poon, A. (1990). Competitive Strategies for a "New Tourism", In: *Progress in Tourism, Recreation & Hospitality Management*, Cooper, C. P., pp. (91-102), Belhaven Press, ISBN: 0471945099, London

Sutherland, R. A.; Bussen, J. O.; Plondke, D. L.; Evans, B. M. & Ziegler, A. D (2001) Hydrophysical Degradation Associated With Hiking Trail Use: A Case Study of Hawai'iloa Ridge Trail, O'Ahu, Hawai'i. *Land Degradation & Development*, Vol. 12, No. 1, (January-February 2001), pp. (71-86), ISSN: 1099-145X.

Wee, Y. C. & Corlett, R. T. (1986). *The City and the Forest: Plant Life in Urban Singapore*, Singapore University Press, ISBN:9971691035, Singapore.

World Tourism Organization. (1999). *Guide for local authorities on developing Sustainable Tourism, (Supplementary Volume on Asia and the Pacific)*, World Tourism Organization, ISBN: 9284402808, Madrid, Spain

8

Conflict and Corollaries on Forest and Indigenous People: Experience from Bangladesh

Nur Muhammed[1], Mohitul Hossain[2], Sheeladitya Chakma[2],
Farhad Hossain Masum[2], Roderich von Detten[1] and Gerhard Oesten[1]
[1]Institute of Forestry Economics, University of Freiburg
[2]Institute of Forestry and Environmental Sciences, University of Chittagong
[1]Germany
[2]Bangladesh

1. Introduction

The South Asian nation of Bangladesh, with a total population of approximately 150 million (mill) and an area of 147,570 km[2], is one of the most densely populated country in the world. The current population density is ~1,127.3 people km[-2] (Food and Agriculture Organization of the United Nations [FAO], 2005), up from 755 people km[-2] in 1991 (Bangladesh Bureau of Statistics [BBS], 1993). The economy is based on agriculture and the society is agrarian, with approximately 75% of the population living in the rural areas (United Nations Population Fund [UNFPA], 2006). Per-capita land holdings are approximately 0.12 ha (Government of the People's Republic of Bangladesh [GOB], 2002). Moist, humid, tropical-monsoon climate, with moderately warm temperatures, high humidity, and a wide seasonal variation in rainfall prevail in Bangladesh (GOB, 2001a). Bangladesh is prone to frequent natural calamities and is perceived as a major climate change victim.

Forest cover is shrinking Worldwide, despite many efforts to halt deforestation. Forest land and resources in many developing countries are serious pressure due to extreme poverty exacerbated by overwhelming increasing population. The forestry situation is even worse in Bangladesh that biotic and abiotic pressure associated with inter and intra competition between different landuses, conversion of forest land into industrial and other non-forest uses resulted in denudation and degradation of the hills, loss of forest areas, biodiversity and wildlife habitat in Bangladesh. Traditional forest management system failed to improve the forestry situation in the country. Large scale participatory social forestry program was introduced in the early eighties of the past century throughout the country's denuded and degraded forests as well as in marginal and newly accreted land.

Forests are the home to more than half of all species living around including human being. Population estimates show that there are about 300 - 400 mill indigenous people worldwide (Hinch, 2001; United Nations, 2009; World Bank, 2000). In developing countries approximately 1.2 billion people rely on agroforestry farming. They are recognized as the inhabitants of the World's most biologically diverse territories, possessor of unique linguistic and cultural diversity as well as they are in possession of huge traditional knowledge. However, they suffer from discrimination, marginalization, poverty, hunger and conflicts. More importantly, their indigenous belief system, cultures, languages and ways of life continue to be threatened, sometimes even vulnerable to extinction (United Nations, 2009). There are about 5,000 such tribes/ethnic races worldwide representing 5% of the World's population (Zeppel, 2006). Indigenous people embody and nurture 80% of the World's cultural and biological diversity, and occupy 20% of the World's land surface (United Nations Commission on Sustainable Development [UNCSD], 2002).

In Bangladesh, except for the mangrove forests of Sundarbans, the other major natural forests are the dwelling place of the most of the indigenous communities. Due to lack of substitute products, people in Bangladesh depend on forests and forest resources especially, for fuelwood and timber. Chittagong Hill Tracts (CHTs) are known to be important reservoirs of forests and forest resources in Bangladesh. Apart from the forest resources, this hilly region of the country is bestowed with magnificent natural landscapes, lakes, hilly streams and rivers. Besides, the indigenous people and their huge cultural diversities and unique handcrafts attract clients from home and abroad. So the CHTs are quite important from both economic and ecological standpoints. The Forest Department, being the State agency responsible for forest management in Bangladesh, considers the indigenous people as a major threat to forest management (Roy, 2004). Therefore, an antagonistic relationship has been in existence between indigenous people and Forest Department since long before. Besides, the indigenous people of the CHTs are in constant conflict with Bangladesh Government with regards to land ownership, resource use, and settlement of non-indigenous migrants and other socio-cultural and political discourses, which has made the area unstable and very sensitive.

The indigenous people of the CHTs are alienated from the mainstream society. Hence, it is not possible to utilize properly the forest resources as well as to accrue the full potential of the CHTs under the prevailing circumstances. It is perceived that if a meaningful solution of the problems of the indigenous people of the CHTs can be achieved, forest and indigenous resources may be properly managed to harbour maximum economic gain that could significantly contribute to the economy of Bangladesh. As both the forests and indigenous people of the CHTs are struggling, it was thought necessary to conduct a study in order to explore the root causes and corollaries of the problems with recommendations for possible solutions through several problem-solving approaches. The specific objectives of this study were as follows:

a. A brief review of the forests and indigenous people in Bangladesh.
b. Historical review of the root causes and corollaries of the conflicts of the indigenous people of the CHTs in Bangladesh.
c. Review and analysis of the on-going efforts of conflict resolution in the CHTs.
d. Suggest recommendations for conflict resolution scrutinizing past as well as on-going efforts.

2. Materials and methods

2.1 Study area

Geographically, Chittagong Hill Tracts of Bangladesh (Fig. 1) lie between 90°54′ and 92°50′ East longitude and 21°25′ and 23°45′ North latitude (Chittagong Hill Tracts Development Facility [CHTDF], 2009). It is bordered with Indian States of Tripura on the North, Mizoram on the East, Myanmar on the South and East, and Chittagong district of Bangladesh on its

West. The CHTs include three hill districts namely, Bandarban, Rangamati, and Khagrachari (Fig.1); covering an area of 13,295 km² (10 % of the country's total landmass (BBS, 2001; CHTDF, 2009) and the CHTs represent nearly 50 % of the total national forest area in Bangladesh (Forestry Master Plan [FMP], 1992).

2.2 Methods

Forest dependent indigenous communities of the CHTs in Bangladesh were the main focus of research. Initially a review on Bangladesh forestry and indigenous people are done discussing contemporary global concern. The CHTs region as a part of the undivided India was ruled by the Mughals (1666-1757) followed by the British rulers (1757-1947). Later this region became a part of Pakistan (1947-1971) and finally, it has become a part of Bangladesh since 1971. History suggests that ignorance and abuses to the indigenous people of the CHTs have a long root of development (Uddin, 2010). In analyzing the conflict, a historical review of the three regimes has been made in order to have clear understanding of the root causes and nature of the conflict.

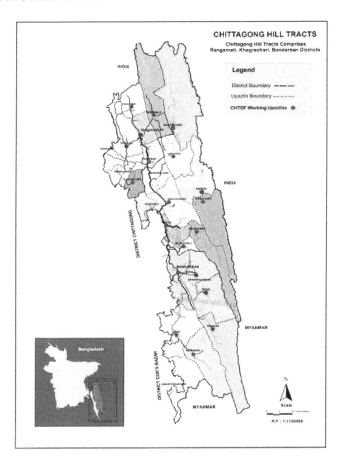

Fig. 1. Study area

Final part of the historical review encompasses an analysis of the Bangladeshi regime. This part of the research was conducted through literature collection and review, empirical analysis based on community survey with open ended and semi - structured questionnaire, key informant interview and focus group discussion. A total of ninety individual respondents were interviewed from nine villages of the three districts viz. Rangamati, Bandarbans and Khagrachari of the CHTs. In order to grasp the problem clearly and to know their opinion, three focus group discussions were held. A total of twelve key informants including three educationists, four political leaders, three cultural activists and two representatives from the Local Administration were interviewed in order to obtain expert opinions about the subject. A qualitative evaluation was done in order to know the respondent's opinion about the factors affecting forest and forest resources in the CHTs. Along with listing the factors, respondents were given a scale of magnitude from 1-5 indicating 1 as very low, 2 as low, 3 as moderate, 4 as high, and 5 as extreme to express the magnitude of effect for each factor.

The survey data encompassed demographic and biophysical data, ethnicity, religion, migration patterns, chronology of conflict, land ownership, etc. An extended and closer stay in the study areas helped particularly to observe, photograph, record the rituals and stories, and also to share experience on the CHTs issues. Additionally, we tried to explore available conflict resolution theories and checked the applicability of such theories in the CHTs context. Current conflict resolution efforts of the Government have been reviewed and analyzed critically to identify its limitations. Based on the analysis, a conflict resolution model has been developed for this region.

3. Bangladesh forestry

Bangladesh – a forest dependent, over-populated country observed a huge loss of forest and forest resources over the past decades (the 1970's deforestation rate of 0.9% rose to 2.7% during 1984 to 1990) (GOB, 2001b). According to Forestry Master Plan (GOB, 1995), of the

Forest types	Area (000 ha)	% of the forest area
Hill forests	551	23.7
Moist deciduous forests	34	1.5
Mangrove forests	436	18.7
Bamboo forests	184	7.9
Long rotation plantation	131	5.6
Short rotation plantation	54	2.3
Mangrove plantation	45	1.9
Rubber plantation	8	0.3
Shrubs/Other wooded land	289	12.4
Wooded land with shifting cultivation	327	14.0
Village forests	270	11.6
Total	**2,329**	**100.0**

Source: GOB, 2009

Table 1. Forest area in Bangladesh by forest types

total area of Bangladesh, agricultural land makes up 64.2%, forest lands account for about 17.8% (2.53 mill ha), whilst urban areas are 8.3%. Water and other land uses account for the remaining 9.9%. However, the last National Forest and Tree Resources Assessment 2005-2007 for Bangladesh (GOB, 2009) indicates a total forest area of about 2.33 mill ha (Table 1) which is about 15.8% (0.02 ha person[-1]) of the total landmass of Bangladesh. This forest area is categorized as natural forests (1.2 mill ha), forest plantations (0.23 mill ha), Shrubs/other wooded land (0.29 mill ha), wooded land with shifting cultivation (0.33 mill ha) and village forests (0.27 mill ha). Of this forest land, the Forest Department (FD) directly controls 1.44 mill ha with the legal status of Reserved Forest (RF) and Protected Forest (PF), and the District Administration controls more than 0.73 million ha (forest management activities entrusted with FD) of Unclassed State Forests (USF).

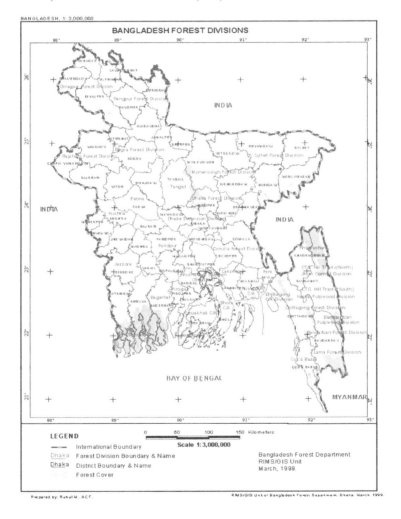

Fig. 2. Map of Bangladesh showing the distribution of forests areas

The distribution of natural forests in Bangladesh is eccentric i.e. the forests are located mostly in the peripheral zones of Bangladesh (Fig. 2). Out of the sixty four districts, there is no forest in twenty eight districts of Bangladesh. Major forest types within Bangladesh include i) tropical evergreen and semi-evergreen hill forests, ii) tropical littoral and mangrove forests, iii) inland moist deciduous Sal forests and iv) freshwater swamp forests (Champion & Seth, 1968). Despite having a century old regulatory forest management, the condition of the forest and forest resources in Bangladesh have been greatly depleted. Almost 50% of the area of Bangladesh has some kind of tree cover. Only 2.3% of the area has a very high tree cover (>70%) and roughly 20% has low tree cover (<5%) (GOB, 2009). The goal of social forestry

Project was to educate, engage and encourage active participation in the management of forest resources, thus creating relevant stakeholders. While traditional forest management resulted in the net loss of forest resource cover, participatory social forestry on the other hand, has the potential in the horizontal expansion of forest cover benefiting thousands of poor people (Salam & Noguchi, 2005; Khan et al., 2004; Muhammed et al., 2008; Muhammed & Koike, 2009). However, the indigenous people in general have been seriously overlooked during the implementation of this program.

4. Indigenous communities in Bangladesh

In Bangladesh, there are about 45 ethnic communities (Costa & Dutta, 2007; GOB, 2008a) with distinct language, culture, heritage and abide by own administrating statutes (Mohsin, 1997). Population statistics on the indigenous people in Bangladesh suffers from reliability and validity. According to 1991 population Census (BBS, 1991), indigenous population was about 1.2 mill (1.13% of the total population) of which the CHTs population was 0.56 mill (Table 2). However, subsequent reports show that indigenous population of the CHTs are about 1.3 mill (CHTDF, 2009), 2 mill (GOB, 2008a) or 2.5 mill (Asian Indigenous and Tribal People Networks [AITPN], 2008). Population Census 1991 reported 12 indigenous groups living in the CHTs. However, the correct number is 11 that have been clarified later by Mohsin (2003) and CHTDF (2009). These indigenous races are *Bawm/Bom, Chak, Chakma, Khumi/Khami, Khyang, Lushai, Marma, Mro/Mru, Murang, Pankhu/Pankho, Tanchangya* and *Tripura. Mro/Mru* and *Murang* categorized into two difference race in the 1991 census which was probably a mistake.

Among the indigenous population of the CHTs *Chakma, Marma* and *Tripura* share about 90%. Dominant indigenous race of the CHTs is *Chakma* with a total population ranging from 252,858 (BBS, 1991) to 382,000 (Joshua Project, 2011). *Marma* race of the CHTs consists of a total population of 157,301 (BBS, 1991) to 210,000 (Gain, 2000). *Tanchangya* is another indigenous race of the CHTs constituting a population size of 21,000 and ranking the 5th among the ethnic communities in Bangladesh (GOB, 2008b). Among the other indigenous races, *Khasi* with variable population statistics [e.g. 12,280 (BBS, 1991) and 81,000 (Joshua Project, 2011)] live within the reserved forests of Sylhet region located in the north - eastern part of Bangladesh. A considerable number of indigenous communities including *Garo, Hazong* and *Koch* live within the fringe of the plain land *Sal* (*Shorea robusta*) forests where *Garo* is the dominant race with a total population of about 64,280 (BBS, 1991).

All the dominant indigenous groups (i.e. *Chakma, Marma and Tripura*) of the region follow the first way of life (Adnan, 2004) that means they have adopted modern life style like the mainstream society in Bangladesh. Majority of the people (i.e. *Chakma, Marma, Tanchangya* and *Mro/Mru*) are Buddhist by religion. *Tripura* follow Hinduism. *Lushai, Pankho, Bawm* and some of the *Mro/Mru* adopted Christianity. Indigenous people of the CHTs have closer ethno - cultural affinities with other Sino -Tibetan people inhibiting in Myanmar and the Indian States of Tripura and Mizoram.

The tribal economy is basically subsistence in nature primarily based on primitive agriculture. As a result the productivity is low. They cultivate their land under input starved conditions due to lack of financial and technical resources. Historically, the indigenous people are dependent on the forests for their livelihood. Indigenous people of the CHTs

Indigenous groups	Bangladesh	CHTs
Bawm/Bom	13,471	13,471
Buna	7,421	-
Chak	2,127	2,127
Chakma	252,858	252,858
Garo	64,280	-
Hajong	11,540	-
Harizon	1,132	-
Khumi/Khami	1,241	1,241
Khasi	12,280	-
Khyang	2,343	2,343
Koch	16,567	-
Lushai	662	662
Mahat	3,534	-
Manipuri	24,882	-
Marma	157,301	157,301
Mro/Mru	126	126
Murang	22,178	22,178
Munda	2,132	-
Oraon	8,216	-
Paharia	1,853	-
Punkhu/Pankho	3,227	3,227
Rajbansi	7,556	-
Rakhaine	16,932	-
Santal	202,162	-
Tanchangya	21,639	21,639
Tripura	81,014	81,014
Urua	5,561	-
Others	261,743	-
Total	1,205,978	558,187

Source: BBS, 1991

Table 2. Indigenous population in Bangladesh and in the study areas

have been practicing shifting cultivation as their principal economic activity in the denuded hillocks with scattered vegetation. It is estimated that at the beginning of the 20th century the tribal people dwelling in forest areas, obtained about 80-90% of their income from minor forest produce (Dasgupta & Ahmed, 1998). But this income has been reduced drastically in recent years. Cultivation of betel leaf (*Piper betle*) on the forest trees is the main economic activity of Khasi people. Similarly, the Garo community has been shaping their life, art and culture based on *Sal* (*Shorea robusta*) forests. Rapid degradation and deforestation of this unique deciduous forests coupled with Government led environmentally adverse development projects are posing immediate threats to this community.

5. Historical review

5.1 British period and Indigenous people of the study areas (1757-1947)

Indigenous people of the CHTs locally known as *Jumma*[1] people were independent before the British colonial period. Bengal region[2] was ruled by the British from 1757 to 1947. The CHTs areas being an important source of raw materials (e.g., timber and cotton) drew the attention of the British ruler (Huq, 2000). Although there was a sharp physiological and cultural gap between the *Pahari*[3] and *Bangali*[4], geographically the CHTs were close to the Bengal region. After occupation, the British annexed the CHTs with Bengal in 1860 as an autonomous administrative district known as 'The Chittagong Hill Tracts' within undivided British Bengal. Few non-indigenous Bangali people co-existed with the indigenous population of the CHTs as original people of the areas. Non-indigenous Bengali people shared only about 1.74% of the total population in the CHTs in 1872 (CHTDF, 2009).

In 1900 Act, British Government kept special regulation to protect the Jumma people from economic exploitation by non-indigenous people, preserve their traditional socio-cultural and political institutions and also ensure their traditional laws and common ownership of land. This 1900 Act safeguarded the Pahari people prohibiting migration and land ownership to non-indigenous people in the CHTs (Asian Cultural Forum on Development [ACFOD], 1997). Respondent's survey could not provide much information about this period (only 3 respondents were able to discuss a little about this period) because of their age and lack of knowledge on the old history. Nine out of the twelve key informants were able to discuss the CHTs during British period. However, we had an effective discussion in all the three group discussions about this period. The result of the discussion as well the available literature concluded that the Jumma people had very peaceful life during this period. Clear felling followed by artificial regeneration was the only management system for the CHTs keeping revenue earning as the major concern of the British Government. This management system even opposed to biodiversity and wildlife conservation did not create any negative concern because of the abundance of huge forests and natural resources in the CHTs.

---

[1] refers to indigenous people or the original population of the CHTs. Jumma people means the groups of people who live on shifting cultivation. Shifting cultivation in its local term is known as Jhum cultivation.

[2] refers to the Bengali speaking part of the undivided part of India.

[3] Indigenous people or the original population of the CHTs. The CHTs are hilly region. Hill in Bengali term means Pahar. So Pahari means the inhabitants of the hill. They also sometimes addressed as Adivasi (original population of the region).

[4] People who speak Bengali as their first language.

5.2 Indigenous people during Pakistan period (1947-1971)

The CHTs remained as a part of Pakistan in 1947, although the CHTs being a non-Muslim populated area were supposed to be a part of India on the basis of the provision of the partition. Despite 98.5% of the CHTs population being Jummas (non-Muslims), the Pakistani leadership ceded the CHTs to the East Pakistan violating the principles of partition (The 2-Nation theory based on religious demographics) and against the desire of the Jumma people. Right after the partition, the Pakistan Government started to ignore the Act and Regulations of 1900 for the CHTs and the Jumma people realized that their life would never be peaceful in Pakistan. In the subsequent years, their anger turned into violence, demanding for an autonomous State of Chittagong Hill Tracts. But the Government adopted more hostile attitude towards the Jumma people of the CHTs annulling the CHTs Police Regulation, 1881 that restricted indigenous people in the police force (Uddin, 2008). Additionally, Jummas were discriminated in jobs, business and education. Besides, the Government amended the 1900 Act several times in order to find a legitimate way for allowing migration of non- indigenous people into the CHTs without consulting the Jumma people (Chowdhury, 2006).

5.2.1 Hydro-electric dam of Kaptai and the indigenous people

In 1960s, the Pakistan Government constructed a hydro-electric dam at Kaptai (popularly known as Kaptai dam). Prior to construction it was neither consulted with the local people nor did it forecast the impact of this dam to the neighbouring lands, resources and people precisely. When the hydroelectric project came into effect in 1962, the water level rose beyond the forecasted one and so most of the rehabilitated (incorrect assumption of the engineers on forecasting the areas requiring rehabilitation before launching the hydro electricity production) areas submerged under water. The resettled people along with thousands of people became homeless loosing their houses, agricultural land, livestock and forests. Many indigenous people of the CHTs (about 100,000) became possibly the first ever environmental refugees in Indian subcontinent due to this huge hydroelectric project (GOB, 1975; Samad, 1994, 1999). As a result of Kaptai dam many Jumma people were moved into sparsely populated regions of Mizoram, Tripura, Assam and Arunachal of India without legal) identity (citizenship rights (Chowdhury, 2002; Uddin, 2008). All the respondents (100%) and key informants were very much aware of the negative impact of Kaptai hydro-electric dam and they considered this dam construction as the starting point of major conflict of the Jumma people with the Government. According to them, this dam not only ruined their life, but also inundated 65,527 ha of land, including 2,590 ha of well stocked reserved forest and about 21,862 ha of cultivable land.

5.3 Bangladesh (1971-) and the indigenous people of the CHTs

The constitution of the sovereign Bangladesh declared Bangladesh as a unitary State and Bengali as the State language. It also declared that citizens of Bangladesh are to be known as Bangali. During that time there was only one representative from the CHTs in the National Parliament who refused to endorse the constitution since it did not recognize the existence of other national communities or sub-national identities (Chowdhury, 2002; Shelly, 1992). Available report suggests that Jumma people remained indifferent to the cause of war against Pakistan (Chowdhury, 2006).

After the liberation, Jumma people demanded for i) autonomy for the CHTs, ii) retention of the CHTs Regulation 1900, iii) recognition of the three kings of the Jummas, and iv) ban on the influx of the non-Jummas into the CHTs (Chowdhury, 2006). The Government rejected these demands and urged the indigenous people to become Bangali, ignoring their ethnic identities (Chowdhury, 2002). This has been considered as the starting point of new conflict of the Jummas with Bangladesh Government. The Jumma people rejected the imposition of Bengali nationalism. According to the summarized result of the focus group discussion and key informant interview, the failure of the Government to recognize the identity of hill people and their political and economic marginalization led them to form an indigenous people's organisation called 'Parbattya Chattagram Jana Samhiti Samiti (PCJSS)' in 1972. A military wing called *Shanti Bahini*[5] was added to PCJSS in the same year (Mohsin, 1997).

5.3.1 State induced military led mass migration

Immediately after the change in 1975 through the assassination of the Father of the Nation and his family, the new Government took more drastic step to militarize the whole CHTs declaring the region a politically special sensitive zone. The Government assumed full military control in CHTs ignoring the local civil administration. In order to earn more control over the region and balance between indigenous and non - indigenous population, the Government adopted a policy of State induced military led migration of non-indigenous poor and destitute folks of other part of Bangladesh to the CHTs without consulting the indigenous people of the CHTs. The establishment of Chittagong Hill Tracts Development Board (CHTDB) in 1976 for the CHTs development deeply strengthened military occupation and military infrastructure in the CHTs furthermore. The CHTDB was formed and administered by the military command and the military was in charge of implementing all development activities in the CHTs.

A close examination of the CHTDB development projects reveals that more than 80% of the CHTDB development budgets were spent on building military infrastructure through construction of military camps, roads and bridges, office buildings, sports complexes, mosques, cluster villages for Pahari and *Bangali* settlers (Bhikkhu, 2007). All the respondents, key informants and groups discussion opined that the strategy of the new Government after 1975 deploying military administration in the CHTs extended the magnitude of conflicts. During this military led administration, counterinsurgency operations were started throughout the CHTs that ruined the scope for accommodation and co-existence of both the hill people and migrated *Bangali* people in the CHTs (B.H. Chowdhury, 2002). Within two years of the new Government (after 1975) more than 80,000 armed forces were deployed in the CHTs for the cause of 'development and security reasons'. In fact, armed forces facilitated the transmigration of non-indigenous *Bangalis* by displacing the *Paharis* (Uddin, 2008).

This military led huge migration changed the population structure and composition in the CHTs. It is found that 1.74% non-indigenous people of the CHTs in 1872 increased to 9.09% in 1951, 19.41% in 1974 and 60% in the current decades (Bhikkhu, 2007; CHTDF, 2009; Uddin, 2008). Fig. 3 shows the pattern of the increase of the non-indigenous people into the CHTs. Actually, as of the advice of the British Government during 1860s, the Circle chiefs brought *Bangali* cultivators to work on this region in order to teach low land farming to the Chakmas and other indigenous races. At that time only the three Circle chiefs were

---

[5] The military wing of the PCJSS formed in 1972 in order to preserve the rights of the indigenous people. '*Shanti*' and '*Bahini*' are two Bengali term meaning peace and armed force respectively.

permitted to own land and the *Bangali* immigrants became sharecroppers. Even knowing that the CHTs land is not arable and life would be very risky, landless floating people accepted migration in order to get a piece of land and a house of their own, along with other financial and food grains support of the Government (Uddin, 2008). During 1979 - 1984 about half a million non-indigenous people have been settled into the CHTs. Bhikkhu (2007) reported that during late 1970s to the early 1980s, more than 400,000 muslims from various plain districts of Bangladesh were systematically migrated to the CHTs under Government sponsored military led settlement programs. It was found that 92% of the respondents and 100%of the key informants viewed this huge migration as demographic invasion or more specifically '*islamization*' of the CHTs. They think that this has been done to diminish their political clout in the muslim dominated Bangladesh. So this State sponsored migration has made the Jumma people a minority in their own homeland. Talukdar (2005) in his study suggested that if this conflict goes on without any resolution, the Jumma people will soon find themselves in a situation of going for unconstitutional struggle.

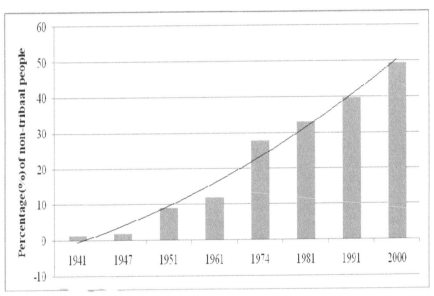

Source: Talukdar, 2005; Uddin, 2010

Fig. 3. Change in population composition in the CHTs

5.3.2 Land crisis and dispossession of Jumma land

According to the available information on site quality, physiognomy and topography, the CHTs do not allow intensive irrigated agriculture except in the limited valley and lowlands (Roy, 2004). According to the soil survey report, only 3.1% of the CHTs lands are suitable for agriculture and 72.9% are suitable for forestry (GOB, 1966). Besides, Kaptai hydro-electric dam further increased the land crisis inundating 21,862 ha of cultivable land (GOB, 1975). In spite of the shortage of fertile farming land and inundation of the available cultivable land, the Government settled thousands of landless non-indigenous people in the CHTs. Besides, each land less settler family was given a legal ownership of 2 ha of hill land or 1.5 ha of mixed land or 1 ha of wet rice land (Chowdhury, 2002). A study of Bangladesh Society for the Enforcement of Human Rights (BSEHR) found that about 61.44% of the indigenous people still face discrimination, 41.86% are victims of corruption and 18.67% have been evicted from their ancestors land (Zaman, 2003). Land and forests are the very basis of life for the Jumma people, so plundering of land is the question of their existence. Respondents (100%) opined that when the newly settled families cannot make a living from their allocated land, they encroach on Jumma owned land. They adopted various ways to occupy the Jumma land and still now the Jumma people are being dispossessed from their own lands (Table 3).

Sl.	Pattern of land grabbing	Actor (s)
1	Expansion of military facilities	Mostly led by the military
2	Expansion of settlements	Mostly led by the military
3	Expulsion in the name of forest protection	Forest Department or local administration
4	Leases of land by the local administration	Forest Department or local administration
5	Attacks	Bengali migrants with direct or indirect support of the military
6	Expulsion through false cases, harassment & other tactics	Bengali migrants with direct or indirect support of the military

Table 3. Patterns of dispossession of Jumma land in the CHTs

5.3.3 Conflict on forest use and indigenous livelihoods

The CHTs are mostly a forested region and the ownership of the forest land lie with District Administration for land and forest management rights lie with Forest Department. Despite of regulatory forest management practices in the CHTs, the forest and forest resources have been depleted greatly over time. Currently, most of the hills are denuded and degraded. Forest Department complains that the indigenous people and their illegal occupancies in the Government forests, illegal logging and their shifting cultivation are the main causes for depletion of forest resources in the CHTs. On the other hand, the respondents (100%) opinion and the result of the group discussion identified that the systemized corruption of the Forest Department, District administration, military administration and associated ministries and political elites, syndicated illegal logging of the timber merchants, poor transit rules, Jhoot[6] permit, inappropriate forest management systems (clear felling in the natural forests followed by artificial regeneration mostly with fast growing exotic species like, *Acacia auriculiformes* and *Eucalyptus camaldulensis*), etc. are the major reasons for forest depletion in this region (Table 4). Qualitative assessment indicates that systemized corruption and syndicated illegal felling by the timber merchants are the major causes of the forest depletion.

From the focus group discussion, it was found that the indigenous people are frequently harassed with false police case against land encroachment and illegal timber cutting. Most of

---

[6] Jhoot means private land. Jhoot permit is the transit rule of the Forest Department which is issued by the Forest Department with necessary approval of the Deputy Commissioner for harvesting, sale and movement of the timber produced in the privately owned land.

the respondents know about the participatory benefit sharing social forestry programs of Forest Department. But they are discriminated to become stakeholders (nine out of the total one hundred two interviewees became the participants). However, the migrated *Bangalis* living in the areas are getting the benefit of such program. Additionally, it was found that most of the saw mills, furniture shops and small scale wood based industries in the CHTs are owned by the non-indigenous migrated population (93%in Ranagamati, 95%in Bandarbans and 94%in Khagrachari).

Sl. No	Factors affecting forests	No. (n) & % of interviewees responded*	**Average magnitude of the effect of each factor in forest degradation				
			1	2	3	4	5
1	Systemized corruption						
a	Forest Department	102 (100)					
b	District administration	78 (76.5)					
c	Military administration	102 (100)					
d	Police Administration	85 (83.3)					
e	Politicians and elites	91 (89.2)					
2	Syndicated illegal logging	96 (94.1)					
3	Poor transit rules	35 (34.3)					
4	Leasing lands to outsiders	25 (24.5)					
5	**Inappropriate management	43 (42.2)					
6	Private forest felling permit	18 (17.6)					

* n=90 (respondents) + 12 (key informants) =102; the figures in the parentheses indicate percentage (%)
** a scale of magnitude from 1-5 indicating 1=very low, 2= low, 3= moderate, 4=high, 5=extreme
*** Inappropriate forest management refers to clear felling followed by artificial regeneration with fast growing species like, *Eucalyptus camaldulensis, Acacia auriculiformes*

Table 4. Major causes of forest depletion in the CHTs

6. Steps towards conflict resolution

Historical review suggests that the Jumma people have been affected largely by the policies of migration, land eviction, cultural assimilation and ethnic discrimination by successive Governments. In their struggle for autonomy, they have been targets of massacres, extra judicial executions, rape, torture and forced relocation. However, the Government took some initiatives in order to calm the situation. Table 5 summarizes the chronological initiatives for peace in the CHTs. Finally after series of meetings and negotiations, Bangladesh Government and the PCJSS came up with a Peace Accord on December 02, 1997. The 68 points Accord deal with variety of subjects ranging from administration to military status, land question, refugee settlement and others. One of the elements of the Peace Accord was to recognize the rights of indigenous communities to land and other sovereign issues. It was agreed that the Ministry of CHTs Affairs will be headed by an indigenous representative. Government also agreed to repatriate and rehabilitate the CHTs refugees in India, resolve land disputes, and cancel illegal leases of land to non-indigenous people. This Accord endorsed a partial release of power to the indigenous authority declaring creation of the 'Regional Council' to look after the entire region. Under the Accord, the militants agreed to surrender and de-commission their arms for general amnesty (Adnan, 2004). However, the Accord was not protected by the constitutional safeguards, and is open to amendment or revocation at any time. It makes no provision for forest and environmental protection and the existing forest act, rules and regulation were not referred. Besides, it makes no provision for social reconciliation between tribal population and migrated Bangali population.

Time period	Steps towards conflict resolution
Early 1977's	Government took first political measure to appease the insurgents in the CHTs, and appointing the mother of the Chakma king.
1977	A Forum formed in the Tribal Convention in order to negotiate at the official level. After initial interest in the process, the PCJSS withdrew them from the Forum due to their internal conflict.
1982	The new Government formed a committee headed by an indigenous leader but this committee failed because the PCJSS did not approve it.
1985	Some announcements were made by the Government that resulted in holding the ever first dialogue with regards to suspension of Bengali settlement, the granting of amnesty to insurgents, and a proposal for direct dialogue with PCJSS leadership. The dialogue proved ineffective to continue the process.
1987-1988	The Government set up a National Committee for the resuming the dialogue with PCJSS. However, every dialogue ended up without any conclusion.
1989	The Parliament enacted the Chittagong Hill Tracts Local Government Council Act, 1989 and the Hill District Act, 1989. The Special Affairs Ministry was constituted in 1990 to look after the CHTs affairs.
1992	Declaration of a general amnesty for the insurgents along with cash rewards for surrendering arms. Some 2,294 insurgents' surrendered and 30,390 indigenous populations came back from the Indian camps. A committee was formed to oversee the most sensitive issue of land ownership.
1992-1993	About 50,000 families affected by the insurgency were provided various relief and rehabilitation support. This created confidence among Jumma people and encouraged many of them to return back from the camps.
1994	The process of the refugee repatriation stopped.
1997	The new Government after a serious of meetings and concerted efforts found a permanent political solution within the framework of the State sovereignty and came up with a historic Peace Accord.

Source: Chowdhury, 2002; Hosena, 1999; Jumma Net, 2009

Table 5. Summary of conflict resolution initiatives the CHTs

6.1 Pace of progress after the Peace Accord

The CHTs Peace Accord was implemented to some extent in the first three years (1998-2000) like, demobilization of the PCJSS, repatriation of Jumma refugees, enactment of the three revised Hill District Council Acts, the Regional Council Act, establishment of the CHTs

Affairs Ministry, etc. But the most important provisions of the Accord, such as the withdrawal of temporary military camps and resolution of land conflicts, remain unimplemented till to date. Since assuming the power by the new elected Government (2001-2006) who kept themselves out of the treaty, implementation of the Peace Accord was quite ignored. During this period the military began to expropriate vast areas of land defying the provisions of the Accord. However, by now various new issues have appeared and the limitations of many provisions of the Accord are hindering the implementation even under the current Government who signed the Accord in 1997. Individual respondent opinion as well the focus group discussion show that indigenous people have again become more sceptical and are loosing their confidence over the Accord. The Peace Accord is facing a number of difficulties regarding implementation as follows.

> "It was better for us when the British were here. We were protected by special rules. Even under Pakistan it was better as we were independent. When the CHTs became part of Bangladesh, every right was taken over. These hills that were once the land of Jumma aren't for us now. Slowly the Bengalis have taken over everything."

Ajit Tanchangya (79)

Box 1. Perception of the conflict by an aged indigenous man

a. Accord is not recognized by the main political opposition (and their allies) of 1997.
b. Implementation Committee can not do their work because of political interferences.
c. Political movement of the indigenous people after the Peace Accord have divided into two groups; the PCJSS - the major group of the Peace Accord and the UPDF (United People's Democratic Front) - a group opposing the Peace Accord.
d. Land crisis, a very delicate issue of conflict, between indigenous and Bangali migrants continue to remain unresolved.

Though the Accord got acclamation from the World community including UNESCO declaring Houpet-Felix Boigny Peace Award in 1999, the CHTs Jumma people still live under duress because of continuous pressure from the civil administration manned by non-indigenous and non-local officials and communal attacks by the Bangali settlers with direct back up from many camps of the Bangladesh security forces (Unrepresented Nations and Peoples Organization [UNPO], 2004). The non-implementation and in some cases violation of vital clauses of the Accord by the government in one hand and extreme Bengali fanatic fronts float against the Jumma people on the other has seriously deteriorated the CHTs situation in recent months (UNPO, 2004). It is found that the Government does not give exact picture of the latest status of the Accord to the countrymen and the concerned international communities.

7. Conflict resolutions models and the case of CHTs

Till date scientists from multiple background and many political analysts contributed notably on developing number of general and case specific tools and instruments on conflict resolution. The famous Thomas-Kilmann Conflict Mode Instrument (Thomas & Kilmann, 1974) is applicable to two basic dimensions viz., assertiveness and cooperativeness. This model proved effective in five different conflicting modes like, competing, accommodating, avoiding, collaborative and compromising. Johan Galtung's seminal thinking on the relationship between conflict, violence and peace which is popularly known as Galtung's model of conflict, violence and peace. Galtung's Conflict triangle works on the assumption that the best way to define peace is to define violence, and its antithesis i.e., this model is effective on both symmetric and asymmetric conflicts (Galtung, 1969, 1996). Ramsbotham and woodhouse's Hourglass model of political conflict resolution combined Galtung model along with escalation (narrowing political space) and de-escalation (widening political space) approaches. Among the other conflict resolution models, Gail Bingham's Environmental Dispute Settlement Process (Bingham, 1986), coercive and non-coercive third party intervention model, Interest-Based Relational Approach, Lederach's model (Lederach, 1997), multi-track conflict resolution model of Ramsbotham et al. (2005), etc. are well known conflict resolution models.

Although theoretically all the models seem sound and effective, practical application is case sensitive because of the multiple magnitudes of conflicts. If we consider the conflict resolution approach of in the CHTs that finally came up with a Peace Accord, it can be said that a combination of the said approaches and models has been applied. But the participation of all possible social actors, all aggrieved parties, political drivers and international bodies were not involved properly here. Therefore, some of the actors for their socio-political interests can not support the accord. Moreover, some sensitive issues like land settlement, forest use and cultural coherence are not truly perceived in the problem solving process; therefore, the implementation of the accord has not advanced duly. Reviewing the conflict resolution theories and models and applying our studied knowledge (based on interview and focus group discussion) on the CHTs, following model has been proposed in order to conflict resolution in the CHTs (Fig. 4).

According to this model both indigenous people and migrated population have been focused with equal attention. Treaty or agreement, whatever is made, should be done based on the mutual trust and consciousness of all aggrieved sides. The Peace Accord of 1997 was an agreement between the military wing of the indigenous people of the CHTs and Government of Bangladesh lacking of consciousness and agreement of the mass indigenous people and migrated people in the CHTs. According to this model, all the relevant parties will sit together through iterative meetings and discussion to bring out a final agreement. Therefore, it will reduce the chance of any group or political party to stand against the agreement or implementation in future. The Peace Accord of 1997 could not make any provision for a third party monitoring and evaluation which is essential in such political discourse resolution process. Our proposed model keeps provision for local, national and international level monitoring and evaluation. As forest, land and military administration are the main issues in the CHTs; therefore, appropriate representatives from these three Departments need to give more chance for discussion and consultation in order to find mutually agreed ways of solution to land and forest. Proper application of this model may help solving the problems of the indigenous people bringing sustainability to the forest management and forest use in this region of Bangladesh.

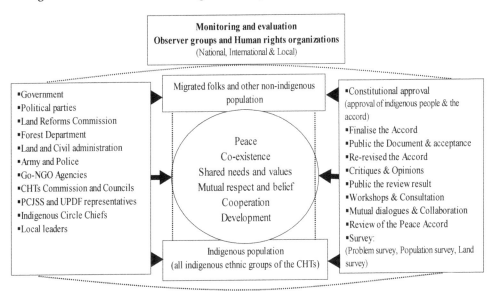

Fig. 4. Proposed model for conflict resolution in CHTs

8. Conclusion

It is understood that the CHTs issue is clearly a political problem created over many decades. The indigenous communities in Bangladesh are left as the most disadvantageous groups by the Government and policy makers. The study shows that the CHTS are one of the resourceful areas with forests, natural landscapes and diversified indigenous people. In order to utilize the economic potential of the areas, both forests and indigenous people are to be treated properly. A strong political will, concerted efforts of all political parties and actors are required to implement the Peace Accord. Land is the one of the major issue of the CHTs conflict which requires more cautious dealing. Empowering indigenous people by resolving the exiting conflicts in the CHTs is essential to help manage forest. Constitutional recognition of the indigenous people in Bangladesh will solve many problems. In order to protect and manage natural forests of the CHTs in a sustainable manner, the cultural and sacred relationship of the indigenous people with the forests must be recognized. Indigenous people should be motivated and be involved in all participatory co-management programs in the CHTs.

We believe that our proposed model can be an effective tool for the CHTs conflict resolution. In the proposed model, it is suggested to start with reviewing the Peace Accord. However, before this review, precise surveys on problem situation, population of the region and land and settlement need to be conducted. According to this model, a meaningful discussion of all relevant actors is needed to finalize the revised Peace Accord before getting it approved by the National Assembly. Local, national and third party led international monitoring are required to evaluate the progress of implementation of all programs in the CHTs.

9. Acknowledgements

This research was conducted with the financial support of the Alexander von Humboldt Foundation, Germany. We are very grateful to them. Our sincere thanks and gratitude are due to many individuals and organizations especially, those who supported us everyway during field data collection.

10. References

ACFOD. (1997). A Conclusion Report: International Peace Conference on Chittagong Hill Tracts (WORKING TOWARD PEACE), Bangkok, Thailand

Adnan, S. (2004). Migration, Land Alienation and Ethnic Conflict: Causes of Poverty in the Chittagong Hill Tracts of Bangladesh, Research and Advisory Services, Dhaka, Bangladesh

AITPN. (2008). Bangladesh: We want the lands, not the Indigenous peoples, New Delhi, India <Retrieved from www.aitpn.org>

BBS. (1991). Bangladesh Population Census 1991: analytical report, Bangladesh Bureau of Statistics, Dhaka, Bangladesh

BBS. (1993). Statistical Year Book of Bangladesh, Bangladesh Bureau of Statistics, Dhaka. Bangladesh

BBS. (2001). Statistical yearbook of Bangladesh. Bangladesh Bureau of Statistics, Ministry of Planning, Government of the People's Republic of Bangladesh

Bhikkhu, P. (2007). CHT on Historical Outline with Special Reference to Its Current Situation, Chittagong, Bangladesh

Bingham, G. (1986). Resolving Environmental Disputes: A Decade of Experience, The Conservation Foundation, Washington D.C.

Champion, H.G., & Seth, S.K. (1968). A Revised Survey of the Forest Types of India, Government of India publication, India

Chowdhury, B.H. (2002). Building Lasting Peace: Issues of the Implementation of the Chittagong Hill Tracts Accord, Program in Arms Control, Disarmament, and International Security (ACDIS) Occasional Paper series, University of Illinois at Urbana Champaign, Available From: http://acdis.illinois.edu/publications/207/publication-BuildingLastingPeaceIssuesoftheImplementationoftheChittagongHillTractsAccord.html

Chowdhury, M.N.H. (2006). Power, law, and history: episodes in consciousness of legality. Asian Affairs, Vol. 28, No. 1, pp. 69-91

CHTDF. (2009). Socio-economic baseline survey of Chittagong Hill Tracts, Dhaka, Bangladesh

Costa, T., & Dutta, A. (2007). The Khasis of Bangladesh, In: A socio-economic survey of the Khasi people, Philip Gain & Aneeka Malik, pp 19-28, Society for Environment and Human Development (SEHD), Dhaka, Bangladesh

Dasgupta, S. & Ahmed, F.U. (1998). Natural resource management by tribal community: a case study of Bangladesh, In: The World Bank/WBI's CBNRM Initiative, access on February, 2011, Available from: < http://srdis.ciesin.org/cases/bangladesh-002.html>

FAO. (2005). State of the World's Forests 1997, Rome, Italy

FMP. (1992). Forestry Master Plan: forest production. Asian Development Bank, Ministry of Environment and Forests, Government of Bangladesh, Dhaka, Bangladesh

Gain, P. (2000). Life at risk, In: The Chittagong Hill Tracts: Life and nature at risk, Roy, R.D.; Guhathakurta, M.; Mohsin, A.; Tripura, P. & Gain, P. (Eds.), 1-41, SEHD, ISBN 984-494-011-7, Dhaka, Bangladesh

Galtung, J. (1969). Violence, peace and peace research. Journal of Peace Research, Vol. 6, No. 3, pp.167–191

Galtung, J. (1996). Peace by peaceful means: peace and conflict, development and civilization, International Peace Research Institute, Oslo and Sage Publication Inc. CA, USA

GOB. (1966). Soil and Land Use Survey report. Chittagong Hill Tracts (Volume 2), Forestal Forestry and Engineering International Limited, Dhaka, Bangladesh

GOB. (1975). Chittagong Hill Tracts, Bangladesh District Gazetteers, Economic Condition, Dhaka, Bangladesh

GOB. (1995). Development Perspectives of the Forestry Sector Master Plan, Ministry of Environment and Forest, Government of the Peoples' Republic of Bangladesh, Dhaka, Bangladesh

GOB. (2001a). Meteorological Report of Bangladesh, Department of Meteorology, Dhaka, Bangladesh

GOB. (2001b). Banglapedia: National Encyclopedia of Bangladesh, <Retrieved from http:/banglapedia.search.com.bd/HT/D_0101.htm>

GOB. (2002). Forest Statistics, Bangladesh Forest Department, Bangladesh, Dhaka

GOB. (2008a). Moving Ahead: National Strategy for Accelerated Poverty Reduction II (FY 2009-2011, Planning Commission, Dhaka, Bangladesh

GOB. (2008b). Banglapedia:Tribal people in Bangladesh, National Encyclopedia of Bangladesh, Government of the Peoples' Republic of Bangladesh, Dhaka <Retrieved from http://banglapedia.search.com.bd/HT/T_0222.htm>

GOB. (2009). National Forest and Tree resources Assessment 2005-2007 Bangladesh, Bangladesh Forest Department, Dhaka, Bangladesh

Hinch, T. (2001). Indigenous Territories, In: The Encyclopedia of Ecotourism, David B. Weaver, pp. 345-358, CABI Publishing

Hosena, S.A. (1999). War and Peace in the Chittagong Hill Tracts: Retrospect and Prospect, Agamee Prakashani, Dhaka, Bangladesh

Huq, M.M. (2000). Government Institutions and Underdevelopment: A study of the tribal peoples of Chittagong Hill Tracts, Bangladesh, Centre for Social Studies, Dhaka, Bangladesh

Joshua Project. (2011). People in Country Profile: Bangladesh, <Retrieved from http://www.joshuaproject.net/people-profile.php?peo3=12654&rog3=BG>

Jumma Net. (2009). Chittagong Hill Tracts White Paper: The Issues of Conflict, Human Rights, Development, and Land of the Indigenous Peoples of the Chittagong Hill Tracts, Bangladesh, Dhaka

Khan, N.A.; Choudhury, J.K.; & Huda, K.S. (2004). An overview of Social Forestry in Bangladesh, Forestry Sector Project, Government of Bangladesh, Dhaka

Lederach, Von J.P. (1997). Building peace: sustainable reconciliation in divided societies, United States Institute of Peace, Washington D.C.

Mohsin, A. (1997). The Politics of Nationalism: The Case of the Chittagong Hill Tracts, Bangladesh, The University Press, Dhaka, Bangladesh

Mohsin, A. (2003). The Chittagong Hill Tracts, Bangladesh: Difficult Road to Peace, Lynne Rienner Publishers

Muhammed, N.; Koike, M.; Haque, F. & Miah, M.D. (2008). Quantitative assessment of people-oriented forestry in Bangladesh: A case study in the Tangail forest division, Journal of Environmental Management, Vol.88, No.1, (July 2008), pp. 81-92

Muhammed, N. & Koike, M. (2009). Policy and Socio-economics of participatory forest management in Bangladesh, Trafford Publishers, ISBN 978-1-4251-8719-4, Canada

Ramsbotham, O., Tom, W., & Hugh, M. (2005). Contemporary Conflict Resolution, Polity Press, USA

Roy, R. D. (2004). The Land Question and the Chittagong Hill Tracts Accord, Dhaka, Bangladesh

Salam, M.A. & Noguchi, T. (2005). On sustainable Development of social forestry in Bangladesh: experiences from Sal (Shorea robusta) forests. Environment, Development and Sustainability, Vol. 7, pp. 209-227.

Samad, S. (1994). Dams Caused Environmental Refugees of Ethnic Minorities. Environmental Refugees of CHT, weekly Holiday, Dhaka

Samad, S. (1999). State of Minorities in Bangladesh, In: Shrinking Space: Minority Rights in South Asia, Banarjee, S. pp. 75-96, South Asia Forum for Human Rights, Kathmandu, Nepal

Shelly, M.R. (1992). The Chittagong Hill Tracts of Bangladesh: The Untold Story, Centre for Development Research, Dhaka, Bangladesh

Talukdar, U. (2005). Chittagong Hill Tracts Issue and Post-Accord Situation, Proceedings of International Conference on Civil Society, Human Rights and Minorities in Bangladesh, Kolkata, India, January 2005

Thomas, K.W., & Kilmann, R.H. (1974). Thomas-Kilmann conflict MODE instrument, Xicom Tuxedo, NY, USA

Uddin, M.A. (2008). Displacement and destruction of ethnic people in Bangladesh. Canadian Social Science, Vol. 4, No. 6, pp. 16-24

Uddin, M.K. (2010). Rights & Demands of the Indigenous People: Perspective from Bangladesh, ARBAN, Bangladesh <Retrieved from www.arban.org>

United Nations. (2009). State of the World's Indigenous peoples, Department of Economic and Social Affairs, NY, USA

UNCSD. (2002). Dialogue Paper by Indigenous People. Economic and Social Council, Addendum No. 3, No. 2-3

UNPO. (2004). Report on the Implementation of the Chittagong Hill Tracts Accord, <Retrieved from http://www.unpo.org/article/568>

UNFPA. (2006). UNFPA Worldwide: Population, Health and Socio - economic Indicators, <Retrieved on January, 20, 2011 from http://www.unfpa.org/profile/bangladesh.cfm>

World Bank. (2000). The Little Green Book 2000, WB, Washington D.C., USA

Zaman, A.K. (2003). Conflicts & People of Chittagong Hill Tracts (CHT) of Bangladesh, Eastern Publications 16, Sylvester House, London

Zeppel, H.D. (2006). Indigenous Ecotourism: Sustainable Development and Management, Cromwell Press, Trowbridge, UK

9

Recent Problems and New Directions for Forest Producer Cooperatives Established in Common Forests in Japan

Koji Matsushita
Kyoto University
Japan

1. Introduction

Property rights for land, including forest land, were introduced in Japan following the Meiji Restoration in 1867. During the Meiji Period (1868–1912), a new land registration system was introduced, and the land-tax system was reformed. Individuals received property rights to forest land at that time. The Meiji Restoration was an important turning point in forest land ownership.

The custom of communal forest management, in which the forest is considered a common forest (*iriairin*), was developed during the Edo Period (1603–1868) in various regions[1]. Generally, forest land had high importance at that time, because of the necessity of various forest resources such as fallen-leaves and tree branches for agricultural production. Livestock management also required grazing land in common forests. Firewood and charcoal were the main energy resources produced from the common forest. Wood and wood-based products were also necessary as construction material. Therefore, the common forest was an essential resource for agricultural-based communities. In this chapter, the community represents the smallest unit of a village, which can be called a hamlet[2].

For a common forest to be sustainable, its users had to manage it carefully (to avoid damaging or destroying it). This included preventing intrusion or utilization of the forest by people from other communities. Thus, the management of a common forest included both internal constraints (e.g., rules) and external exclusions. During the Edo Period, forest land was one of the major sources of conflict between communities, and in some cases, struggles continued over several decades. As the security of the common forest is directly linked to the livelihood and agricultural production of the entire community, all community members united to protect their common forest. The boundaries of common forests gradually became clear through such struggles, and the solidarity among community members strengthened. As a result, a sense of equality developed among common forest members (Takasu, 1966). A unanimity rule became important within such communities for making decisions on various matters, including utilization of the common forest. For example, when forest management practices commenced, unanimity rule determined who used the specific forest site and how the profit from the common forests was used. A unanimity rule on the common forest is also thought to be important today[3].

During the Meiji Period, the property rights system was introduced to law in Japan, and how the common forest was organized and managed needed to change to fit into this system. This problem was extremely difficult from the beginning of land reform[4]. This is discussed further in this chapter. However, due to space constraints, only the most important points are explained. The most important right pertaining to the common forest was that people living in a specific area had a conventional right. It must be paid attention to that the right is based on the convention, which is, in most cases, originated from the Edo Period. According to Articles 263 and 294 of the Civil Law of 1896 (Act No. 89), any common forest rights must first consider the customs of a given area.

Common forests with no property rights in modern law gradually decreased in Japan after the Meiji Period. During the early Meiji Period, part of common forest was considered national forest land. In 1889, the concept of the municipality was introduced, and part of common forest became the property of newly founded municipalities. In 1910, the Public Forest Reorganization and Unification Project started and continued until 1939. Under the policy program, the common forest was considered municipal property. However, some people with rights to the common forest took various countermeasures to protect their rights, in some cases, registering the area under his/her name or that of other members, or as a shrine or temple. The management of conventional rights to common forests in the modern legal system has proved challenging. Regardless of whether the land was registered as national, municipal, or private property, rights for common forest may still exist. In such an instance, the important legal point would be the historical fact regarding conventional forest management, including utilization.

After World War II, the common forest faced another economic problem. In 1964, the Forestry Basic Act (Act No. 161 of 1964) was enforced. According to Article 2 of the Act, the main objectives of forestry policy are to increase timber production, timber productivity, and the income of forestry workers. Under the Act, the Forestry Agency strongly promoted the change in species composition throughout Japan, from broad-leaved trees or natural forests to planted forests of coniferous trees such as *Cryptomeria japonica* and *Chamaecyparis obtusa* in the 1960s and 1970s. As the main species of the common forest were broad-leaved trees, the forest was considered a good site for the change in species composition.

In 1966, a new act related to the common forest was enacted, and many forest producer cooperatives, which are called *Seisan Shinrin Kumiai*, were established and started forest management practices. This type of cooperative organization is the main topic of this article. In section 2, the contents of the act of 1966, the policy programs under the act, the founding of forest producer cooperatives, and the current management problems of these cooperatives is explained. Their management is now facing extreme difficulties, and some of the cooperatives have undergone liquidation or soon will do so. However, this has sparked a new movement among some cooperatives. In section 3, three such cooperatives (located in Hyogo, Mie, and Fukui Prefectures) are described, including a brief summary of each cooperative, the major problems they are now facing, and the new movement they are involved in. Section 4 discusses the differences and similarities among these three examples, and section 5 presents our conclusions.

2. Modernization of rights for common forest

2.1 Act on modernization of rights for common forest

In 1966, the Act on Advancement of Modernization of Rights in Relation to Forests Subject to Rights of Common (Act No. 126 of 1966) was enacted. The main contents of the Act are as follows. In Article 1, the objective is to increase agriculture and forestry in the common forest[5]. At the time the Act was enacted, the total area of common forest was estimated to be approximately 2 million ha, which is equal to 13% of the total area of the non-national forest (Iriai Rinya Kindaika Kenkyukai, 1971). Generally, utilization of the common forest was extensive and did not contribute to an increase in income for the residents of mountainous areas, which they obtained from agriculture or forestry. For a long time, broad-leaved trees in the common forest were used as fuel-wood, so the common forest was necessary, but this use of the common forest as an energy source ended in the 1950s and 1960s. Thus, it was important to modernize the rules governing the common forest to promote its utilization. In this Act, modernization meant abolishment of common forest rights, and the granting of property rights[6].

Article 3 of the Act ordained the implementation procedures for modernizing the common forest rights. It was necessary for all rights holders to agree with the abolishment of rights. To this end, they had to be convinced of the importance of the common forest as a custom. To plan the modernization of the system, it was also necessary to record the locations and precise areas of each common forest. These two procedures — getting all rights holders to agree on relinquishing their personal common rights, and the measurement of land — were particularly complicated and difficult.

After the abolishment of common rights, new property rights were granted, in two different ways: land was either divided equally among all rights holders, or a cooperative organization was established. Both methods were available, but the Forestry Agency recommended establishing cooperatives such as forest producer cooperatives or agricultural producer cooperatives (*noji kumiai hojin*), because equal division allowed for subdivision of the forest land. Generally, in Japan, the unit size of private forest land is so small that management costs are relatively high. Among the cooperative organizations related to agriculture and forestry, the Forestry Agency strongly recommended forest producer cooperatives, under the Forestry Cooperative Act (Act No. 36 of 1978)[7]. This cooperative organization was actually introduced in the 1951 Forest Act, and the main objective was not related to the modernization of common forest policy.

The modernization of common forest policy was also affected by Article 12 of the Forestry Basic Act. This article included various policy measures, including the modernization of rights to the common forest, to realize the expansion of small-scale private forest management in Japan. As the area of common forest was generally broader than that of private forest, the transformation from common forest to cooperative organization conformed to the policy direction subject to Article 12 of the Forestry Basic Act. After the 1960s, the planting policy changed to increase coniferous trees to secure future domestic forest resources for industrial round wood; the land owned by the forest producer cooperatives was adapted to this planting policy.

2.2 Establishment of forest producer cooperatives

Under the Act for modernizing common forest rights, subsidy programs to promote the modernization of rights started in 1967. The following is a statistical summary of this 43-year program, which ran from fiscal year 1967 and to fiscal year 2009 (Forestry Agency, 2011b): 6,651 places were given permission to modernize rights. The total number of common forest rights holders before modernization was 432,906, and 423,618 were granted property rights after modernization. The total acreage[8] of common forest whose policies were modernized was 575,125 ha. After the modernization procedure, 52.4% of the land was classified as forest producer cooperative, 41.0% was divided equally among common forest rights holders, 5.5% became jointly-owned private forest, and 1.0% was classified as agricultural producer cooperative (Forestry Agency, 2011b). According to these statistics, 59.0% of the land was classified into some form of cooperative, and more than half of the modernized area became forest owned by forest producer cooperatives. The main utilizations of the land after modernization were forestry (97.8%), agriculture (1.9%), and other activities (0.3%). As already mentioned, the main objective of the Act was to increase agriculture and forestry use in the common forest, which it achieved.

The annual areas of forest that were modernized over time are shown in Figure 1. The modernization procedure was conducted most intensively during the latter half of 1960s through the 1980s, with a peak of approximately 53,000 ha processed in 1974. After the 1980s it tended to decrease, and just 950 ha were processed in 2009. The proportion of land converted to forest producer cooperative of the total land area that was modernized from 1967 to 2008 was > 50%. However, this proportion has decreased gradually. By contrast, the percentage of private ownership increased.

There were 3,459, 3,364, and 3,224 forest producer cooperatives in all of Japan at the end of fiscal years 1999, 2004, and 2009, respectively. The number is now decreasing due to several problems, which are discussed below. Figure 2 shows the number of forest producer cooperatives. Since about 1996, the annual change has been negative (with one exception); that is, more cooperatives have been dissolved than have been newly founded[9].

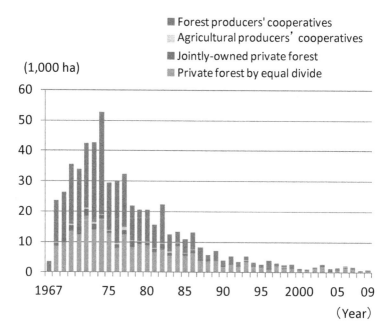

Fig. 1. The areas of common forest modernized over time (1967-2009)
Source: Forestry Agency (1992, 2005, 2006–2011)
Note: Fiscal years are shown.

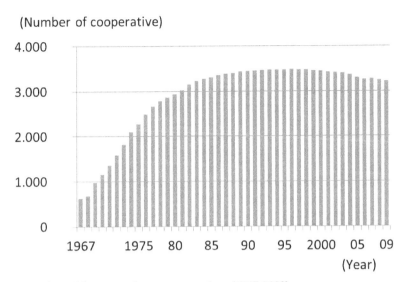

Fig. 2. The number of forest producer cooperatives (1967-2009)
Source: Forestry Agency (1969-80, 1992, 2005, 2006–2011)
Note: End of fiscal year is shown. The figures for the period of 1967-71 represent the number of forest producer cooperatives, which were required to answer the survey conducted by the Forestry Agency in order to compile the Forestry Cooperative Statistics. Thus, the data before 1971 do not completely connect to that after 1972.

2.3 Management problems of the forest producer cooperatives

Based on the statistics[10] for fiscal year 2009, 67.9% of the forest producer cooperatives was founded by the modernization procedure. There were about 247,000 cooperative members, including 7,000 living outside of the community. The common forest rights, which were relinquished at modernization, were strictly limited to the households living onsite. Only about 1% of the cooperatives had full-time executives; thus, nearly all cooperatives were organizationally weak. The total forest area owned by forest producer cooperatives was approximately 357,000 ha, including 10.8% of profit-sharing forest[11]. The percentage of planted forest to the total forest area, of which the main species were *Cryptomeria japonica* and *Chamaecyparis obtusa*, was 50.3%. In fiscal year 2009, total newly planted area was 147 ha, the total area for tending operations was 5,045 ha, the total area for final cutting was 109 ha, and the total area for thinning was 1,007 ha, representing 0.04%, 1.41%, 0.09%, and 0.28% of the total area, respectively. As the area for final cutting and thinning was small, the income from cutting was generally low. The forest producer cooperatives are now facing several issues. Three such challenges are highlighted below.

The first problem is due to the age-class distribution of the planted forests. As planting activities were generally conducted just after the cooperative started from the mid-1960s to mid-1980s, the current average age is 30 or 40 years. Before the modernization procedure, the main species of the common forest were broad-leaved trees for the production of fuel-wood or charcoal; thus, the area of coniferous trees tended to be small. Accordingly, most of the planted forest could not produce income via, for example, timber sale, through final cutting. In addition, certain age classes of tree require thinning. However, at the current domestic price for logs, it is difficult for many cooperatives to make money from thinning. Thus, the income from cutting is almost zero in many cooperatives. In addition, there are costs to maintain the organization, including taxes[12], which many cooperatives have difficulty paying, which can lead to debt[13].

The second problem is related to cooperative member labor. In a forest producer cooperative, it is ideal for forest practices to be conducted by the cooperative members themselves. Indeed, one common rule of cooperatives is that more than half of the members must engage in forest management. However, these principles are difficult to realize under the present circumstances. For one, the demographic composition of communities has changed since modernization took place. When the Act was enacted, there was an outflow of the rural population in almost all mountainous communities, and this trend continues today. In addition, the average age in these populations has increased considerably. As a result, some mountainous settlements have been forced to close, and the sustainability of associated cooperatives has become questionable. Problems related to ageing of the population are common throughout Japan. Due to the increasing population of elderly persons, the number of cooperative members that can engage in forest practice has been decreasing. It may be possible for aged members to planting trees or weed, which was the main work in 1960s and 1970s, but it is impossible for them to conduct thinning of 30- or 40-year-old coniferous trees. Hence, there is currently an extreme shortage of labor power[14].

The third problem is a lack of organizational management. Many forest producer cooperatives are founded on administrative advice based on the national and prefectural forestry policies, including modernization policy for common forest rights, planting promotion policy for coniferous trees, and policy on profit-sharing forest. The percentage of planted forest increased to approximately 50% in fiscal year 2009, which seems to indicate that the planting program was a success. However, there seem to be no forest management practices being undertaken, such as improved cutting and thinning, after tree planting is finished, and any practices that are undertaken tend to be performed just after planting. As a result, forestry engineers were not included in the cooperatives[15]. In some cases, the executive of the cooperative was determined by rotation, and the members had no experience in, or concern with, forestry. There are almost no forestry technicians, managers, or other staff, and most cooperatives have never created a specific forest-management plan. Without a long-term forest-management plan, cooperatives are unable to conduct appropriate forest practices or construct road systems, and without an appropriate forest road system, for example, it is difficult or impossible to harvest trees.

These three problems are complicated, and correlated. The aging of cooperative members is occurring before many planted forests reach the final cutting age. Such rapid demographic changes are affecting almost all social and economic activities in mountainous areas, including forestry. In addition, younger people are leaving such areas, resulting in not only a decrease in labor but also a lack of managers for the future.

Thus, it is difficult to find a clear reason to continue cooperatives: income from the forest is lacking, there is no future vision for forest management, the tax payments are difficult, and debt is increasing. As a result, the number of established forest producer cooperatives has been decreasing, and the number of dissolved cooperatives has been increasing. Future research should investigate how forest lands are being managed after dissolution of forest producer cooperatives[16].

At the time the Act was enacted, there were approximately 2 million ha of common forest. As mentioned above, the total area of modernized forest during the 43 years from 1967 to 2009 was approximately 575,000 ha. As of December 1, 2007, there were 618,000 ha of common forest remaining, and the modernization procedure began on approximately 109,000 ha among these 618,000 ha. Among the 618,000 ha, only 32,000 ha, which is equal to approximately 5% of the total common forest area, is continuing to undergo modernization. The modernized procedure was discontinued for the remaining 95% of common forest and there has been no will to modernize common forest rights for this area. The deteriorating economic and social conditions described above have also occurred in areas in which modernization has not taken place, and several small settlements are now facing dissolution in the near future. In such areas, the common forest rights will become more ambiguous and complex and it is possible that such areas will be removed from forestry policy and thus problems related to non-timber functions may occur.

3. Case studies

Most forest producer cooperatives are facing management difficulties. They planted coniferous trees. If cutting activities are not conducted, the cooperatives cannot obtain income from the selling of logs. As a result, some cooperatives have chosen to dissolve the organization. However, a small number of cooperatives are persisting. Here, three such cooperatives, located in Hyogo, Mie, and Fukui Prefectures, are described[17].

3.1 Case 1 (Hyogo Prefecture)[18]
3.1.1 Description
This cooperative was founded in 1971. In fiscal year 2009, there were 184 members and the cooperative was managing 563 ha of forest, including 466 ha of planted forest (Hyogo Prefecture, 2011b). This cooperative has planted a substantial number of coniferous trees; currently 82.8% of its land is planted forest, which is extremely high compared to the national average of 41.7% in non-national forests in this prefecture (Hyogo Prefecture, 2011a). *Cryptomeria japonica* and *Chamaecyparis obtusa* are the two main species in the planted forest. The age-class distribution of the forest is as follows[19]: 1.6% (1–10 years), 2.4% (11–20 years), 12.2% (21–30 years), 31.8% (31–40 years), 46.7% (41–50 years), 2.7% (51–60 years), and 2.6% (≥61 years). Thus, there is a large area of 30- to 50-year-old trees. As this cooperative was founded in 1971, trees that are about 40 or more years old were planted before the modernization of rights. Approximately half of the planted forest was already planted during the period when the forest was legally common forest. Generally, common forest is not utilized intensively, but in this case, coniferous trees had been planted in a large area.

3.1.2 Largest problem and background
The biggest problem that the cooperative is now facing is managing the large area of the planted forest, in particular, thinning it. Generally, in Japan, the difference between the selling price of a log and the actual logging cost is small or even negative for many forest sites. Accordingly, thinning activities tend to be put off.
Final cutting was last conducted in 2001. The total forest area for thinning between 1992 and 1995 was 45 ha, whereas that in 2003 was 2 ha and that between 2007 and 2009 was 33 ha. These values are extremely low compared to the area of planted forest.

Thinning is necessary in a planted forest, particularly in a dense plantation, to direct the current forest stand to the targeted forest stand, and in this sense, a delay in thinning or changing the method are generally permitted. However, problems[20] related to thinning have become more political and more complex since the Kyoto Protocol in 2008. The government promised a 6% reduction in greenhouse gasses compared to the base year, including a 3.8% contribution to the carbon sink of domestic forest resources. Thinning must be conducted in the planted forest to fulfill the national target. Without thinning, the planted forests are not considered well managed. Hence, thinning activity is now under political and social pressures related to the Kyoto Protocol, which is unconcerned with forestry activities. The Forestry Agency is now promoting thinning throughout Japan. In 2008, a special measure promoting thinning was enacted[21]. This is a typical problem being experienced by forest producer cooperatives.

3.1.3 New activity and background

This area, including the cooperative forest site, is among the model areas for promoting rapid and intensive timber production. In 2009, an association of which the forest producer cooperative is a member began a project to stabilize timber production.

The new initiative of the cooperative included the introduction of a Japan Verified Emission Reduction (J-VER) Scheme, which is a public certification scheme for carbon offsetting managed by the Ministry of the Environment. In July 2011, there were four J-VER projects related to forest management in progress in Hyogo Prefecture. The representative business operator and the holder of the offset credits was the Hyogo Prefectural Federation of the Forest Owners' Cooperative Association[22]. This association founded a company that provides offset credits in 2008. This prefectural association can sell the offset credit to private companies by way of the offset provider company.

The total area of forest for offset is 97 ha, and the credit period is from 2008 to 2012. The total CO_2 sink for emission reduction is assumed to be 2,657 tons in CO_2 over 5 years, which is 531 tons CO_2 per year. The project was already registered and a field investigation was conducted in 2011. Due to declines in the log price and other reasons, forestry profitability is not expected to be regained, and the CO_2 sink funds are expected to be utilized as part of the future cost of forest management and logging. Under these difficult economic circumstances, the cooperative expects income from their CO_2 business.

The J-VER Scheme, started in 2008, certifies the greenhouse gas reduction or carbon sink realized by domestic projects as a credit. In March 2009, with collaboration between the Ministry of the Environment and the Forestry Agency, the forest management project, in which project CO_2 sink by doing forest practices such as thinning and planting, was added to the list of domestic projects which will act as a carbon sink. The forest management projects[23] in J-VER Scheme are divided into two types, those that will promote thinning, and those that will promote sustainable forest management. This case is the former type.

3.1.4 Characteristics of and evaluation of the new activities

At the time of our investigation, the carbon credit sales produced from the forest owned by the forest producer cooperative was not determined; the direct financial contribution is unclear. As the cooperative expects income from carbon credits, how these profits are to be distributed between the offset provider company, the cooperative, and other related organizations will be determined is the most important. The new relationships that the cooperative wants to build are evaluated from the following three points of view.

First, this cooperative is the first among 3,224 cooperatives that challenged joining the J-VER Scheme[24]. The forerunner of many forest producer cooperatives is common forest originating from the Edo Period, and the common forest had been utilized among a limited number of residents. Generally, the management policy of such forests seems to be limited, thus, the cooperative is isolated from global environmental issues. As there are many forest producer cooperatives, future expansion of the J-VER Scheme might be expected, but it is difficult to rapidly increase forest for carbon offset in cooperatives, due to the current complex and cumbersome procedures and the fact that most cooperatives do not have full-time staff.

Second, the introduction of the carbon offset program is related to the relationship between the younger generation in the community and forest owned by the cooperative. The profits from timber production and the direct or indirect distribution of the profit are tangible results to residents, including the younger generation. The past common forest had brought various tangible results to the rights holders of the forest. However, younger people tended to work outside of their community, and were far removed from agriculture and forestry. As a result, their concern for the forest owned by the cooperative has been decreasing. One important challenge is to renew the involvement of younger people in the forest. Despite the stagnation in timber production, the forest resources owned by the cooperative will contribute to the carbon offset program under the J-VER Scheme, and thus younger people might turn to the forest owned by the cooperative.

A third activity is to enhance the partnership between forest owners among districts. Carbon credits based on the forest owned by the cooperative and jointly owned private forest located next to the cooperative's forest are for sale as one credit. After their sale, the profit will be distributed among those involved[25]. The cooperative and the jointly owned private forest participate in the same forestry promotion group and recently planned the construction of a new skidding road that passes both forests. The jointly owned forest was certified by the Sustainable Green Ecosystem Council (SGEC)[26], which is a Japanese forest certification organization, and now the cooperative is considering obtaining forest certification. This new movement could lead to collaboration with the credit purchaser for the purpose of carbon offset in city areas. According to a newspaper report[27], the profit from carbon credit sales would be used for forest management in the area where the cooperative is located. Thus, the number of people concerned with the forest is expected to grow.

3.2 Case 2 (Mie Prefecture) [28]
3.2.1 Description
This cooperative was founded in 1960, to consolidate the municipalities at that time and to prevent common forest from being absorbed into the new municipal government after consolidation. Thus, all of its members are residents of a village, which was ultimately consolidated into a new municipality. In 2009, the total number of members was 505 (Mie Prefecture, 2010). The area of forest is 23.8 ha, including 21.4 ha of planted *Cryptomeria japonica* and *Chamaecyparis obtusa* forest. The age of the planted forest is approximately 40 years. The cooperative is holding lands other than forest, of which the largest and most important is land for a golf course. The management is generally good, as the cooperative can collect rent from a golf company every year. The main work of the cooperative is management of their forested and non-forested land.

3.2.2 Largest problem and background

Residents can work in the city, as this cooperative is located in an urban neighborhood. It is not a wood-producing area, and activities related to forestry are generally sluggish. There are a few saw mills, in which the logs purchased from outside the area are processed. This cooperative has no forestry technicians. Thus, forest practices have not been introduced, although the cooperative has money for forest management. As thinning has never been conducted, the light intensity in the forest is low and future growth is not expected.

The planted forest was intensively managed in the 1960s and 1970s, but various problems relating to thinning are now occurring. In Case 1 described above with similar problems, a lack of funding for forest practices including thinning was a problem, but not in this case. Rather, an almost total lack of forestry knowledge and experience is the most important problem in this cooperative[29]. Such a situation is common in forest producer cooperatives located in urban neighborhoods, and seem to be related to the following three points.

First, almost all residents, including the cooperative members, have lost a relationship to the forest. Thus, they cannot judge when and how the forestry practices must be conducted. By the 1960s, some connections remained to the forest, such as using it as a source of fuel-wood utilized in daily life. However, wood and fuel-wood are not necessary for residents today. Thus, interest in forestry has dissipated, and along with it, the necessary knowledge, experience, and techniques to manage a forest.

Second, the age-class of the planted trees is increasing. In the 1960s and 1970s, planted trees were short when harvested and special techniques or large machines were not necessary for planting or weeding. Although this was certainly hard work, it was not dangerous or impossible. Today, however, specialized experience and forestry machines are necessary to thin 40-year-old trees. The cut trees must be moved to specific points or a forest road. Such practices cannot be conducted by cooperative members who have no forestry experience.

Third, even if the cooperative members cannot develop and conduct necessary forest practices, there is no problem if there is a forestry-related organization around for consultation. However, there is no such organization near this cooperative. Even if there were, most forest owners' cooperatives are generally weak in urban neighborhoods. In most cases, there are no forestry workers or machines, and their main work is paperwork.

3.2.3 New activity and background

This cooperative started thinning using funds sponsored by an automobile-related enterprise located in the same municipality. Mie Prefecture started a system in which private enterprises can allocate money for sustainable private forest management. When the board of the cooperative discussed thinning with the administrative office, the enterprises were searching for an appropriate site by chance. In 2006, the forest sites for thinning and supporting funds were identified. A tree-planting ceremony was held in 2008. A thinning and tree-planting ceremony was planned to utilize the money over 5 years. Many from the wider community, including elementary school children, attended the ceremony in April 2010. This is important because awareness regarding the forest within the community, and forestry activities generally, spread to the residents, including children and their families.

The chief of the cooperative pointed out that the development of consciousness for forest can be found in the cooperative board members as a result of the conduct of thinning and tree-planting ceremony. Although their major previous thought was that forest can be left unmanaged without cutting grass, but they begin to think that leaving it uncontrolled was not good practice. When the cooperative conducted thinning, it developed a relationship with the logging company located in the same area.

Various private companies in Japan have recently started supporting management and forest conservation, including both national forests and non-national forests. In the case of national forest policy programs, companies can allocate money for future management costs for a specific forest site, which consists of coniferous trees in most cases, and the company can obtain some of the profits when the forest is finally cut[30]. During a contract with the national forest, the company can utilize the forest site without cutting. For example, the company can place a sign indicating that the company is contributing to the national forest environment, or it can utilize the forest for employee or customer events. Of course, companies can highlight their contribution to the national forest in their annual report on corporate social responsibility. Companies can easily fund national forest management as a public contribution as it is governed by a department in the national government.

By contrast, sustainable forest management in the privately owned non-national forest is supported mainly by subsidies. Recently, many prefectures have developed schemes by which private companies can support non-national forest[31]. The support of forest producer cooperatives by companies, such as Case 2, is a recent phenomenon. Forests owned by cooperatives are non-national forests in which the management area is generally larger than that of forests owned by individuals. Moreover, as the cooperatives are legally incorporated organizations, it seems that they have an advantage in making such contracts with private companies.

3.2.4 Characteristics of and evaluation of the new activities

This contract resulted in good forest practices such as thinning of the planted forest to some extent, which is strictly necessary, but the area that can be supported by a company is limited. More significant is whether the connection between the cooperative and the private company will result in sustainable forest management for the cooperative in the future. Expansion of forest practices in the regional forest might be expected. As the contract has recently just started, an evaluation from these points of view cannot be clearly shown, but several new possibilities are suggested by Case 2.

One of the important characteristics of Case 2 is the strong relationship between the forest producer cooperative and the residents' association. In this case, the forest producer cooperative, as owner of the forest, and the residents' association, as an organization that promotes regional activities in all fields of daily life, are working together. This co-sponsored fund resulted in a new relationship between the cooperative and the residents' association.

As mentioned above, some of the forest producer cooperatives are facing management difficulties such as financial problems, and one of the solutions is dissolution of the cooperative and the transfer of the forest from the cooperative to the residents' association. In these cases, forest management is impossible for the cooperative but may be possible for the association. In such cases, the major role of forest management is changing from timber production to land management, because the main objectives of the residents' association do not include forest management. However, Case 2 shows that the appropriate divisional cooperation will result in profits for both organizations.

In the case of a forest producer cooperative that was founded in a common forest originating during the Edo Period, the cooperative members were originally residents who were living in a limited area at the time the cooperative was created. However, as years pass since the creation of the cooperative, the members' profile changes, because new residents move there from outside the area and some people, mainly younger people, move away to cities to get jobs. After a long period of time, the area may include a large proportion of people who moved there after the cooperative's founding, and who thus do not have the right to become a cooperative member. However, they can become members of the residents' association. Only in this light, dissolution of the cooperative and transfer of the forest to the residents' association can benefit the community.

Second, when a cooperative is located in an urban neighborhood, various efforts are necessary to promote sustainable forest management, from the point of human resources. In most cases, the chair of the cooperative has no forestry knowledge or experience with forestry. In this case, the board members hold 3-year terms and there may not be a chance to learn forestry at all during their term, or their terms might expire after they have learned some forestry methods.

In this cooperative, thinning activity is conducted by a fund from the company. The board members of the cooperative had no idea that thinning even had to be done. Considering that the cooperative was founded under a national forest policy that ensured the site for planting *Cryptomeria japonica* and *Chamaecyparis obtusa*, the administrative sector also had to show how to sustainably manage the coniferous tree plantation forest, particularly the necessity for thinning. Even if the cooperative board members notice the need for thinning, they cannot progress without any support from forestry-related organization including administrative sector.

In this case, the cooperative had to conduct the thinning and tree-planting ceremony with support from the company, according to the contract between the cooperative and the company. The board member of the cooperative had new relationships with the officer of the municipality, officers of the department of forestry, local officers of the prefectural government, and the local logging company. Thus, the cooperative was able to identify people and organizations involved in forestry in the community and learn about the differences in the forest stand before and after thinning.

Third, it must be pointed out that the contribution of human resources from the cooperative is very important—not only the efforts of the cooperative board members but also the existence of a key person who is willing to participate actively outside the cooperative. In Case 2, the person responsible in this company was enthusiastic about this activity. He spent more time than expected on the project, created various documents for people both inside and outside the company, and carefully prepared for the ceremony.

Generally, the role of the local officer of the prefectural government is to promote forest practices and to expand forest techniques. In this case, his work became more difficult and complex; for example, he often visited the forest site undergoing thinning and generated many official documents. His contribution was also important because he played a coordinating role.

3.3 Case 3 (Fukui Prefecture)[32]
3.3.1 Description
The cooperative was founded in 1968. The total area of forest is large at approximately 2,500 ha, including approximately 200 ha of planted forest[33]. The remaining area is covered by natural forest including secondary broadleaved forest that had been used for the production of fuel-wood or charcoal for a long period of time. When the cooperative was founded, a profit-sharing reforestation program was introduced to plant coniferous trees. In the 1960s and 1970s, the Forestry Agency and the prefectural government promoted the establishment of this type of forest producer cooperatives.

3.3.2 Largest problem and background

The largest problem that this cooperative is now facing is the same as that for Case 1, but to a greater extent. That is, the population has decreased and aged dramatically. In the area where the cooperative is located, there were 93 households and 535 persons, respectively, in 1891. In 1950, the population decreased to 163 persons, and today is down to only two people (MAFF, 2010). The main jobs in the area were forestry and sericulture, and both have fallen into decline. The remaining residents are now living in a neighboring area. Unlike Cases 1 and 2, this cooperative is not located near a city with jobs, which in those other cases has helped to prevent the collapse of the community. Thus, the largest problem of this cooperative is how to reorganize itself under the current circumstances.

3.3.3 New activity and background

With only two people remaining in the community, the cooperative cannot move ownership to a residents' association. As two people can do very little, the cooperative, a large part of the membership of which are now living in a neighboring community, is attempting to revive local interest in exploiting the forest for a wide range of activities. In the annual meeting of the cooperative in 2005, various important problems were discussed, including how to improve degraded forest resources, protect the natural environment, address a problem of illegal dumping, enhance the low morale of mountain climbers, spark community regeneration and activation, and further utilize forest resources (Kuniyoshi, 2008). The group ultimately decided to develop ecotourism in their forest.

The cooperative set up an organization, together with community members, which included previous community members, the staff and students in an architectural course at a local private college, a local private railway company, a non-profit organization related to the mountains, and individuals, and they are actively recruiting sponsors and volunteers (Kuniyoshi, 2008). Various ecoprojects are now in progress (Kuniyoshi, 2008). The organization helps to maintain trails in Hakusan National Park, with the aim of running ecotourism project there as well. Many tours are already available, and some hunting is allowed. Some old traditional houses are being renovated with help from college staff and students. These renovated houses are expected to become major ecotourism draws. The renovations are also viewed as a starting point for community regeneration. Restoration of the local waterway is also being conducted as part of landscape management. A tea-growing area has been introduced into an abandoned cultivation area, which is also expected to be a major attraction for ecotourists in the future.

It is interesting that the cooperative clearly identified the development of ecotourism as their main forest-management goal, given that ecotourism has never before been a major objective of a forest producer cooperative[34].

3.3.4 Characteristics of and evaluation of the new activities

In this case, the final results of the cooperative's initiative will not be realized until the future, but this case study illustrates several new ways to manage a forest producer cooperative.

First, linking the cooperative to ecotourism is a thought-provoking concept. Ecotourism is related to forest resources and many national forests managed by the Forestry Agency, such as Yakushima Island, Iriomoteshima Island, the Shirakami Mountain Range, and Shiretoko, are famous in Japan[35]. In these cases, the forests have been classified as Forest Ecosystem

Protected Areas and almost all forest-related activities are prohibited within them. In contrast, this cooperative owns the forestland, which is non-national forest[36], and thus can open it up for ecotourism.

The second is the basic policy that the existing facilities are important. For example, the renovation of traditional houses, which were damaged by extraordinarily heavy snowfall, is in progress. The first projects are in cooperation with volunteers, and the next project will be developed to effectively utilize the houses. Such a basic idea of utilizing existing facilities is actually related to funding problems. In the past, public work projects for tourism in forested areas required building new facilities, which were paid for by subsidies from the national or prefectural governments. However, in the case of this cooperative, the costs are covered or mitigated using volunteers and sponsors and involving the private sector in planning and finance.

Third is the fact that the ecoprojects were conducted directly by forest producer cooperative that own the forestland used for the ecoprojects, with the cooperation of individuals and organizations that have had no previous relationship with forestry. In contrast, the involvement of the forest owners' cooperative, prefectural federation of forest owners' cooperative association, and logging companies, which played an important role in the new activities of the forest producer cooperative in Cases 1 and 2, seems to be small in Case 3. There are many tourism-related sites and various activities in the national forest, but the Forestry Agency and its regional offices are not involved directly in tourism management. Their role is only that of being the owner of the forestland. The fact that the forest producer cooperative participate directly in the ecoprojects as the forest owner seems to be a new direction in forest resource management.

Fourth is that this cooperative's initiative could lead to substantial earnings. The community is located at the entrance to Hakusan National Park. The cooperative pointed out that the environment, history, and culture of the community are being lost due to depopulation and the increase in the number of climbers in the national park[37]. Particularly for the latter reason, a cooperation fund for protecting the local environment, which is a toll fare system[38], was introduced in 2007, and toll gates were constructed at the entrance to the forest road. Although the income from toll fare is limited, the cooperative is expecting more income from the other activities it has planned.

In Japan, the national park system is based on a zoning system, and the government does not generally own the land within a national park. Thus, the role of private land is important in regions where privately owned land is dominant. However, the Ministry of the Environment, which is responsible for managing the national parks, has done almost nothing to protect and maintain the natural environment in privately owned forests, except to regulate cutting activities. Simultaneously, residents living within park areas do not understand the characteristics of the region in general. Under these circumstances, the new projects of the cooperative have great significance for the national park system, because there are many national parks in which the percentage of privately owned forest is high. The cooperative is contributing to the maintenance of the natural environment around the national park.

Finally, this cooperative has shelved timber production and forest practices, the major objectives of all cooperatives at their founding, in favor of preserving the community's history, culture, and lifestyle. There are other cooperatives and communities in similar situations throughout Japan, particularly in mountainous areas. Still, many challenges remain. If, for example, the remaining two residents leave the community, the long history of the community will be completely lost. Furthermore, while it is possible to repair the old traditional Japanese houses damaged by heavy snow, it will become more difficult in time. Some previous residents, who are mostly elderly, are still living in surrounding communities, and when the generation changes, a major part of history will disappear. The remaining time is strictly limited.

4. Discussion

Three case studies were introduced in this study to illustrate new trends in the management of forest producer cooperatives. The cooperative in Case 1 reestablished timber production and increased their income by trading carbon credits; that in Case 2 developed a novel relationship between the cooperative and forestry-related organizations; and that in Case 3 has started a number of initiatives aimed at regenerating the community, increasing the population, and embracing ecotourism activities and principles. In all cases, the cooperatives introduced a new social and economic system. Importantly, a small number of cooperatives throughout Japan have begun to apply similar ideas to their cooperatives.

As the management of forest producer cooperatives becomes increasingly difficult and timber prices continue to drop, fewer cooperatives are being formed, and those that were founded long ago are increasingly being dissolved. Yet, these three cooperatives have found new ways to manage their lands in ways that are completely different than previous management styles. The three cooperatives have some common characteristics, which may help inform more effective management of other cooperatives in the future.

First, each has embraced outside organizations that have no connection to forestry, from the private companies purchasing CO_2 credits in Case 1 to the automobile enterprise involved in Case 2 and the college involved in Case 3. Concern about global environmental problems has been gradually increasing in Japan, and the first commitment period for the Kyoto Protocol started in 2008. Environmentally related activities of private companies or organizations have been increasing, and, at the same time, the content has been changing. Thus, in some cases, a forest producer cooperative can open up their lands for such environmental activities. The cooperatives and other forestry-related organizations should explain their difficulties regarding sustainable forest management to the public.

Second, each made substantial administration changes, particularly in Cases 1 and 2. Forestry policy programs have traditionally been run mostly by the national and prefectural government, and in a different manner[39]. However, in Case 1, the carbon credit program is mainly operated by the Hyogo Prefectural Federation of the Forest Owners' Cooperative Association. In addition, the carbon credits produced from the forest are held by the prefectural association. The carbon offset provider is also closely related to the prefectural association. Similarly, in Case 2, many initiatives are sponsored by private companies, but prepared and managed by the prefectural government (a major reason why the companies feel safe supporting the activities). For example, a company paid for thinning and tree planting via the Mie Prefectural Federation of the Forest Owners' Cooperative Association, which is an extra-departmental body of the prefectural government.

Third, each cooperative has found new ways to be funded. In many ways, this is natural consequence of the two previous points. For example, a change in administration means a change in subsidy, and providing benefits to outside organizations can lead to earnings from those outside sources. For instance, in Case 2, the forest producer cooperative received rent from non-forest land every year.

Finally, each cooperative underwent a change in leadership[40]. The original presidents of these cooperatives did not come up with these novel ideas. It took a new generation of presidents, who decided to act in a broader area than just forest practice and timber production. In Case 1, the president had a long-term vision for joint regional timber production and sustainability of the community. In Case 2, the president agreed to participate in an in-forest event for residents and students, and in Case 3, the president was finding a way to regenerate the community. They are acting not only for the business of the cooperative but also for the community and its members.

5. Conclusions

The origin of the forest producer cooperative was the common forest starting during the Edo Period. At that time, forests were an essential resource for agriculture, energy, and daily life. A limited number of people living in a specific area conducted sustainable forest management by following local rules and excluding outsiders. Since 1966, some of such forests have been changed to forest producer cooperatives. These cooperatives have endured many hardships and continue to face major financial and other challenges. However, some novel solutions have recently been applied to revive cooperatives. All of these have included expanding the cooperative's business outside of the forest owned by the cooperative. This has resulted in the formation of new business networks, acceptance of new types of administrative services, and new sources of funding. Moreover, a new generation of leaders, with fresh ideas and different job experiences compared to past presidents, appears to be having an impact, taking the cooperatives into new, unexplored directions.

6. Acknowledgments

The author thanks the staff of the department of forestry of the prefectural government, board members of the forest producer cooperative and related organizations for the case studies in the three prefectures. The author also thanks anonymous reviewers for their useful comments.

7. Notes

1. See Handa (1988) and McKean (1992) for English language accounts of the common forest in Japan.
2. A term that means the smallest unit relating to the rights of the common forest is difficult. In Handa (1988, 2001), "hamlet" is used. "community forest" sometimes means another forest other than the smallest unit of the village. For example, it means forest owned by the city, town, and village in GHQ (1951).
3. Itoh (2009) pointed out the current decision by the Supreme Court.
4. See Totman (2007) for more on land reform during the Meiji Period.
5. This act was not only for the common forest but was also for common land among the municipality forest. As referred to in section 1, part of the common forest became public forest at the time when the municipality system was introduced in 1889. At that time, conventional utilization, which continued from the Edo Period, was permitted in some cases of newly established public forest that originated from the common forest. Namely, there was common land among municipality forests. The important point is

that the conventional utilization in the municipal forest was not the right for the common forest and was permitted by the town or village assembly. However, for simplicity, we refer only to "common forest" in this chapter.

6. Nakao (1969, p.56–69) reported five characteristics of common forest rights. (1) The rights were based on the customs of the area. (2) People who were living in a definite hamlet had rights. (3) Households had rights. (4) It was impossible to inherit rights. (5) It was also impossible to assign or sell rights to anyone. These five points make up the difference between common forest rights and property rights.

7. Under the Forestry Cooperative Act, there were two types of forestry-related cooperative organizations at the local level: forest owners' cooperative and forest producer cooperative. See Forestry Agency (1955), Matsushita & Hirata (2002), and Ota (2009) for more information about the forest owners' cooperative in Japan.

8. Total forest area for modernization of rights was divided by the total number of holders of common forest rights, resulting in 1.33 ha. This figure is just for reference, but it shows the approximate size of forest land that each person would have received if the land was divided equally.

9. The statistics by the Forestry Agency do not include the annual numbers of established cooperatives and dissolved ones, only the number of existing cooperative (Matsushita, 2012).

10. These statistics were compiled annually for all forest producer cooperatives. In the fiscal year 2009 survey, survey sheets were sent to 3,224 cooperatives, and the number of respondents and the rate of respondents was 2,723 and 84.5%, respectively. The figures shown here are for the 2,723 responding cooperatives.

11. Various combinations of profit-sharing forestation systems have been developed in Japan since the Edo Period. After World War II, the Act on Special Measures concerning Shared Forest (Act No. 57 of 1958) was enforced, and planting of coniferous trees was promoted. Most forest producer cooperatives prepared forest land for planting. The Prefectural Forestation Corporation or Forest Development Corporation allocated funds for planting trees, and the profit was divided as a constant percentage, which was determined at planting.

12. A fixed asset tax is common between private forest and forest owned by forest producer cooperatives. A corporate inhabitant tax is required regardless of income.

13. From the Forest Cooperative Statistics of fiscal year 2009, the percentage of forest producer cooperatives with short-term debt and long-term debt is 14.5% and 22.6%, respectively.

14. When labor power is lacking, forest producer cooperatives must outsource forest practices to the forestry company or employ forestry workers. This may conflict with the principles of independent business. Plus, it is important for cooperatives to balance log price and labor cost. Generally in Japan, labor costs have been increasing, and log prices have been decreasing. In 2009, the average wage for male loggers was 12,898 yen per day and the average price for a medium-sized log of *Cryptomeria japonica* was 10,900 yen per 1 m^3 (Forestry Agency, 2011b). In 1980, when the average log price peaked, these figures were 8,550 yen and 39,700 yen, respectively (Forestry Agency, 1992). Although a clear comparison is not possible due to the difference in survey methods between the years, it is clear that the relative log price to the logging wage has decreased drastically.

15. Today, forest management practice of forest producer cooperatives is financially subsidized and generally undertaken by forest owners' cooperatives (Kawamura, 2010).

16. Recently, there has been a rise in the number of forest producer cooperatives that have dissolved their cooperative in order to transfer the forest into a regional organization such as a residents' association (Sakai, 2005; Yamashita 2006). This was unexpected, but is legal, and will likely lead to many future problems, for example, conflicts over how to distribute profit after harvesting trees.

17. These three cases were picked from the annual meeting of the Middle Japan Common Forest Conference.

18. Part of the explanation of this section is based on the personal interview by the author with the board members of the forest producers cooperative in July, 2011, and presentation by Fukuda at the 32th Annual Meeting of the Middle Japan Common Forest Society in September 1, 2011.

19. These data are based on a forest planning system summary table managed by the prefectural government; not all planted forests are included in the system.

20. See Matsushita and Taguchi (2011) for more information about global warming and forest policy in Japan.

21. In 2008, the Act on Special Measures concerning Advancement of Implementation of Forest Thinning, etc. (Act No. 32 of 2008) was enacted. Thinning above the usual levels was promoted by this Act. During fiscal year 2007 and fiscal year 2012, the usual thinning target area is 350,000 ha per year, and the additional area is 200,000 ha per year. In total, 3.3 million ha of planted forest is planned to be thinned as a result of this Act. Additionally, planting in non-reforested land is promoted.

22. Ishimaru (2011) explained the current situation and problems regarding the introduction of the J-VER Scheme to the forest owners' cooperatives, using the examples of Osaka and Hyogo Prefectures.

23. On December 2010, the total number of registered forest management J-VER projects and the quantity of credit was 60 and 74,038 tons CO_2, respectively, and among these registered projects, the total number of entirely certificated projects was 26 and 34,993 tons CO_2, respectively (Forestry Agency, 2011a).

24. The number of cooperatives was based on the Forestry Cooperative Statistics for fiscal 2009.

25. Part of the carbon credit from the forest owned by the cooperative and jointly owned private forest was purchased by a private railway company to offset the carbon dioxide emitted from a railway station, which was newly opened on March 14, 2009. This is the first case of such a project related to a railway station (The Kobe Newspaper, March 12, 2010). The railway company's homepage indicated that the quantity of CO_2 offset during March 14, 2010 and March 13, 2011 was 37 tons in CO_2 for operating the station, 267 tons CO_2 for the train, 3 tons CO_2 for the station stand and automatic vending machines, or a total of 308 tons CO_2.
(http://holdings.hankyu-hanshin.co.jp/eco/information/information_110314.html, 2011/09/06)

26. This organization was founded in 2003. As of April 5, 2011, the total number of certified forests by SGEC was 116, and the total certified forest area was 864,351 ha. More certified forests in Japan have been certified by SGEC than any other certification organization.

27. Kobe Newspaper, March 12, 2010.

28. Part of the explanation of this section is based on Kawasugi (2009), Matsushita (2009), and Nagata (2009).

29. In Case 1, the area in which the forest producer cooperative is located is among the most important forestry areas in Hyogo Prefecture. The Prefectural government has introduced various forestry policy programs in this area, and there are relatively many forestry workers.

30. In fiscal year 2009, 486 sites in national forest were used for activities sponsored by private companies (Forestry Agency, 2011a).

31. In fiscal year 2009, 638 sites in non-national forest were used for activities sponsored by private companies (Forestry Agency, 2011a).

32. Part of the explanation of this section is based on Kuniyoshi (2007, 2008) and Matsushita (2008).

33. The holding area of forest is larger than that of the private forest owned by individuals. This is an important characteristic of the forest producer cooperative, and Handa (2001) and Hirata (2008) have pointed out that this characteristic has to be utilized more often.

34. The Government enacted the Act on Promotion of Ecotourism (Act No. 105 of 2007) in 2007 and enforced it in 2008. Article 3 of the Act indicated a philosophy of ecotourism, including protection of the natural environment, promotion of tourism, regional development, and environmental education. The national government created basic ecotourism policy and the municipal offices made the regional master plan. Ecotourism started before enforcement of the Act, but enacting ecotourism into law was recent. In case 3, some activities related to ecotourism started before enforcement of the Act.

35. Yakushima Island, Shirakami Mountain, and Shiretoko were listed as World Natural Heritage Sites by UNESCO in 1993, 1993, and 2005, respectively.

36. Sacred Sites and Pilgrimage Routes in the Kii Mountain Range were listed as World Cultural Heritage Sites by UNESCO in 2004. Most of the forest is non-national forest.

37. Based on a presentation by Kuniyoshi (2008) at the 28th Annual Meeting of the Middle Japan Common Forest Society.

38. The major contents of the leaflet for the visitors are as follows. There are problems such as damage in forest roads and climbing trails, illegal picking of natural plants in private land, and illegal waste dumping due to the increase in visitors and climbers. The fee is 300 yen for adults and 100 yen for children of elementary school age or younger. The use of the fund is limited to improvement of climbing trails, maintenance, operation, and activities related to nature protection, and so on.

39. The basic policy of the past 45 years on the common forest rights is facing limits. When the act was enforced in 1966, the Forestry Agency and the departments of forestry of prefectural governments did not have ideas such as these three cases. One of the reasons is that these three examples may conflict with the basic principles of independent businesses, which are required for the forest producer cooperative. However, in these three cases, the department of forestry of the prefectural government gave silent approval.

40. In the three cases, the former job of the president was not forestry. Thus, it is possible that their experience in fields other than forestry might have influenced their decisions. New ideas may continue to develop as the ageing population grows in rural areas, particularly

in mountainous areas, and generations change, and new presidents or board members with different experiences and fresh ideas take over the cooperative's business.

8. References

Forestry Agency (1955). *Forestry of Japan*, pp.177–184, Tokyo, Japan

Forestry Agency (1969-1980). *Shinrin kumiai tokei (Forestry Cooperative Statistics): fiscal year 1967-1978*, Tokyo, Japan

Forestry Agency (1992). *Ringyo tokei yoran: jikeiretsu ban (Annual statistics on forestry: time-series version)*, p.62, p.90, Rinya Kosaikai, Tokyo, Japan

Forestry Agency (2005). *Shinrin ringyo tokei yoran: jikeiretsu ban (Annual statistics on forest and forestry: time-series version)*, p.55, p.148, ISSN 1344-543X, Rinya Kosaikai, Tokyo, Japan

Forestry Agency (2006-2011). *Shinrin ringyo tokei yoran (Annual statistics on forest and forestry)*, Tokyo, Japan

Forestry Agency (2011a). *Shinrin ringyo hakusyo (White Paper on Forest and Forestry)*, p.48, p.61, ISBN 978-4-88138-259-2, Zenkoku Ringyo Kairyo Fukyu Kyokai, Tokyo, Japan

Forestry Agency (2011b). *Shinrin ringyo tokei yoran (Annual statistics on forests and forestry)*, p.87, p.168, pp.232–235, ISSN 1344-543X, Nihon Shinrin Ringyo Shinkokai, Tokyo, Japan

GHQ, General Headquarters, Supreme Commander for the Allied Powers (1951). *Forestry in Japan*, p.16, Tokyo, Japan

Handa, R. (1988). Policy for the administration of public forests and common forests, In: *Forest Policy in Japan*, Handa, R. (ed.), pp.211–225, Nippon Ringyo Chosakai, ISBN 4-88965-001-6, Tokyo, Japan

Handa, R. (2001). A 50-year history of forest producer's cooperatives and hamlet forest, *Forest Economy*, No.637, (November 2001), pp.1–13, ISSN 0388-8614

Hirata, K. (2008). Seisan shinrin kumiai no keiei doko (Management of the forest producer cooperative), *Sonraku to Kankyo (Rural community and the environment)*, No.5, (August 2009), pp.3–8

Hyogo Prefecture (2011a). *Hyogo-ken ringyo tokeisyo (Annual statistics on forestry in Hyogo Prefecture)*, p.4, Hyogo Prefecture, Kobe, Japan

Hyogo Prefecture (2011b). *Hyogo-ken shinrin kumiai tokeisyo (Annual statistics on forestry cooperatives in Hyogo Prefecture)*, pp.92–93, Hyogo Prefecture, Kobe, Japan

Iriai Rinya Kindaika Kenkyukai (1971). *Iriai rinya kindaika mondoshu (Questions and answers of modernization of common forest)*, Rinya Kosaikai,Tokyo, Japan

Ishimaru, K. (2011). *Shinrin kumiai ni okeru offset credit (J-VER) seido fukyu no genjyo to kadai (Current situations and problems of the diffusion of carbon credit scheme (J-VER) in forest owners' cooperative)*, Graduation Thesis of Kyoto University, Kyoto, Japan

Itoh, H. (2009). Iriaiken to zennin icchi gensoku: Kaminoseki iriaiken sosyo saikosai hanketsu wo megutte (The rights for the common forest and the unanimity rule: Supreme Court decision on the rights for common forest in Kaminoseki), *Journal of Middle Japan Common Forest Society*, No.29, (March 2009), pp.6–14, ISSN 1349-8584

Kawamura, M. (2010). Iriai rinya keiei to Seisan shinrin kumiai (Management of common forest and forest producer cooperative), *Journal of Middle Japan Common Forest Society*, No.30, (March 2010), pp.26–38, ISSN 1349-8584

Kawasugi, S. (2009). Iriai rinya no aratana kanosei (New possibility of common forest), *Journal of Middle Japan Common Forest Society*, No.29, (March 2009), pp.17–18, ISSN 1349-8584

Kuniyoshi, K. (2007). Genkai syuraku no saisei to kankyo hogo kyoryokukin no donyu (Regeneration of the marginal community and introduction of the cooperation fund for the protection of environment), *Quarterly Ecotourism*, (August 2007), No.36, p.12

Kuniyoshi, K. (2008). Eco-green tourism ni yoru iriai rinya no kanosei (Potential of common forest by eco-green tourism), *Journal of Middle Japan Common Forest Society*, No.28, (March 2008), pp.18–20, ISSN 1349-8584

Matsushita, K. (2008). Seisan shinrin kumiai ni yoru ecotourism (Ecotourism by forest producer cooperative), *Journal of Middle Japan Common Forest Society*, No.28, (March 2008), pp.44–46, ISSN 1349-8584

Matsushita, K. (2009). Seisan shinrin kumiai to kigyo no mori (Forest producer cooperative and forests supported by private company), *Journal of Middle Japan Common Forest Society*, No.29, (March 2009), pp.37–38, ISSN 1349-8584

Matsushita, K. (2012). Gendai ni okeru seisan shinrin kumiai no jittai haaku (Proposals for survey of forest producer cooperative), *Journal of Common Forest*, No.32, (March 2012), pp.24-35, ISSN 2186-036X

Matsushita, K. & Taguchi, K. (2011). The Kyoto Protocol and the private forest policy at local governments in Japan, *Small-scale Forestry*, Vol.10, (March 2011), pp.19–35, ISSN 1873-7617

Matsushita, K. & Hirata, K. (2002). Forest owners' associations, In: *Forestry and the forest industry in Japan*, Iwai, Y., (ed.), pp.41–66, UBC Press, ISBN 0-7748-0883-7, Vancouver, Canada

McKean, M. A. (1992). Management of traditional common lands (Iriaichi) in Japan, In: *Making the commons work*, Bromley, D. W. (ed.), ICS Press, pp.63-98, ISBN 1-55815-198-2, California, U.S.A.

Mie Prefecture (2010). *2009 Shinrin kumiai tokei (Statistics on Forest Cooperative)*, p.108, Mie Prefecture, Tsu, Japan

MAFF, Ministry of Agriculture, Forestry, and Fisheries (2010). *Syokuryo nogyo noson hakusyo (White paper on food, agriculture, and agricultural community)*, Saiki Printing, p.238, ISBN 978-4-903729-76-3, Tokyo, Japan

Nagata, M. (2009). *Seisan shinrin kumiai ni okeru kigyo no mori seido no genjyo to kadai (Current situation and problems in the "Forest sponsored by private company" system applied to the forest producer cooperative)*, Graduation Thesis of Kyoto University, Kyoto, Japan

Nakao, H. (1969). *Iriai rinya no horitsu mondai (Legal problem of common forest)*, Keiso Shobo, Tokyo, Japan

Ota, I. (2009). Activities and significances of forest owners' cooperative in Japan, In: *Legal aspects of European forest sustainable development, Proceedings of the 10th international symposium in Sarajevo*, Avdibegovic, M., Herbst, P., & Schmithüsen, F., (eds.), pp.101–108, Sarajevo, Bosnia and Herzegovina

Sakai, M. (2005). Seisan shinrin kumiai wo meguru futatsu no mondai (Two problems of the forest producer cooperative), *Sonraku to Kankyo* (*Rural community and the environment*), No.1, (March 2005), pp.25–38

Takasu, G. (1966). *Iriai rinya kindaika no shihyo* (*Index of modernization of common forest*), Nikkan Ringyo Shinbunsha, Tokyo, Japan

Totman, C. (2007). *Japan's imperial forest, Goryorin, 1889–1946: with a supporting study of the kan/min division of woodland in early Meiji Japan, 1871-76*, Global Oriental, pp.7–14, ISBN 978-1-905246-30-4, Folkestone, U.K.

Yamashita, U. (2006). Effect of instituting "authorized neighborhood associations" in the iriai forest: A case study on Iiyama City and Sakae Village in Nagano, *Forest Economy*, No.697, (November 2006), pp.17–32, ISSN 0388-8614

10

Setting Up Locally Appropriate Ecological Criteria and Indicators to Evaluate Sustainable Forest Management in Dinh Hoa District (Northern Vietnam)

Anna Stier, Jutta Lax and Joachim Krug

Johann Heinrich von Thünen-Institute (vTI), Federal Research Institute for Rural Areas
Forestry and Fisheries, Institute for World Forestry
Germany

1. Introduction

The forests of Vietnam provide a high conservation value considering habitat diversity despite massive forest destruction within the last decades (World Bank, 2010). Following recent studies Vietnam is one of the 34 biodiversity hotspots in the world (Indo-Burma hotspot) (Myers et al., 2000; Brooks et al., 2002; World Bank, 2010; Werger and Nghia, 2006), but at the same time one out of eight tropical forest hotspots which will lose the largest number of species by cause of deforestation (Brooks et al., 2002). The impacts of the forest environment – and its ongoing degradation – on local socio-economic factors cannot be neglected. Actually, Vietnam has been defined as an archetypal case for a positive correlation between a high forest cover and a high poverty rate combined with a low poverty density (Sunderlin et al., 2007). In other words, regions with high forest cover are often sparsely populated but after all are among the poorest of the country. Forests are populated by the poor, but it is nowadays also an evidence that it is the poorest households which generally depend more on forests (Cavendish, 2003; Wunder, 2001), deriving several goods, income and services from them (Arnold, 2001; Dubois, 2002).

Human and ecological factors in Vietnam make it a candidate for the implementation of sustainable forest management (SFM) with the objective of win / win solutions for both human well-being as well as forest ecosystems (Sunderlin and Ba, 2005) The recently implemented forest management types defined by the government had to face wide criticism concerning their success in reaching such win / win objectives, such as the existing gap between state intentions and local applications of policies, the poor involvement of households in the forestry sector and their insufficient payment for protection activities, or the disturbance of traditional land-use systems (Clement and Amezaga, 2008; Boissière et al., 2009; Sunderlin and Ba, 2005; Wunder et al. 2005). In the course of national decentralisation processes the former state organised forest enterprises were fragmented and land / forest was reallocated to communities and private stakeholders. It is essential to record and compare the different stakeholder perceptions concerning SFM to elaborate adequate criteria and indicator (C&I) sets to be able to measure the sustainability of current forest management regimes (Karjala et al., 2004; Sherry et al., 2005; Ritchie et al., 2000).

The national Vietnamese set, based on Forest Stewardship Council standards, has not been finalized and accepted yet because of the lack of local consultation (anonymous personal communication). The mostly used sets, based on expert consultations, give results which often differ from local needs (Pokharel and Larsen, 2007; Purnomo et al., 2005; Adam and Kneeshaw, 2008; Sherry et al., 2005). By experience the ecological elements demonstrated the highest similarity among C&I frameworks (Purnomo et al., 2005; Sherry et al., 2005; Adam and Kneeshaw, 2008). It has still to be tested how far local perceptions differ from institutional ecological C&I sets in the case of Dinh Hoa, and how far they differ among different local communities depending on different forest management types. Accordingly an ecological C&I template that is appropriate to Dinh Hoa District for SFM assessment was set up, by:

1. Comparing local perceptions of SFM with those from institutional top-down approaches (comparing the sets from local communities with the sets resulting from national, province and district level workshops); and
2. comparing local perceptions between forest use type categories (comparing the sets from communities which hold high proportions of special use, protection and production forests).

2. Material and methods

2.1 Study site

Dinh Hoa is a district of the Thai Nguyen province in Northern Vietnam where forest land represents 68,7 % of the total area in 2005, with 33,0 % of it being classified as planted forest (Data provided by the Agriculture and Rural Development Department of Dinh Hoa). The high forest cover, combined with a high population density (189 habitants / km^2) results in high pressure on the forest resources.

All Vietnamese forest use types are represented in the district: special use forest (8 728 ha, 24 % of the forest area), protection forest (7050 ha, 20 % of the forest area) and production forest (20 009 ha, 56 % of the forest area) (The Prime Minister, 2008).

Following the law on forest protection and development of December 3rd 2004 (The President of the Socialist Republic of Vietnam, Tran Duc Luong, 2004) and by declaration of the department of policy and rural development (FAO and RECOFTC, 2000), the objectives of the forest use types / categories are defined as followed:

Special-use forest is predominantly related to the conservation of nature, scientific research and protection of landscapes and historical / cultural relics. Management boards directly manage these forests. Contracts are made on long-term basis with households for ecological restoration, afforestation and protection. Households are entitled to collect dead wood for self consumption.

Protection forest mainly fulfils protection purposes as to protect water sources and land, to prevent erosion and desertification, restrict natural calamities and regulate the local climate. Management boards make contracts with households, communities, individuals or organizations to protect and regenerate forestland. Contracted stakeholders possess some restricted utilization rights.

Production forest is managed mainly for the production and trading of timber and non-timber forest products. It includes natural and planted production forests.

2.2 Set development

Criteria and Indicator sets were built up through three workshops with forest management experts (top-down method where a generic set was modified using multi-criteria decision making) and group discussions with 12 local villages (bottom-up method where sets were elaborated from local visions) (Fig. 1), resulting in 6 criteria and 27 indicator (see Tab. 3a and 3b). These sets were then compared and compiled to a final set for all forest use types of the Dinh Hoa District, and the differences between the sets were analyzed.

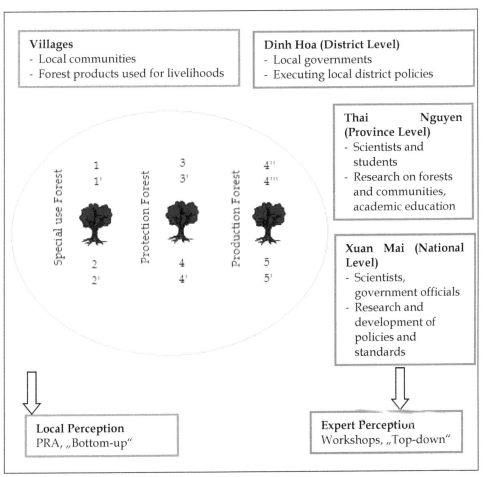

Fig. 1. Conceptual Framework of Dinh Hoa forests. The figure displays the connections of the forest concerned stakeholders to each other and towards the forest. The Bottom-up approach was implemented in twelve villages from five communes, meaning four villages from two different communes per forest use type. The Top-down approach was implemented through three workshops at national, provincial and district level. 1 and 1', 2 and 2', etc.: two villages belonging to the same commune.

2.2.1 Top-down approach

A generic set was built up through the combination of already existing templates:
-   The **CIFOR** (Centre for International Forestry Research) generic template (CIFOR, 1999b);
-   the **ITTO** set (ITTO, 2005); as the set was designed for tropical forests and focus on South-East Asia;
-   **Lepaterique** (Anon., 1997) and **Tarapoto** (ACT, 1995) ; for further inputs concerning tropical forests;
-   **Regional Initiative dry Forests Asia** (Anon., 1999) ; for further directions concerning Asia in general, inter alia because it is applied in countries close to Vietnam like Thailand;
-   **MCPFE** proposal (MCPFE, 2003) ; an external example showing efficient methods to work towards a relevant C&I set; and
-   **a local set of C&I** elaborated in Dinh Hoa in 2009 for elements and formulations that had already demonstrated their effectiveness for the region.

This generic set was modified during three workshops from May to June 2010, at national (Xuan Mai University of Hanoi), province (Thai Nguyen University of Agriculture and Forestry) and district level (Cho Chu, the Dinh Hoa District centre). The participants could eliminate, modify and add elements to the set.

Three tools were used during these workshops: rankings, ratings and pairwise comparisons, following the Multi-Criteria Decision Making (MCDM) application guidelines from CIFOR (CIFOR, 1999a), and its applications (Mendoza and Prabhu, 2000a; Mendoza and Prabhu, 2000b; Andrada II and Calderon, 2008; Gomontean et al., 2008; Ritchie et al., 2000). The proceeding lead to one final C&I set per workshop.

2.2.2 Bottom up approach

Sampling design:

The communes and the villages for the Participatory Rural Appraisal (PRA) were chosen following the decisional framework presented in Fig. 2: For each forest use type, two communes with territorial dominance of the concerned were chosen. Only those villages were chosen where the forest area is managed solely by households (Criterion 2.1) and belong exclusively to the forest function of interest (Criterion 2.2). Then, the two villages with the largest forest area were chosen for the Participatory Rural Appraisal (PRA) (Criterion 2.3).

In each village, 15 participants were chosen for the PRA method, including five people representing the village organizations and 10 households owning forest. The five representative villagers included the head of the village as well as representatives of the old soldier union, the farmer association, the women association and the youth union. The 10 households per village were chosen by respecting equity in gender, age, well-being and educational level.

Participatory methods and tools:

PRA tools were used at village level during half day group discussions. These discussions took place in the Fig. 2 mentioned villages, during July 2010. The tools included participatory mapping, open ended questions, semi-structured questionnaires and brain storming sessions, leading to one set of ecological C&I per village.

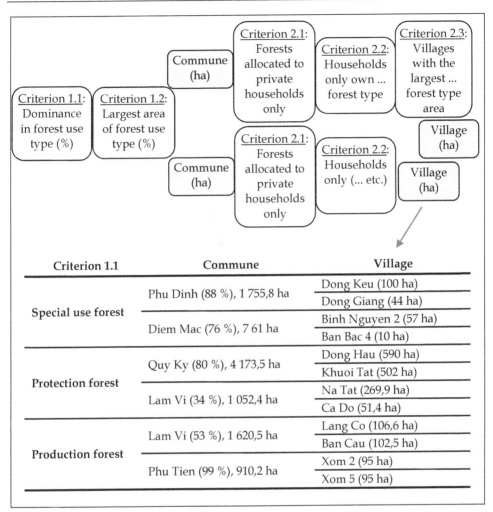

Fig. 2. Decisional framework for village selection. For each forest use type, two communes with territorial dominance of the concerned type (in % of the commune's forest area and ha) were chosen.

2.3 Analyzing methods
2.3.1 Final set elaboration
A consolidated list of indicators was generated with all proposed elements from all the workshops and PRA sessions. This list was used as a basis for comparison, so that the presence of an element in both consolidated and stakeholder list was coded as "1" whereas the absence of element analogy was coded as "0" (Tab. 1).

The final C&I set should have both expert and local population acceptance, and be applicable to all forest use types. Regarding the single forest use type, only elements accepted by more than 50 % of the workshops and villages were accepted. The same counted for new elements proposed by villages which got incorporated if they were proposed by more than 50 % of villages under each forest use type (Fig. 3a and 3b).

Stakeholder	Indicator				
	1	2	3	...	27
Workshop 1	1	1	1	...	0
Workshop 2	1	0	0	...	0
Workshop 3	0	1	0	...	1
Village 1	1	1	0	...	1
Village 2	0	0	1	...	1
...	...	...	...	...	...
Village 12	1	0	0	...	1

Table 1. Binary representation of the stakeholders' perceptions (note: This table is made up and does not contain data of the study).

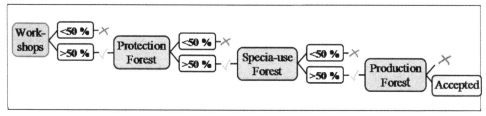

Fig. 3. Decisional Framework for the final set elaboration of indicators. To be accepted in the final list, an indicator had to be accepted by minimum 50 % of the expert workshops and 50 % of the villages for each forest use type.

2.3.2 Cluster analysis

One key objective was the identification of commonalities or differences within the participants' perceptions concerning the local value of ecological indicators suitable for the assessment of sustainability of forest management. Cluster analysis allows the identification of uniform groups of data within a data set (called clusters), meaning groups of data that have sufficient similarities. Cluster analysis has already been applied to analyze perceptions of different stakeholders concerning C&I (Purnomo et al., 2005). Following Gower and Legendre (1986), the simple matching coefficient is used as similarity coefficient. This coefficient is calculated as follows (Gower and Legendre, 1986):

$$S = (a+d)/(a+b+c+d)$$

with S the similarity coefficient and the standard notation a for the number of (+, +) matches, b for (+, -), c for (-, +) and d for (-, -).

Considering the made up data for workshop 1 and 2 and the example indicators 1, 2, 3 and 27 in Tab. 4, the similarity coefficient would be $S = (1+1)/(1+2+0+1) = 2/4 = 0.5$.

Setting Up Locally Appropriate Ecological Criteria and Indicators to Evaluate Sustainable Forest Management
in Dinh Hoa District (Northern Vietnam)

209

Given these similarity measures for all possible pairs of stakeholders, the data was organized into useful / meaningful groups, so that those within each group (cluster) were more closely related to one another than subjects in different clusters. Hierarchical clustering can either follow *agglomerative* or *divisive* methods (Janssen and Laatz, 2010; Manning et al., 2008). The output can be illustrated by a so called dendrogram (Fig. 4).

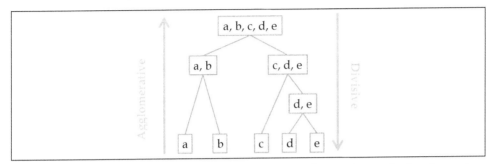

Fig. 4. Example of a dendrogram with fictive data. The agglomerative method makes series of fusions of the n objects into groups whereas the divisive method separates n objects into finer groupings.

As a result of various ways of calculating the distance between the clusters (Janssen and Laatz, 2010; Manning et al., 2008), different fusion procedures exist for the agglomerative method. In *single linkage*, the distance between two clusters is given by the value of the shortest link between two objects of the two clusters. In *complete linkage*, the distance between two clusters is given by the value of the longest link between two objects of the two clusters. In *group average linkage*, the distance between two clusters is defined as the average of distances between all pairs of objects, where each pair is made up of one object from each cluster (Fig. 5). This type of linkage appears to be the most useful for this study, because it takes into account all the possible pairs of distances between the C&I sets.

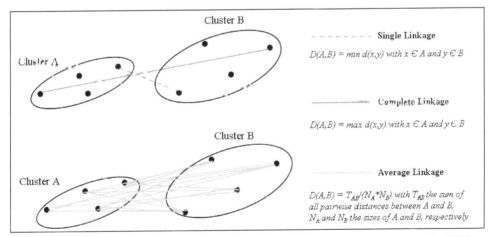

Fig. 5. Examples of three linkage calculation methods (adapted from Manning et al., 2008). The average linkage method is used in this study for its use of all possible pairs of elements.

2.3.3 Hypotheses testing

The cluster analysis evaluates the similarities of perceptions among the stakeholders based on all pairs at once, but it does not allow drawing conclusions about the analogy of the perceptions between stakeholders. In fact, a cluster can indicate a high similarity within its subjects, and still display significant differences among them when tested pair wise. Finally two hypotheses were tested. The first hypothesis was the similarity of perceptions between the villages and the experts. This test is relevant because experts play a decisional role in law and policy making although the results are implemented at local level. The second test concerned the similarity of perceptions between villages managing different forest use types. The data indicating the absence or presence of an indicator against a consolidated reference list of indicators is of binary character. Thus, the Phi coefficient [mean square contingency coefficient] was calculated (Janssen and Laatz, 2010). To test the significance of this Phi coefficient, the Pearson Chi-square was applied if the expected cell frequencies were all ≥ 5, otherwise the Fisher exact probability test was used (Janssen and Laatz, 2010; Sachs, 2002), both with a significance level of $\alpha = 0,05$. The hypotheses were:

H0: Perceptions of X and Y are not associated   *H1: Perceptions of X and Y are associated*

If there is no association between variables, the answers of stakeholders are independent, meaning that the C&I sets are NOT similar: Perceptions of X ≠ Perceptions of Y.
Accordingly, if there is an association between variables, the answers are dependent, meaning that the C&I sets ARE similar: Perceptions of X = Perceptions of Y.

3. Results

3.1 Final set of C&I

During the workshops, the generic set was left almost unmodified. Strictly speaking no elements were added and two indicators were eliminated at the provincial level workshop. In all group discussions the villagers eliminated the same 12 indicators, including the 5 genetic indicators (compare Tab. 3a and 3b). The group discussions led to the addition of 2 new indicators. Generally the villagers had the same perceptions for eliminating and keeping indicators: 86 % of the indicators had over 80 % of similar answers (either 0 or 1) in the villages (Tab. 2).

Part of indicators (%)	Number of different perceptions	Similarity (%)
67	0	100
15	1	92
4	2	83
11	3	75
4	5	58
Total = 100		

Table 2. Similarity in perceptions between villages.

The final set was composed out of 15 indicators under 6 criteria and 2 principles (Tab. 3a and 3b).

One special case was decided to remain included in the final version of the C&I set, Indicator *"1.3.4 Minimization of soil degradation"*: This indicator was accepted by 100 % of experts, special use and production forest villages, and rejected by three out of four protection forest villages. Dinh Hoa District is part of the mountainous regions of Vietnam which cover 3/4th of the country, having a complex topography and steep slopes (Werger and Nghia, 2006). Soil degradation and erosion is generally a great risk in the northern mountainous regions of Vietnam (Pomel et al., 2007; Thao, 2001). Land erosion has been identified to be a key point impacting many elements which influence farming systems (like water quality / quantity and soil fertility) thus causing crop yield reduction leading to a general income loss (Thao, 2001; Pomel et al., 2007). Though forests are the main subject of this study plus forests represent the land use option with the smallest erosion rate in Northern Vietnam (Pomel et al., 2007) the indicator was kept.

Principle 1 : Ecosystem integrity is maintained

Criterion		Indicator	Work-shops	Protection Forest	Special use Forest	Production Forest	Indicator Acceptance (Yes/No)
			% of agreement on the acceptance of concerned indicators				
1.1 Extent of forests	1.1.1	Maintain/Improve the forest area	100	100	100	100	Y
	1.1.2	Control of forest area loss	100	0	0	0	N
1.2 Forest ecosystem health	1.2.1	No chemical contamination	67	50	50	75	N
	1.2.2	No natural degradation	100	100	100	100	Y
	1.2.3	No human degradation	100	100	100	100	Y
	1.2.4	Regeneration and forest structure	100	100	100	100	Y
	1.2.5	Soil/Decomposition	100	100	75	100	Y
1.3 Forest ecosystem services	1.3.1	Product provision for local people	100	100	100	100	Y
	1.3.2	Protection of riparian forests	100	0	0	25	N
	1.3.3	Maintain the water quality/quantity	100	100	100	100	Y
	1.3.4	Minimize soil degradation	100	25	100	100	Y
	1.3.5	Valuation of Carbon sequestration	100	0	0	0	N
	1.3.6	Forest protection/valorisation for tourism	100	75	50	100	Y
	1.3.7	Minimize floods	0 (new)	75	100	100	Y
	1.3.8	Pleasantness of environment	0 (new)	100	100	100	Y

Table 3. a. Final Indicator selection. Principle 1: Ecosystem integrity is maintained.

3.2 Cluster analysis

As described in section 2.3.2, cluster analysis operates in successive stages of fusions, based on the calculation of similarity coefficients. The final structure of the grouping is determined by the desired number of clusters, or by a previously fixed level of similarity that is considered as "acceptable". There is no general rule for the minimal similarity level. In order to make the results comparable with a previous similar study implemented in Indonesia (Purnomo et al., 2005), 80 % of similarity are specified here as acceptable. For instance, the pairs of clusters (12, 13), (7, 12), (10, 11), (7, 10) and (1, 3) show 100 % similarity (Tab. 4),

Principle 2 : Socio-economic & cultural benefits are linked to ecosystem integrity or are of prime importance

Criterion		Indicator	Work-shops	Protection Forest	Special use Forest	Production Forest	Indicator Acceptance (Yes/No)
			% of agreement on the acceptance of concerned indicators				
2.1 Forest ecosystem diversity	2.1.1	Maintain the forest landscape	100	100	75	75	Y
	2.1.2	No human habitat diversity destruction	100	0	0	0	N
	2.1.3	Presence of corridors	100	75	50	100	Y
2.2 Forest species diversity	2.2.1	Preserve species with key functions	67	0	0	0	N
	2.2.2	Preserve species diversity of animals/plants	100	100	100	100	Y
	2.2.3	Population sizes and demographic structures	100	0	0	0	N
	2.2.4	Protection of rare/endangered species	100	100	75	100	Y
2.3 Forest genetic diversity	2.3.1	Genetic diversity is preserved in rare/commercial species	100	0	0	0	N
	2.3.2	Implementation of measures for genetic diversity conservation	100	0	0	0	N
	2.3.3	All phenotypes are preserved	100	0	0	0	N
	2.3.4	Gene flow is maintained	100	0	0	0	N
	2.3.5	Mating system doesn't change	100	0	0	0	N

Table 3. b. Final Indicator selection. Principle 2: Socio-economic & cultural benefits are linked to ecosystem integrity or are of prime importance.

meaning the valuation of the indicators were similar for these groups of stakeholders. Moreover, about 93 % of the stakeholders (14 out of 15) display at least 80 % similarity (i.e. all stakeholders excepted number 8).

The 80 % threshold can be used to calculate the final number of clusters from the Agglomeration Schedule Table (Tab. 4) (Janssen and Laatz, 2010; Manning et al., 2008):

Number of clusters = Number of subjects – Value of the last stage over 80% similarity
=> 15-12 = 3 clusters

This makes it possible to draw a line across the dendrogram to specify the final grouping which is considered as meaningful with a fixed minimum level of 80 % similarity (Fig. 6). According to this final partitioning, 3 clusters were considered as reasonable:

1.   The cluster (1, 2, 3) containing the three expert workshops;
2.   the cluster containing all villages except village number 8; and
3.   village number 8 alone, corresponding to the special use forest village Ban Bac 4.

The subgroups in cluster 1 and 2 are based on minimal differences (similarity between 87 and 100 %), showing a strong homogeneity in the stakeholder's perceptions. Noteworthy, the cluster formation is detached of the forest use types, all forest types appear in all sub-clusters (Fig. 6).

	Agglomeration Schedule					
Stage	Cluster Combined		Coefficients	Stage Cluster First Appears		Next Stage
	Cluster 1	Cluster 2		Cluster 1	Cluster 2	
1	12	13	1,000	0	0	2
2	7	12	1,000	0	1	4
3	10	11	1,000	0	0	4
4	7	10	1,000	2	3	6
5	1	3	1,000	0	0	9
6	7	14	0,963	4	0	7
7	6	7	0,957	0	6	10
8	4	9	0,926	0	0	11
9	1	2	0,926	5	0	14
10	6	15	0,915	7	0	12
11	4	5	0,889	8	0	12
12	4	6	0,873	11	10	13
13	4	8	0,791	12	0	14
14	1	4	0,479	9	13	0

Table 4. Cluster analysis using group average linkage. Cluster 1 and Cluster 2 (under cluster combined) display the membership of the stakeholders (3 workshops and 12 villages) towards the cluster. The bold line represents the 80% similarity threshold.

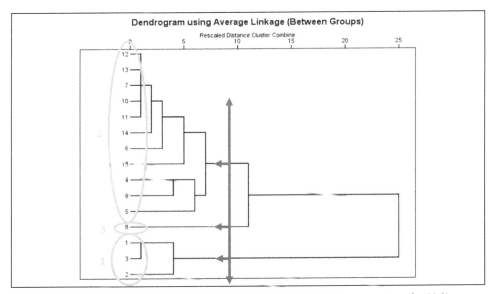

Fig. 6. Cluster Dendrogram of stakeholder perceptions. The red line represents the 80 % similarity threshold, delimitating three meaningful clusters in green. The numbers at the end of each cluster arm represent the concerning stakeholder: 1, 2, 3 = expert workshops; 4, 5, 6, 7 = protection forest, 8, 9, 10, 11 = special use forest and 12, 13, 14, 15 = production forest.

3.3 Hypotheses testing

As described in section 2.3.3, the null and alternative hypotheses are associated to similar or different perceptions between stakeholders. The results show that the null hypothesis cannot be rejected for any of the pairs of experts and local populations (all p-values > 0, 05). Thus, the local populations have different perceptions of what they consider to be important ecological elements for a SFM compared to national and province experts, as well as local authorities represented by the district workshop. The p-value is at the limit of significance (0, 055) for the national and district workshops compared to the village Ban Bac 4. This is in accordance with the cluster dendrogram results (Fig. 6).

There are similar perceptions about SFM among villages, irrespective of the forest use type they manage (all p-values are < 0, 006) (Fig. 6).

4. Discussion

4.1 Local and expert perceptions of sustainable forest management

Alike this study indicates, previous studies equally show that local and expert perceptions of indicators for sustainable forest management often differ, but that these differences decrease while concerning ecological indicators (Karjala et al., 2004; Purnomo et al., 2005; Pokharel and Larsen, 2007; Sherry et al., 2005; Adam and Kneeshaw, 2008). This has been explained by the fact that C&I processes always largely focus on environmental (not socio-economic) issues, so that there is less disagreement in what should be included in a meaningful set of indicators for ecological sustainability: ecosystem condition, biodiversity and ecosystem services are nearly always included (Adam and Kneeshaw, 2008). Rural populations often not only depend on natural resources, they also inherit a thorough traditional environmental knowledge (TEK) (Karjala et al., 2004) about their surrounding environment which is often in accordance with expert formulations, even if the vocabulary is different. Requirements like water and soil protection, critical habitat preservation or productive functions of forests are the concerns of both local people and experts.

In fact, the differences in perception between experts and villages in this study are based for a non-negligible part on the genetic diversity concept. About 42 % of eliminated indicators (5 out of 12) concern the criterion of genetic diversity. The neglect of genetic issues by local populations can be attributed to a lack of knowledge and difficulties to understand the concept.

On the other hand, villagers added some elements which were considered as relevant, independently from their scientific importance. For instance, the elements leading to the addition of *"1.3.8 Pleasantness of environment"* as an indicator included the beauty of the landscape, air quality, temperature and provision of shade. Aesthetic issues have been identified in previous studies to be typical local requirements which are not integrated at the expert levels (Adam and Kneeshaw, 2008). Bottom-up approaches and TEK can thus be seen as a way to integrate and connect ecological issues with cultural and communal aspects. This integration of connections / interlinkages in some indicators could be an answer to recent critics about the strict structure and isolation of elements into ecological, social and economic issues (Adam and Kneeshaw, 2008; Mendoza and Prabhu, 2003; Requardt 2007). Further this result also confirms that expert sets fail to address particular values and needs of local populations. Elements generated by local communities can complement expert sets by adding valuable knowledge. Moreover, they can increase the legitimacy of those sets, facilitate their implementation and the acceptance of the results, they can contribute to the conservation and recognition of TEK and reduce hierarchical conflicts.

4.2 Perceptions among local communities

Local populations of the villages where PRA was implemented had almost the same perceptions of ecological sustainability of forest management as shown in section 4.0 and there were no measurable differences resulting from the forest use type the villagers were managing. The existing discrepancy of one indicator (Tab. 3a and 3b) dealt with the significance of forests for protecting or mitigating soil degradation. Villages surrounded by protection forests never experienced landslides or soil degradation, ergo could not make the link between the presence of forests and the absence of soil degradation. This does not mean that the indicator is not applicable; it even makes it a suitable element, showing that protection forests actually really protect the soil. It can thus be discussed if future research has to consider forest use types as a meaningful subdivision in the sampling design or not.

The village Ban Bac 4, representing the outlier in all data analyses, was the village with the smallest forested area (10 ha), entirely young *Acacia spp.* plantations planted in 2006 and decimated by a disease in 2008. The government did not support new plantations since then, resulting in a general disinterest in forests. Therefore, forest area could be a key element influencing the correct implementation of PRA and the resulting lists.

4.3 Pertinence of Multi-Criteria Decision Making (MCDM) for top-down approaches

MCDM has been identified in previous studies to be a pertinent method to use with experts (Mendoza and Prabhu, 2000b; Mendoza and Prabhu, 2000a; Andrada II and Calderon, 2008; Gomontean et al., 2008). It can save time to begin from a generic set and to modify it afterwards with local experts instead of generating new lists of indicators from scratch. MCDM is a method which easily helps to reach an agreement among all participants. The fact that in this study nearly no indicators were modified from the existent generic set could be explained in two ways. The first could be that the generic template, already resulting from several international processes and expert consultations, covered all requirements of the workshop members. Anyway, ecological elements have often been those where the most agreement appeared among stakeholders worldwide (Purnomo et al., 2005; Sherry et al., 2005). The second explanation could be that the method does not allow easy modifications of the generic template, for the following reasons. Providing a generic set resulting from several international consultations may make local experts hesitant to freely reject / modify elements. Moreover, even if MCDM allows the elimination of elements, the method is hardly adaptable to the addition of new topics.

5. Conclusion

This study reveals that local and expert perceptions differ in their perceptions of ecological sustainability of forest management. However, among experts and among local communities the perceptions were relatively uniform.

It can thus be recommended to combine expert consultations to ensure the scientific validity with local perceptions to ensure the recognition of local values and perceptions. To use only one of the two approaches may reduce the acceptance of the representativeness of the resulting set, leading to conflicts causing difficulties of implementation, and finally to change forest management practices if necessary.

6. Acknowledgements

We wish to thank Dr. Aljoscha Requardt (head of the Observatory for European Forests (OEF) of the European Forest Institute (EFI), Nancy, France) for his help during the preparation of the study. We as well wish to thank the Thai Nguyen University (Vietnam), the Department of International Relations and the Thai Nguyen University of Agriculture and Forestry for their support. Our special thanks go to Dr. Do Anh Tai (Thai Nguyen University, associate professor), Dr. Dai Tran Nghia (Thai Nguyen University, head of the international relations) and Dr. Tran Quoc Hung (Thai Nguyen University, Dean of the Forestry Faculty).
Our strong consideration goes to Nguyen Thi Thanh Ha, Nguyen Huy Hoang and Prem Raj Neupane for their assistance in the field and organisational help. As well we would like to thank Andre Iost for the statistic review of the analysis.

7. References

ACT, 1995. Proposal of Criteria and indicators for sustainability of the Amazon Forest. Pro Tempore Secretariat, Lima, Peru.,

Adam, M.C. and Kneeshaw, D., 2008. Local level criteria and indicator frameworks: A tool used to assess aboriginal forest ecosystem values. Forest Ecology and Management, 255, pp.2024-2037.

Andrada II, R.T. and Calderon, M.M., 2008. Sustainability criteria and indicators for the Makiling Botanic Gardens in Los Banos, Laguna, Philippines: An application of the CIFOR C&I toolbox. Ecological Indicators, 8, pp.141-148.

Anon, 1997. Results of the FAO-CCAB-AP experts' meeting on criteria and indicators for sustainable forest management in Central America. Tegucigalpa, Honduras, 20-24/01/97,

Anon., 1999. Regional Initiative for the Development and Implementation of National Level Criteria and Indicators for the Sustainable Management of Dry Forests in Asia. "Workshop on National-Level Criteria and Indicators for Sustainable Management of Dry Forests in Asia/South Asia" organised by FAO/UNEP/ITTO Indian Institute of Forest Management, Bhopal, India. 30 November - 3 December 1999.,

Arnold, J.E.M., 2001. Forestry, Poverty and Aid. CIFOR Occasional Paper, (33).

Boissière, M. et al., 2009. Can engaging local people's interests reduce forest degradation in Central Vietnam? Biodiversity and Conservation, 18(10), pp.2743-2757.

Brooks, T.M. et al., 2002. Habitat Loss and Extinction in the Hotspots of Biodiversity. Conservation Biology, 16(4), pp.909-923.

Cavendish, W., 2003. How do forests support, insure and improve the livelihoods of the rural poor? CIFOR A research note.

CIFOR, 1999a. Guidelines for Applying Multi-Criteria Analyses to the Assessment of Criteria and Indicators,

CIFOR, 1999b. The CIFOR Criteria and Indicators Generic Template,

Clement, F. and Amezaga, J.M., 2008. Afforestation and forest land allocation in Northern Vietnam: Analysing the gap between policy intentions and outcomes. Land Use Policy, 26, pp.458-470.

Setting Up Locally Appropriate Ecological Criteria and Indicators to Evaluate Sustainable Forest Management
in Dinh Hoa District (Northern Vietnam)

217

Dubois, O., 2002. Forest-Based Poverty Reduction: A Brief Review of Facts, Figures, Challenges and Possible Ways Forward. In International Workshop on "Forests in poverty reduction strategies: Capturing the potential". Tuusula, Finland: FAO.

FAO and RECOFTC, 2000. Decentralization and Devolution of Forest Management in Asia and the Pacific.

Gomontean, B. et al., 2008. The development of appropriate ecological criteria and indicators for community forest conservation using participatory methods a case study in northeastern Thailand. Ecological Indicators, 8, pp.614-624.

Gower, J.C. and Legendre, P., 1986. Metric and Euclidean Properties of Dissimilarity Coefficients. Journal of Classification, 3, pp.5-48.

ITTO, 2005. Revised ITTO criteria and indicators for the sustainable management of tropical forests including reporting format,

Janssen, J. and Laatz, W., 2010. Statistische Datenanalyse mit SPSS 7 ed., Springer.

Karjala, M.K. et al., 2004. Criteria and Indicators for Sustainable Forest Planning: a framework for recording Aboriginal resource and social values. Forest Policy and Economics, 6, pp.95-110.

Manning, C.D. et al., 2008. Hierarchical clustering. In Introduction to Information Retrieval. Cambridge University Press.

MCPFE, 2003. Improved Pan-European Indicators for Sustainable Forest Management as adopted by the MCPFE Expert Level Meeting 7-8 October 2002, Vienna, Austria,

Mendoza, G.A. and Prabhu, R., 2000a. Development of a Methodology for Selectin Criteria and Indicators of Sustainable Forest Management: A Case Study on Participatory Assessment. Environmental Management, 26, pp.659-673.

Mendoza, G.A. and Prabhu, R., 2000b. Multiple crietria decision making approaches to assessing forest sustainability using criteria and indicators: a case study. Forest Ecology and Management, 131, pp.107-126.

Mendoza, G.A. and Prabhu, R., 2003. Qualitative multi-criteria approaches to assessing indicators of sustainable forest resource management. Forest Ecology and Management, 174, pp.329-343.

Myers, N. et al., 2000. Biodiversity hotspots for conservation priorities. Nature, 403.

Pokharel, R.K. and Larsen, H.O., 2007. Local vs official criteria and indicators for evaluating community forest management. Forestry, 80, pp.183-192.

Pomel, S., Pham, Q.H. and Nguyen, V.T., 2007. Les états de surface des sols au Nord Vietnam: une méthode pour estimer et cartographier les risques d'érosion. In Actes des JSIRAUF: Journées Scientifiques Inter-Réseaux de l'Agence Universitaire de la Francophonie. Hanoi.

Purnomo, H. et al., 2005. Analysis of local perspectives on sustainable forest management an Indonesian case study. Journal of Environmental Management, 74, pp.111-126.

Requardt, A., 2007. Pan European criteria and indicators for sustainable forest management: Networking Structures and Data Potentials of International Data Sources. Hamburg: Universität Hamburg.

Ritchie, B. et al., 2000. Criteria and Indicators of Sustainability in Community managed Forest Landscapes CIFOR.,

Sachs, L., 2002. Angewandte Statistik: Anwendung statistischer Methoden Zehnte, überarbeitete und aktualisierte Auflage., Springer.

Sherry, E. et al., 2005. Local-level criteria and indicators: an Aboriginal perspective on sustainable forest management. Forestry, 78, pp.513-539.

Sunderlin, W.D. and Ba, H.T., 2005. Poverty Alleviation and Forests in Vietnam.

Sunderlin, W.D. et al., 2007. Poverty and forests: Multi-country analysis of spatial association and proposed policy solutions. CIFOR Occasional Paper, (47).

Thao, T.D., 2001. On-site Costs and Benefits of Soil Conservation in Mountainous Regions of Northern Vietnam. EEPSEA Research Reports, (13).

The President of the Socialist Republic of Vietnam,Tran Duc Luong, 2004. The Law on Forest Protection and Development passed by the XIth National Assembly of the Socialist Republic of Vietnam at its 6th session on December 3, 2004.

The Prime Minister, 2008. Decisions No. 1134/QD-TTg of August 21, 2008, approving the scheme on forest protection and development in Dinh Hoa safety zone, Thai Nguyen Province, in the 2008-2020 Period.

Werger, M.J.A. and Nghia, N.H., 2006. Vietnamese forestry, biodiversity and threatened tree species,

World Bank, 2010. Country databank for Vietnam, Available at: http://data.worldbank.org/country/vietnam. Last access the 31/09/2010

Wunder, S., 2001. Poverty Alleviation and Tropical Forests - What Scope for Synergies? World Development, 29(11), pp.1817-1833.

Wunder, S. et al., 2005. Payment is good, control is better: Why payments for forest environmental services in Vietnam have so far remained incipient.

Part 4

Europe

11

Sustainable Forest Management in Galicia (Spain): Lessons Learned

Edward Robak[1], Jacobo Aboal[2] and Juan Picos[3]
[1]University of New Brunswick
[2]Dirección Xeral de Montes, Xunta de Galicia
[3]University of Vigo
[1]Canada
[2,3]Spain

1. Introduction

Galicia is an autonomous region of Spain that produces more than 8 million cubic metres (m³) of timber, with the Galician forestry sector currently providing 12% of industrial employment in the region (Monte Industria, 2010a). However, Galicia's potential as a forest product producer can be considered to be under-developed since both the amount and unit value of forest production could be greatly increased (Xunta de Galicia, 2001). Given that the Galician agriculture and fishery sectors (traditional bases for economic activity, especially in rural communities) are declining, a healthy forestry sector can be seen as an engine for regional and rural economic development.

One of the primary causes for forest sector under-development is the high degree of private forest ownership in small, scattered holdings. According to Ambrosio et al. (2003), private forests comprise approximately 97% of Galician forestlands, with about two thirds of those in holdings of less than 2 hectares (often in several non-contiguous parcels). Approximately 30% of private forests are owned by communities, but even these average only several hundred hectares in size. About half of the community forest area is managed by an agency of the regional government as a result of agreements signed in the second half of the twentieth century.

This fragmented ownership pattern has made it difficult to promote sustainable forest management (SFM) and the development of the sector. Only a small portion of the forested land is managed in a patently sustainable manner, which does not bode well for the future of industrial forestry given the pressure for certified SFM from governments, the general public and the forest product marketplace. This makes it difficult to justify public and private investment in forestry, which in turn impedes investment in forest industry modernization. If the industry is not modernized, the degree of "value-added" processing will remain low, with most raw production sent to other regions for processing.

Given the situation, the government department primarily responsible for forest management (Dirección Xeral de Montes, or DXM, of the Galician Rural Development Ministry) recognized that it was necessary to formulate new strategies, policies and processes aimed at the development of the forestry sector based upon the principles of SFM.

The goal of this chapter is to describe the Galician SFM strategy framework as initially envisaged, assess its evolution and implementation to date, describe important initiatives that have been undertaken by the private sector itself, and conclude with a summary of what we believe that should be learned from the entire process.

2. Development of the initial strategy

In order to develop the strategy framework, the DXM relied upon guiding principles and foundations of SFM gleaned from international, European Union (EU) and Spanish policies and agreements. Although a wide range of documents were reviewed, the following agreements and resolutions were seen as being most relevant:

- The Rio Declaration on Environment and Development in 1992; (UN General Assembly, 1992)
- Resolutions of the Ministerial Conference for the Protection of Forests (MCPFE, 2011);
- Pan-European Criteria, Indicators and Operational Level Guidelines for Sustainable Forest Management (MCPFE, 1998)
- The Proposals for Action of the United Nations Intergovernmental Panel of Forests and Intergovernmental Forum of Forests (IPF/IFF, 1998)
- The EU Forestry Strategy (Council of European Union, 1999)
- The Spanish Forestry Strategy (Ministerio de Medio Ambiente, 2000)
- The European Environment Action Programmes; (Council of European Union, 1998);
- The Spanish Forestry Plan (Ministerio de Medio Ambiente, 2002);

Furthermore, in order to ensure that a new strategy and its related programs would be consistent with forestry and environmental policy of the EU, the DXM reviewed and summarized documentation concerning relevant Community programs and initiatives.

Further inspiration for the development of a new forest strategy was drawn from the concepts of hierarchical forest management (HFM) and integrated forest management (IFM). HFM is based upon the tenets of hierarchical production planning as described by such authors as Hax and Candea (1984). Explanations of the hierarchical approach to forest management can be found in various documents, but a paper by Weintraub and Davis (1996) is especially recommended.

The term IFM has been used to describe several distinct (though related) concepts, but in the case of the strategy development effort in Galicia, IFM was taken to mean the integration of management processes and systems to ensure that the objectives of HFM are achieved (Gallis & Robak, 1997; Robak, 1996).

Based upon these foundations, a Strategy for Sustainable Forest Management for Galicia was developed and unveiled in the spring of 2002. (DXM, 2002)

3. The SFM framework for Galicia

The SFM framework proposed by the DXM represented a new approach to managing the forests of the autonomous region according to the principles and norms of sustainable forestry. The eight lines of action to implement of the framework included:

1. **Development of the legal framework for sustainable forest management.** SFM requires the formulation and enactment of integrated and coherent sets of policies, laws and regulations.
2. **Establishment of integrated management structures and processes for sustainable forestry.** Based on the principles of HFM, new integrated planning, monitoring and

control structures would be implemented at the regional, district[1] and forest management unit levels to ensure the continuity of strategic, tactical and operational decision processes.

3. **Development of the criteria and indicators of sustainability**. Forest management processes in Galicia should be developed that are consistent with regional criteria and indicators of sustainability, which will be, in turn, based upon those affirmed in the 3rd MCPFE in Lisbon (MCPFE, 1998).

4. **Establishment of an accurate and reliable system of forestry information**. Good information is essential for forest management planning, monitoring and control, and to evaluate and document actions and results in relation to accepted criteria.

5. **Promote increased research into forest sustainability**. Forest management should be based upon scientific knowledge, and research directed by management needs.

6. **Foster public forestry education to facilitate understanding and participation**. Informed participation of the public and the forestry sector of Galicia are critical to the success of the Strategy.

7. **Foster and support the economic development of the forest sector of Galicia**. Priorities include promoting timber and non-timber forest products as renewable resources, enhancing the role of the forestry sector in rural development, and supporting cooperation amongst forest owners and forest owners associations.

8. **Promotion of forest certification**. Forest certification initiatives that lend credibility and transparency to the forest management process should be fostered, especially those that enable certification by small forest owners.

The DXM believed that a new integrated process should be a critical component of the new strategy. The following sections of this paper focus on the new SFM process and supporting information infrastructure proposed by the DXM.

4. The proposed SFM process in Galicia

While many actions and programs would be required to implement the new strategy as envisaged, a key component would involve the implementation of a new forest management process. This new process, which is illustrated schematically in Figure 1, would be aimed at integrating and coordinating forest management at the regional, district and forest ownership levels while at the same time fostering the active participation of forestry stakeholders and Galician society at all levels.

The following are brief descriptions of the three major sub-processes of the proposed new SFM process for Galicia, followed by a description of how they are to interact (Fig. 1.)

4.1 Regional management sub-process

The first major goal of Regional Management would be to develop a revision to the current Plan Forestal de Galicia – PFG (Xunta de Galicia, 1992) based upon principles of sustainability, input from the public and the forest sector, and the best current forestry knowledge. A Regional Committee for Sustainable Forestry representing all regional stakeholders would endorse a Declaration of Regional Principles of Sustainability based

---

[1] Galicia is divided into 19 Forest Districts that have common physical, biological, economic and social characteristics. It can be argued that it is only at the level of the Forest District that it is possible to manage for critical landscape, territorial and community objectives and constraints.

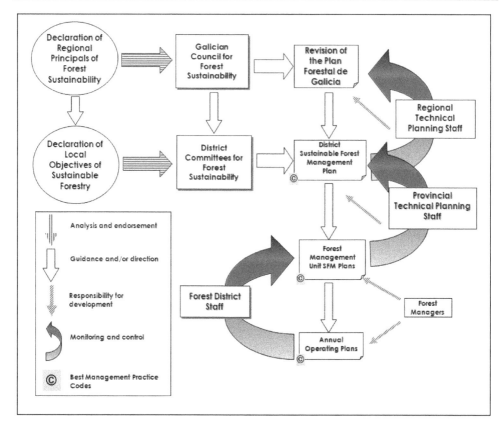

Fig. 1. Proposed SFM Framework for Galicia

upon international, EU and Spanish principles of forest sustainability, but recognizing the specific goals and constraints of Galicia. The revised PFG, developed by regional technical staff within the integrated process described below, would make explicit the long-term forest management goals of the region, the actions required to achieve goals, and terms of the "*co-responsibility contract*" that define responsibilities of forest sector players with respect to actions.

The second goal would involve the monitoring and control of results and actions to ensure that the specific objectives of SFM for Galicia are being achieved.

4.2 District management sub-process

Forest District Management Plans (FDMPs) are intended to bridge the gap between the PFG and the management plans and actions of Forest Management Units (FMUs). According to the most recent draft Plan Forestal de España-PFE (Ministerio de Medio Ambiente 2002), this is the level in the management hierarchy where it is most appropriate to accommodate strategic social and economic development objectives defined by local communities and group and also consider landscape-level environmental constraints and objectives that require planning across forest ownership boundaries (Ministerio de Medio Ambiente, 2002).

It should be noted that the PFE established general goals and guidelines for lower levels of planning in order ensure the fulfillment of the international commitments assumed by the Spanish government.

4.3 Forest management unit sub-process

Since most individual forests in Galicia are privately owned, the government does not have direct control of the forest management undertaken in them. This is particularly true of the very small ownerships that comprise almost 70% of Galician forests, where forest management (if there is any) is up to the individual owner. Even in the 30% of private forests owned by communities and managed by government foresters, community objectives may be at odds with those of the region or district. However, the government is not without tools to influence the management of private forests. The government controls subsidies for management activities and has the right to regulate some forest activities such as harvesting in certain forest types or authorizing the plantation of certain species (such as *Eucalyptus globulus*). Furthermore, the government could be seen as the sole organization capable of implementing a management infrastructure that would be capable of facilitating regional forest certification, which could be seen as the most viable approach to certification given ownership patterns. This infrastructure would include the planning and control mechanisms, Best Management Practice (or BMP) Codes and a series of silvicultural models (SMs) for major forest species.

In the absence of clear regional and district plans, it was difficult for the government to justify the use of such tools in any focused manner, and advancing credible forest certification was seen as difficult. With regional and district forest management guidelines and plans in place, the government would be able to give priority for subsidies and harvest approvals to forest owners who followed the district (and, thereby, regional) plans. As well, the implementation of the planning, monitoring and control systems envisioned in the new SFM process would facilitate the certification of even small forest ownerships, as long as they are managed in a way that conforms to the local FDMP, BMPs and SMs.

4.4 Management integration processes

The application of HFM requires that problems be decomposed and the various elements handled at the appropriate level of the management hierarchy. However, since it is usually impossible to solve all parts of the problem simultaneously, it is necessary to use an iterative approach to planning and control. This means that, although higher levels of the hierarchy give direction to (or constrain) the lower-level management processes, the lower-level processes should provide feedback to the upper levels so that the plans and decisions at higher levels can be refined and improved. Depending upon the complexity and importance of the problem and the time involved and available, several iterations of this process might occur. In the case of the proposed SFM process for Galicia, the main steps of this iterative approach were expected to be:

1. A draft regional plan for Galicia would be developed by the regional DXM technical staff based upon the declaration of principles, state of the forest, knowledge of forest processes, forecasts of forest products markets and other economic data and forecasts. The plan would be general, but would define specific goals and constraints.

2. This rough first version of the plan would then be passed down to districts as (generally) aspatial guidelines and constraints to Forest District planners. It might

indicate, for example, district targets related to reforestation, wildlife habitat, timber production, production of non-timber products over a long time horizon.

3. Initial district objectives and constraints would then be defined by the District Committees by way of the Declaration of Local Objectives of Sustainable Forestry in conjunction with provincial and Forest District staff that would provide data, information and guidance to ensure that regional guidelines are properly interpreted and that objectives are realistic given the geophysical and cultural characteristics of the district.

4. The provincial technical staff would then develop a long-term, strategic, generally aspatial, multi-objective forest management plan for the entire district **as if all of these forests were being managed by the forest service**. Again, forecasting and modeling tools would be required that use knowledge and information concerning forests and economic and social factors in the district. These district plans would also take into account the expected probabilities of success in convincing forest owners to follow the district plan. While it is recognized that there is no assurance that individual forest owners would follow the plan, it should be possible to model the probability of compliance for given levels of incentives and regulation. The process of objective setting and plan development would itself be iterative, since it is certain that the planning process would uncover problems or opportunities that require modification of the initial objectives and constraints. The goal would be to produce a district plan indicating actions to be undertaken, results expected and resources (and policies) required for implementation. The district plan would also describe the actions and investments promised by other "co-responsible" parties involved in implementation, such as a community that promises to find investment for a value-added plant to process a certain kind of product that is or can be made available from the forest.

5. When a district plan that is acceptable to the entire district (and provincial staff) has been developed, it would be passed back up to the regional staff for evaluation and possible approval. The evaluation process would involve ensuring that the guidelines have been followed, the regional objectives have been met, and the resources (and policies) required by the district are appropriate. For example, although each individual district plan could be reasonable, the budget requirements of all the districts together might not be able to be satisfied. It would be up to the regional staff to use their own information and models, along with the information from the districts, to produce a rational distribution of resources. Thus, budget rationing may require that some district plans be revised, taking into account the budgetary constraints for that district. It is also likely that information from district plans would prompt revisions of the regional plan.

6. Once final versions of the district plans have been completed and approved, these would be used to produce a final version of the Plan Forestal de Galicia since most of the actions, results and resource requirements necessary to carry out the regional plan are in the district plans.

7. During the period of implementation of the regional and district plans, it would be necessary to monitor the actions and results in individual FMUs to ensure that the plans are being followed and that the results are as expected. Besides acting as a control mechanism, the monitoring processes would help to provide the data and information necessary for subsequent iterations of the planning cycle. Such a monitoring system, if

properly designed, could also support regional forest certification for any forest owner who follows the district plans (BMPs and SMs).

The implementation of the proposed SFM process for Galicia would require much greater availability of reliable data and information for planning, and much better monitoring systems than existed in 2002. The following section of this paper gives a brief description of the Information Technology (IT) infrastructure envisaged at that time for the planning and control functions of the process.

5. Data and information infrastructure

A great deal of time and effort went into the design and documentation of the information systems and data structures required to support the new SFM strategy and process. For the purposes of this paper, these are summarized as:

Spatial Forest Data Infrastructure (SFDI): The SFDI would supply basic spatial and attribute forestry data to be used by all levels of management and, eventually, the public. Based upon such concepts as Open GIS (promoted by the international Open GIS Consortium) and web-enabled designs, it would foster standardized gathering and storage of data required for planning and control of forest management, as well as for the development and evaluation of forest policies and programs.

Integrated Forest Management System (IFMS): An integrated system of management tools would be required to ensure that plans at all levels of the management hierarchy are consistent, that actions and outcomes are monitored and controlled, and that decisions are justified and documented. These would consist of planning decision support systems (including forest modeling and forecasting tools to enable sensitivity and trade-off analysis) and monitoring and control tools to ensure that plans are being followed, and that objectives are being achieved.

Monitoring and Control Systems: Although specific monitoring and control tools would be part of IFMS, others were expected to be required to implement the SFM Strategy. In particular, these would include systems to compare outcomes to criteria and indicators of sustainability, to support regional forest certification initiatives, and to enable reporting of results to the public and to national, EU and international agencies.

6. Implementation and revisions to the strategy

As can be imagined, the implementation of such a great change was not without its problems. Perhaps the greatest obstacles that had to be overcome were the lack of knowledge concerning the proposed new management processes on the part of key players, the lack of information concerning the forests and other key factors, and the great difficulty in coordinating the design, development and implementation of so many interrelated programs and actions. In the nine years following the development of the strategy framework, the following pieces have been put in place:

- Education sessions related to SFM, forest certification and information systems for forest management have been provided to forest service personnel;
- SFDI, which provides web access to forestry spatial and attribute data to forest service staff, is now in place;
  - http://rimax.xunta.es/VisorRIMAX/Default.aspx
- Preliminary designs for the IFM system and monitoring and control systems have been developed;

- New instructions for forest management planning have been instituted that are more consistent with the principles of SFM as outlined in the Strategy;
- New guidelines for the submission of standardized forest management plan data (consistent with criteria and indicators of sustainability in the EU) have been developed and put in place.
- The Galician Council for Sustainable Forestry has been created;
  - As described in the Galician government document "Decreto 306/2004" (http://www.xunta.es/dog/Publicados/2004/20041229/Anuncio2558E_es.html) and in subsequent legislation and regulations.
- A manual of best management practices (BMPs) has been published for landowners, forest services companies and forest harvesting companies;
  - http://mediorural.xunta.es/areas/forestal/xestion_sustentabel/boas_practicas/
- Preliminary steps have been undertaken to establish the Regional Declaration of Principles of Sustainability, such as its endorsement by the "Mesa de la Madera" (Galician Wood Council") in 2008.
  - http://mediorural.xunta.es/areas/forestal/producion_e_industrias/mesa_da_madeira/

Some of the changes to the implementation of the strategy were the result of new regional, national and international reports, protocols, guidelines and proposals that informed and refocused the strategy framework. These included:

- The reports of the 4th and 5th Ministerial Conferences on the Protection of Forests in Europe (UNECE, 2003 and UNECE, 2007) held in Vienna and Warsaw, respectively.
- The EU Forest Action Plan which was adopted in June, 2006 (European Commission, 2006), described at http://ec.europa.eu/agriculture/fore/action_plan/index_en.htm.
- The Biomass Action Plan (COM(2005) 628 final – Official Journal C 49 of 28.02.2005) which is described in this European Union document: http://europa.eu/legislation_summaries/energy/renewable_energy/l27014_en.htm.
- Other EU and regional directives and initiatives, such as those related to good governance and public participation supported the direction and approach taken in the new strategy framework. The EU directive on good governance, for example, (http://ec.europa.eu/governance/governance_eu/index_en.htm) insists that public participation must be part of the development of national policies, while public participation in environmental plans has been enforced with the endorsement of law 27/2006 in Galicia (http://noticias.juridicas.com/base_datos/Admin/l27-2006.t3.html).
- Marey et al (2007) provide a description of the proposed content, structure and processes related to district plans in Galicia.

One of the critical pieces of the SFM strategy framework that have not been modified (or developed) and tested for the Galician situation, as was originally envisaged, is the IFM system. While this is still considered a critical element of the strategy framework and pieces that feed into it (see Data Infrastructure above) are mostly in place, the development of the IFM itself continues to be delayed.

Finally, though not enacted, a new Forest Law for Galicia is being developed, which would require more formal and effective public participation in the forest management process in the autonomous region. While this was envisaged in the original strategy (see Regional Management Sub-Process above), the fact that such participation was not mandated and institutionalized at the outset of is a major reason why the implementation of the strategy framework has been slow and incomplete. This is more fully explained in the Conclusions and Lessons Learned section, below.

7. SFM and forest certification in the private sector

During the same period the aforementioned efforts were being promoted by the government, the Galician private forest sector was under growing pressure to demonstrate SFM and due diligence with respect to legal source procurement. The growing importance of certification in the forest product marketplace and legislative initiatives such as European Regulation (EC) No 1024/2008 (European Commission, 2008), (EU) No 995/2010 (European Parliament, 2010) or Spanish Order Pre/116/2008, (Government of Spain, 2008), have been instrumental in increasing pressure on the private sector in this respect. In addition, the global economic downturn that followed the collapse of major US financial institutions significantly reduced demand for forest products (UNECE/FAO, 2010). SFM, previously considered by many players as a "tool for reaching new markets", suddenly turned into something "compulsory for maintaining declining core markets".

Despite growing pressures, forest certification schemes have not been very effectively implemented in Spanish forestry in general and in Galician forestry in particular. The share of certified area in Galicia is far below that of most of European countries, as is shown in Figure 2. In 2009, 9% of European forests were certified under PEFC or FSC, but if the Russian Federation is not included then this figure rises to an average of 46%, with several major wood-producing countries having certification rates of 60% or higher.

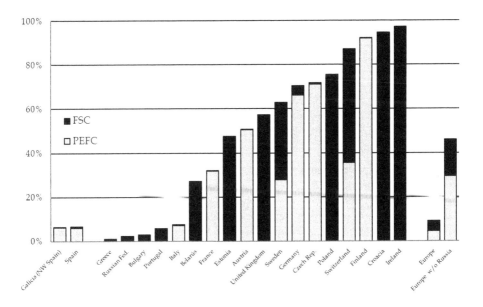

Fig. 2. Share of Forest Area under SFM certification in different European countries in 2009. Sources PEFC (2011), FSC (2011), FAO (2009). Note that due to difficulties in cross checking the FSC and PEFC databases, areas certified under both schemes are counted twice.

As of 2011, only 6.9% of the forest land in Galicia was certified under internationally recognized schemes. Regarding the type of certification scheme, 97% of the certified area is PEFC-certified and only 3% FSC-Certified. In Figure 3, contrary to what would be normally expected for region where a small private ownership predominates, it can be stated that despite some group and regional initiatives being recently launched, individual certification is most common (72%). This is probably due to the fact that 62% of the certified area is managed by DXM (Figure 4) and 10% by industries, while certification processes and SFM initiatives have not been implemented in areas managed by small private non industrial owners.

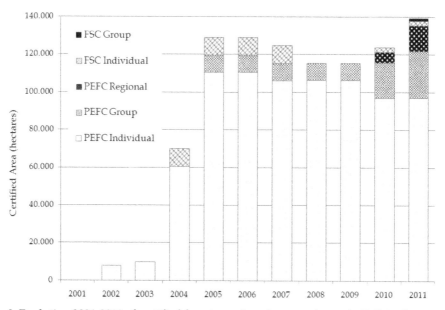

Fig. 3. Evolution 2001-2011 of certified forest area by scheme and type in Galicia. Source: Calculated from certificates public reports in registries of certification schemes. PEFC (2011), FSC (2011),

There are several possible reasons of the slow pace of forest certification in Galicia. Some authors, such as Ambrosio (2006), for example, postulate that the small size of private forest holdings implies relatively huge certification implementation and auditing costs. As well, such an ownership pattern implies significant traceability costs. For example, in 2010 there were more that 33.000 timber harvesting operations with an average of only 210 m³ obtained from each harvest (Monte Industria, 2010a).

Beyond the matter of scale, Picos (2009) suggests that the requirements of the PEFC and FSC are more stringent than those of forest certification schemes in place in other wood-producing countries. This fact represents a commercial disadvantage to industries that depend upon the local Galician wood supply. Local certified wood cannot compete with imported products on price, quantity or certainty of continuous supply. The paradox is that this situation favors operations audited according to less stringent standards and more distant suppliers, thereby requiring more long distance transport.

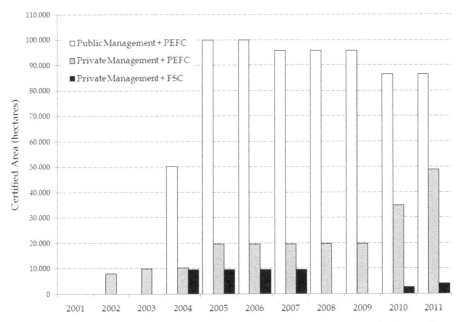

Fig. 4. Evolution 2001-2011 of certified forest area by management type and scheme. Source: Calculated from certificate public reports in registries of certification schemes. PEFC (2011), FSC (2011),

However, in the past five years, some of the Galician forest industry has been making significant investments in Chain-of-Custody certification. From indications provided by PEFC (2011), FSC (2011) and Monte Industria et al (2008a), there is a currently a potential annual demand for 3.5 million m³ of certified roundwood in Galicia alone, regardless if it is PEFC or FSC certified. In addition, the industry in neighboring regions (mainly Asturias and Portugal) would demand close to an additional million m³ from Galician forests. According to calculations based on public statistics on timber sales (Xunta de Galicia, 2010) and public reports on certified companies registered in PEFC (2011) and FSC (2011), as of 2009, forests in Galicia only produced approximately 275,000 m³ of PEFC-certified production, while no certified production at all came from FSC-certified forests. As well, it should be recognized that there a large number of small forest contractors that buy standing timber but do not have Chain-of-Custody certification. This means that some of the timber that was purchased from certified forests by such contractors lost its certification on its way to sawmills, wood-based panel factories or pulp mills. According to the summaries of certified timber auctions in public managed forests published by Xunta de Galicia (2010), this would have further reduced the supply of certified timber to industry by 60%.

The overall situation regarding difficulties in implement forest certification schemes may also be due to the slow and incomplete progress in the implementation of the proposed SFM strategy in Galicia. Ambrosio (2006) refers to complaints by the private sector about the lack of speed of the region's forest administration in formulating strategies, policies, processes or cost-effective methods aimed at helping the private forestry sector adopt principles of SFM and certify its performance under internationally recognized schemes.

Concerned by the situation regarding Forest and Chain-of-Custody certification, in 2009 the Galician forest industry, in cooperation with some forest owners associations, founded a Forest Certification and Chain-of-Custody Group, (Grupo Galego de Certificación Forestal e Cadea de Custodia or CFCCGA) aimed at achieving certification for small forest owners (under PEFC and/or FSC) and designing a due-diligence system that would comply with the imminent introduction of the Commission Regulation (EC) No 1024/2008. According to CFCCGA (2009a), in less than two years, more than 4,500 forest holdings (with an average size of 0.7 ha) have joined the group, making this initiative the first and only one to date which has been able to implement forest certification for small forest owners in Galicia. The aims and actions of this group are relevant to this paper because they can be seen as an attempt by a stressed private sector to take more responsibility for and ownership of broad SFM initiatives despite the fact that the SFM strategy framework developed by DXM has not been fully implemented. It is notable that some of the tools and processes developed by this group, namely Grouped Management Plans (Picos, 2010a), Best Management Practice codes (CFCCGA, 2009b) and Silvicultural Models for major forest species (CFCCGA, 2009b), are quite similar to the FDMPs, BMPs and SMs proposed in the Sustainable Forest Management Framework for Galicia.

8. Lessons learned

Almost a decade after the development of the SFM strategy framework for Galicia, while it can be said that much has been accomplished with regard to its implementation, it must be admitted that much more needs to be done. Although the complex political and jurisdictional situation in Galicia (as is described below) has contributed to slow and incomplete progress in the implementation of a SFM strategy in Galicia, in the authors' opinions the major direct contributing reasons for this state of affairs has most to do with three major deficiencies in the strategy framework development and/or implementation processes: the insufficient public participation, the lack of clarity with respect to the appropriate roles of major players, and insufficient legislative underpinnings. All of these are described below as "lessons learned", though it should be recognized that our lessons are likely far from complete.

Lesson One – Earlier and More Complete Public Participation: Originally, the intent was to present an initial, but comprehensive, draft strategy as a "straw man" or "white paper" to be revised based upon the suggestions and comments, but this has not worked since many stakeholders felt that they were being presented with a "fait accompli". While the DXM forest administration may have the broadest understanding of the situation, it cannot assume that it knows, a priori, all of the major problems in the region's forest sector. The principal forest stakeholders demand that government officials first ask them their opinions concerning the main issues, problems (or potential market opportunities) before beginning to design and develop systems, instruments, tools to satisfy them. Although it is the responsibility of the government to define policies, it is critical that the principal stakeholders are consulted at the outset in order to avoid subsequent rejections or negative political influences. Therefore, it is now felt that it was a mistake not to work with major stakeholders (perhaps by means of workshops and advisory groups) to identify issues and directions BEFORE drawing up any draft document.

Lesson Two – Clarification of Roles: While there are several levels of government and government administration that impact SFM in Galicia (the Spanish national government, the regional government, the provincial government and the 315 local municipalities), all of which could lead to confusion and problems, the authors believe that it is the lack of clarity regarding the roles of local municipalities, in particular, that impeded the implementation of the proposed strategy framework. For example, local municipalities regularly endorse plans, laws, decrees and regulations that regulate (ban or allow) harvesting operations, plantations and/or specific forest species in ways that contradict the Galician government's plans and laws. On the other hand, there is resistance at higher levels of government to the creation of district SFM committees for SFM due to fears that some local representatives may use these committees to advance political platforms or pressure the government for funding that is not related to SFM objectives. Therefore, before undertaking this process, the roles and responsibilities of the various levels of government (and the reasons for these as they relate to critical competencies with respect to SFM decision-making) should have been clarified and made explicit in a legal framework – preferably one that would advance sustainable management and reduce political manipulation.

Lesson Three – Legislative Support: Given the legislative changes required to ensure advancement of such a significant and politically sensitive initiative, it is necessary to receive full political support by the government in power right from the beginning of the process, and continuing support over the implementation period. Unfortunately, the large number of significant actions required to implement the strategy meant that inadequate progress was made before the elections of 2005, when there was a change in the Galician government, at which time the new government had to be educated regarding the strategy framework and the details of its proposed implementation. While progress continued to be made, it was slow and was then again slowed when another new government was elected in 2009. For these reasons, we believe that it would have been advisable to begin the strategy development and implementation process at the beginning of a political mandate and obtain broad political support so that any changes in government would be less likely to impede progress.

Lesson Four – Operational Priorities and Organizational Gaps: There are practical problems that impeded the DXM from making a sustained effort to manage and control the continued development and implementation of the new SFM strategy. The DXM has two main responsibilities: to develop and implement forest policies, and to prevent and fight forest fires. The principal forest stakeholders have persistently claimed (Monte Industria et al., 2008a; Monte Industria 2010b; Picos, 2010b) - that 95% of the time of a district director is taken up with organizing and managing forest fire fighting brigades during fire season, while forest services company associations have complained that all the technical staff of forest districts are fully occupied with fire detection and fire-fighting responsibilities from at least July 1st to September 30th, and in bad weather conditions the main fire season period may be further lengthened significantly.

While these complaints may be somewhat exaggerated, it must be recognized that the individuals and units tasked with implementing forest management practices at the district level are preoccupied with fire prevention and fighting for up to six months of the year. This significantly reduces the time and attention they have available for overseeing the implementation of new forest management strategies. It should be recalled that, within the proposed strategy (and generally), district-level staff are expected to:

a. control and monitor the degree of accomplishment of means proposed in District plans;

b. provide information and guidance to landowners and forest managers about regarding district and regional priorities, funding applications, BMPs, SMs and so on;

c.   monitor and control the progress of on-the-ground management plans and related actions.

Instead, however, technical staff members in districts are almost completely dedicated to fighting fires for up to half of the year. Therefore, in order to implement any new forest strategy in a coherent manner, it is necessary that responsibilities be changed so that some technical staff would be completely dedicated to forest management regardless of the severity of the fire season.

Lesson Five – Advantages for Small Ownerships: The implementation of a well-designed and supported SFM framework would boost forest certification in small non-industrial forestry settings. In Galicia, group and regional certifications are underdeveloped, in part as a result of the high degree of small, fragmented private ownership. It is important to note that the certification initiatives that have been most successful for small non-industrial private forest owners use documentation and processes that are very similar that those designed for the SFM strategy framework by DXM. This may indicate that a complete framework would aid to develop smallholder certification initiatives throughout the region and would reduce their implementation costs, which are a major constraint. Moreover, a SFM framework with a public participation may help auditors and group managers to verify and register the achievement of indicators and ease some of the required procedures. In addition, it also may be possible that strong public participation in the SFM framework development could help in objectively defining the certification requirements in Galicia and comparing them to what is required in other countries. This process could facilitate reviews and modifications to both FSC and PEFC national standards or, perhaps, enable the development of regional ones.

9. Conclusion

In conclusion, although the initial stages of its development seemed to progress rapidly, the continued development and implementation of the SFM strategy framework has been delayed and is far from complete, , with some major elements progressing only slowly, if at all. Greater time and effort spent obtaining clarity of roles and support from all stakeholders and actors at the initial stages of the process, while slowing the early stages, would likely have led to more progress by this time.

Meanwhile, the forest industry in Galicia is facing increasing pressure by markets and public opinion to demonstrate that it has adopted sustainable practices. The Galician private forest sector, which is moving in this direction despite the difficulties in Galicia caused by the very small holdings and fragmented ownership, has begun to realize that the full implementation of the SFM strategy framework could mitigate some of forest sector's problems in this regard and thus might be critical to its long-term survival. Given that some positive results that correlate very well with the government's strategy framework have been achieved by private initiatives in a short period of time, there is reason to be optimistic that more and faster progress may be made in the near future.

10. References

Ambrosio, Y. (2006). La certificación forestal en Galicia. *CIS madeira: Revista do Centro de Innovación e Servizos Tecnolóxicos da Madeira de Galicia*, ISSN 1886-3752, N°. 14, 2006 , p. 70-83

Ambrosio, Y. ; Picos, J. & Valero, E. (2003) Small Non Industrial Forest Owners Cooperation examples in Galicia (NW Spain). In Celic, K. (ed.); Robek, R. (ed.); Arzberger, U. (ed.) Proceedings of FAO/ECE/ILO Workshop on Forest Operation Improvements in Farm Forests. Logarska Dolina (Slovenia) 9 – 14 September 2003. Date of Access: July 19, 2011. Available from:
<http://webs.uvigo.es/jpicos/slovenia_ambrosio_picos_valero%20COMPLETO.pdf>

CFCCGA (2009a) Grupo Galego de certificacion Forestal e Cadea de Custodia. Certificados y Auditorias. Date of Access: July 26, 2011. Available from:
<http://certificacionforestal.blogspot.com/2009/12/certificados.html>

CFCCGA (2009b) Grupo Galego de certificacion Forestal e Cadea de Custodia. Documentos del Sistema. Date of Access: July 26, 2011. Available from:
<http://certificacionforestal.blogspot.com/2009/11/documentos-del-sistema.html>

Council of European Union (1998) Decision of the European Parliament and of the Council on the review of the European Community programme of policy and action in relation to the environment and sustainable development 'Towards sustainability`. Decision No 2179/98/EC of 24 September 1998

Council of European Union (1999) Council Resolution of 15 December 1998 on a forestry strategy for the European Union. CR 1999/C 56/01; Access: August 26, 2011. Available from:
<http://eur-lex.europa.eu/LexUriServ/LexUriServ.do?uri= OJ:C:1999:056:0001:0004:EN:PDF>

DXM (2002). Estratexia Galega de Xestión Forestal Sostible. Dirección Xeral de Montes e Industrias Forestais. Consellería de Medio Ambiente.

European Commission. (2006). Communication From The Commission To The Council And The European Parliament On An EU Forest Action Plan. Date of Access: July 19, 2011. Available from:
<http://ec.europa.eu/agriculture/fore/action_plan/index_en.htm>

European Commission. (2008). Regulation (EC) No 1024/2008 of 17 October 2008 laying down detailed measures for the implementation of Council Regulation (EC) No 2173/2005 on the establishment of a FLEGT licensing scheme for imports of timber into the European Community Date of Access: July 26, 2011. Available from:
<http://eur-lex.europa.eu/LexUriServ/LexUriServ.do?uri=OJ:L:2008:277:0023:0029:EN:PDF>

European Parliament. (2010). Regulation (EU) No 995/2010 of the European Parliament and of the Council of 20 October 2010 laying down the obligations of operators who place timber and timber products on the market Date of Access: July 26, 2011. Available from:
<http://ec.europa.eu/development/icenter/repository/TimberRegulation%20JO %2012_Nov_2010%20EN.pdf>

FAO. (2009). State of the World's Forests 2009. Rome: U.N. Food and Agricultural Organization, 2009. Date of Access: July 26, 2011. Available from:
<http://www.fao.org/docrep/011/i0350e/i0350e00.htm>

FSC. (2011). Global Forest Stewardship Council certificates: type and distribution. Date of Access: July 26, 2011. Available from: <http://www.fsc.org>

Gallis, C. & Robak, E. (1997). The Proposed Design for an Integrated Forest Management System for Greek Forestry, *Proceedings of the IUFRO 3.04 conference on Planning and Control of Forest Operations for Sustainable Forest Management, Madrid.* ETSI de Montes. Fundación Conde del Valle de Salazar. Pp. 213

Government of Spain. (2008). Orden PRE/116/2008, de 21 de enero, por la que se publica el Acuerdo de Consejo de Ministros por el que se aprueba el Plan de Contratación Pública Verde de la Administración General del Estado y sus Organismos Públicos, y las Entidades Gestoras de la Seguridad Social. Date of Access: July 26, 2011. Available from:
<http://www.boe.es/boe/dias/2008/01/31/pdfs/A05706-05710.pdf>.

Hax, A. & Candea, D. (1984). *Production and Inventory Management,* Prentice-Hall, Inc., ISBN-13: 978-0137248803, Englewood Cliffs, New Jersey

IPF/IFF (1998) IPF/IFF Proposals for Action. The Intergovernmental Panel on Forests (IPF) Intergovernmental Forum on Forests (IFF). Access: August 26, 2011. Available from: <http://www.un.org/esa/forests/pdf/ipf-iff-proposalsforaction.pdf>

Marey, M.; Aboal, J. & Bruña, X. (2007). Planes Ordenación De Recursos Forestales (Porf): Aplicación A Galicia Con Metodología Project Management, *Xi Congreso Internacional De Ingeniería De Proyectos, Lugo.* Available from:
<Http://www.aeipro.com/index.php/remository/func-startdown/523/>

MCPFE (1998). Resolution L2. Pan-European Criteria, Indicators and Operational Level Guidelines for Sustainable Forest Management. Third Ministerial Conference on the Protection of Forests in Europe 2-4 June 1998, Lisbon/Portugal. Access: August 26, 2011. Available from:
<http://www.foresteurope.org/filestore/foresteurope/Conferences/Lisbon/lisbo n_resolution_l2.pdf>

MCPFE (2011). Forest Europe Commitments. Includes the nineteen resolutions from the Ministerial Conferences on the Protection of Forests in Europe 1990, 1993, 1998, 2003,and 2007. Access: August 26, 2011. Available from:
<http://www.foresteurope.org/eng/Commitments/Commitments/>

Ministerio de Medio Ambiente. (2000). Estrategia Forestal Española. Madrid. ed. Organismo Autónomo de Parques Nacionales, 240 p.

Ministerio de Medio Ambiente. (2002). Plan Forestal de España. Date of Access: July 19, 2011. Available from:
<http://www.madrimasd.org/cienciaysociedad/ateneo/dossier/plan_forestal/m ma/pfe.pdf>

Monte Industria, Fearmaga y CMA. (2008a). 13 Medidas imprescindibles. Executive Briefing Date of Access: July 26, 2011. Available from:
<http://monteindustria2.blogspot.com/2008/04/informe-13-medidas.html>

Monte Industria, Fearmaga, CMA y FECEG. (2008b). Una estrategia para la organización de la Administración Forestal en Galicia Executive Briefing. Date of Access: July 26, 2011. Available from: <http://monteindustria2.blogspot.com/2009/03/una-estrategia-para-la-organizacion-de.html>

Monte Industria. (2010a). Presentación de Resultados Industria de la Madera de Galicia 2010. Date of Access: July 19, 2011. Available from:
<http://www.box.net/shared/vppiivpra7>

Monte Industria. (2010b). Proxecto de Orzamentos 2011 e o sector forestal Internal Report. Presentation. Date of Access: July 26, 2011. Available from: <http://monteindustria2.blogspot.com/2010/10/proxecto-de-orzamentos-2011-e-o-sector.html>

PEFC. (2011). Programme for the Endorsement of Forest Certification Schemes Database Date of Access: July 26, 2011. Available from: <http://register.pefc.cz/statistics.asp>

Picos J. (2010b). Impact of forest fires in Galician forest economy conference in FIREMAN Project Workshop FIREMAN- Fire Management for Maintain Biodiversity and Mitigate Economic Loss Oleiros 30.09.2010. Date of Access: July 26, 2011. Available from: <http://monteindustria2.blogspot.com/2010/10/impacto-dos-lumes-na-economia-forestal.html>

Picos, J. (2009). FSC España. Problemática de su situación actual. *Reunión de la Cámara Económica de FSC España con representantes de FSC Internacional* Date of Access: July 26, 2011. Available from: <http://monteindustria2.blogspot.com/2009/10/fscespana-situacion-y-problematica.html>

Picos, J. (2010a). Plan Conxunto de Xestión Forestal Sostible dos adscritos ó Grupo Galego de Certificación Forestal na Comarca Costa Norte. CFCCGA. Date of Access: July 26, 2011. Available from: <http://certificacionforestal.blogspot.com>

Robak, E. (1996). " IFMS Designs for North American Forest Product Companies", *Proceedings of Planning and Implementing Forest Operations to Achieve Sustainable Forests, Joint Meeting of COFE/IUFRO, Marquette, Michigan* (August, 1996). USDA General Technical Report NC-186

UN General Assembly (1992). Rio Declaration on Environment and Development, Date of Access: August 26, 2011. Available from: <http://www.un.org/documents/ga/conf151/aconf15126-1annex1.htm>

UNECE (United Nations Economic Commission for Europe). (2003). State of Europe's Forests 2003: The MCPFE Report on Sustainable Forest Management in Europe. Date of Access: July 19, 2011, Available from: <http://www.foresteurope.org/filestore/foresteurope/Publications/pdf/forests_2003.pdf>

UNECE (United Nations Economic Commission for Europe). (2007). State Of Europe's Forests 2007: The MCPFE report on sustainable forest management in Europe. Date of Access: July 19, 2011. Available from: <http://www.foresteurope.org/filestore/foresteurope/Publications/pdf/state_of_europes_forests_2007.pdf >

UNECE/FAO. (2010). Forest Products Annual Market Review, 2009-2010. 188 p. Date of Access: July 26, 2011. Available from: <http://live.unece.org/fileadmin/DAM/timber/publications/sp-25.pdf>

Weintraub, A. & Davis, L. (1996). Hierarchical planning in forest resource management: Defining the dimensions of the subject area, *Hierarchical Approaches to Forest Management in public and private organizations*. Petawawa National Forestry Institute Information Report PI-X-124: Canadian Forest Service. Edited by Martell, D., Davis, L.S, Weintraub, A.

Xunta de Galicia. (1992). Plan Forestal de Galicia. Consellería de Agricultura, Gandería e
     Montes. Dirección. Xeral de Montes e Medio Ambiente Natural. Santiago de
     Compostela, España.
Xunta de Galicia. (2001). *O Monte Galego En Cifras*. Dirección Xeral de Montes e Industrias
     Forestais, Consellería de Medio Ambiente, Xunta de Galicia
Xunta de Galicia. (2010). Produción e Industrias. Subastas. Date of Access: July 26, 2011.
     Available from:
     <http://mediorural.xunta.es/es/areas/forestal/produccion_e_industrias/subastas/>

12

Can Forest Management in Protected Areas Produce New Risk Situations? A Mixed-Motive Perspective from the Dadia-Soufli-Lefkimi Forest National Park, Greece

Tasos Hovardas
University of Thessaly
Greece

1. Introduction

Sustainable forest management is conceptualized as a management system which attempts to conciliate economic, ecological, and social dimensions (Maes et al., 2011; Vierikko et al., 2008). In this direction, sustainable forest management will have to foresee mechanisms and institutional arrangements for resolving trade-offs among multiple actors with conflicting interests (Bernués et al., 2005; Van Gossum et al., 2011). Decision-making schemes that involve a variety of interest groups might lead to increased complexity due to the societal demand for rational and transparent deliberation processes (Brechin et al., 2002; Wolfslehner & Seidl, 2010). A major challenge to sustainable forest management is to handle this complexity as well as confront uncertainty, which originates from changes in positions of stakeholders, scientific theories, and social institutions (Foster et al., 2010).

One response to complexity is to employ adaptive management, which presupposes feedback mechanisms and reflexive learning processes as necessary prerequisites to cope with an undetermined future and the core issues of risk and uncertainty (Plummer & Fennell, 2009; Von Detten, 2011). However, social aspects of sustainability are frequently downplayed even in adaptive management configurations. Previous studies showed that conventional methods of social science research and widely implemented management policies might underestimate the social heterogeneity at the local level to a substantial extent (Berninger et al., 2010; Sugimura & Howard, 2008). This inadequate handling of the social context might be expressed in matters of environmental governance. For instance, when local communities are represented by spokespersons adhering to majorities, minority positions are excluded (Hovardas et al., 2009; Hovik et al., 2010).

In the present paper we will approach the notion of sustainable forest management critically by examining the case study of the Dadia-Lefkimi-Soufli Forest National Park (DNP) in Greece. Our basic objective is to exemplify how forest management in a protected area can succeed in guaranteeing viable population sizes of endangered species but, at the same time, how it can eventuate in new risk situations regarding fire suppression and ecotourism development. Next, we will present a mixed-motive perspective formulated by local residents, who take advantage of the new risk situations encountered in the area and aim at the renegotiation of forest management and building a new social consensus. Finally, we will attempt a philosophical grounding of the mixed-motive perspective and we will discuss its potential contribution in promoting a new conceptualization of sustainability in forest management.

2. The Dadia-Lefkimi-Soufli Forest National Park

The study area is a public forest situated in north-eastern Greece next to the Greek–Turkish border (Fig. 1). The forest is dominated by Aegean pine (*Pinus brutia*), black pine (*Pinus nigra*), and oaks (*Quercus frainetto, Q. Cerris, Q. pubescens*). A development project funded by the World Bank in the late 1970s was about to intensify forest production, promote the clearing of oaks and reforestation with fast-growing pines. Based on a report of IUCN and WWF, the Greek Government established the Dadia Forest Reserve in 1980 by a Presidential Decree. The reserve was later included in the Natura 2000 sites proposed in Greece. The status of the reserve was further upgraded in 2006, when the protected area was recognized as a 'national park'. The park includes two strictly protected core areas (7290 ha), where all human activities are prohibited apart from those which are considered necessary for biodiversity conservation and scientific research. In the buffer zone (35170 ha), forestry is the main activity.

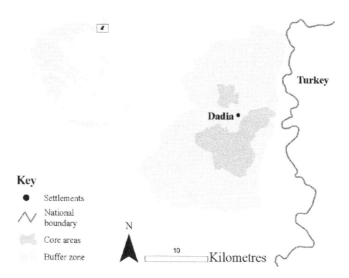

Fig. 1. The Dadia-Lefkimi-Soufli Forest National Park

Dadia is most known for its raptor fauna; thirty-six out of thirty-eight European raptor species can be observed in the park. The conservation of the black vulture (*Aegypius monachus*) is the central subject of forest management in the region (Adamakopoulos et al., 1995; Poirazidis et al., 2004), since DNP hosts the only breeding population of the species in

Can Forest Management in Protected Areas Produce New Risk Situations? A Mixed-Motive
Perspective from the Dadia-Soufli-Lefkimi Forest National Park, Greece

241

the Balkans (Skartsi et al. 2010). The two cores have 85% cover of pinewoods and mixed pine-oak woods and are crucial as nesting sites for the black vulture (Poirazidis et al., 2004). Maps of probability of occurrence for the nest sites of black vultures showed that nests are very likely to fall within the cores of DNP (Poirazidis et al., 2004). While the population in 1979 amounted to no more than 5 breeding pairs and a total of 26 individuals, the number of birds counted today in the park may range between 70 and 80 individuals, including about 20 breeding pairs (Skartsi et al., 2010). A number of projects implemented by WWF-Greece supported the increase of the population of the black vulture (Adamakopoulos et al., 1995; Poirazidis et al., 2002), for instance, vulture food supplement is provided in a feeding table that has been landscaped in the big core of the study area. Together with vulture food supplement, prohibition of hunting as well as strict control of forestry activities and road access that may lead to breeding failure have enabled the recovery of the black vulture (Poirazidis et al., 2004; Skartsi et al., 2008).

When the protected area was established in 1980, local people opposed fiercely the suspension of logging in the core zones of the forest reserve, since the loggers' cooperative in Dadia included more than 70 members, namely, about one-tenth of the total population of the village (Catsadorakis, 2010). This negative stance gradually shifted to an acceptance of the environmental conservation regime principally due to ecotourism development (Hovardas & Stamou, 2006a; Svoronou & Holden, 2005). The primary sector is currently decreasing in terms of employment, while the tertiary sector, including services and tourism, is gaining importance in the local economy (Liarikos, 2010). Ecotourism has been developed around wildlife viewing. Visitors are transferred to a Bird Observatory in full view of the vulture feeding table, where they can watch vultures feeding on carcasses (Hovardas & Poirazidis, 2006). The annual number of visitors has risen from less than 2000 in 1994 to more than 40000 today and infrastructure was gradually expanded to cover the rising demand (Hovardas, 2005). Apart from supporting ecotourism development, WWF-Greece has launched since the 1980s a series of environmental awareness campaigns including the celebration of the Annual Birds' Day.

2.1 Fire suppression in DNP

The biodiversity in DNP has to be attributed to a highly heterogeneous patchwork of habitats which are the result of traditional activities such as livestock grazing, small-scale agriculture and logging, as well as the use of fire (Grill & Cleary, 2003; Kati & Sekercioglu, 2006). However, many of these anthropogenic activities have either diminished or are undertaken rather infrequently (Schindler, 2010). Analogous socio-environmental trends have been reported for other Mediterranean mountainous regions (Álvarez Martínez et al., 2011; Badia et al., 2002; Bernués et al., 2005; Tàbara et al., 2003). In accord with the regulation of logging activities after the designation of DNP, the developments that have been described above have resulted in forest expansion and a decrease of forest clearings (Triantakonstantis et al., 2006), which comprise a crucial structural component of the foraging habitat of vultures (Catsadorakis et al., 2010; Gavashelishvili & McGrady, 2006). Therefore, protected area managers in DNP call for an urgent restoration of traditional activities with special reference made to grazing, which can halt the accumulation of dry plant biomass and reduce the risk of a wildfire (Catsadorakis et al., 2010; Poirazidis et al., 2004; Vasilakis et al., 2008).

Despite the fact that fire has been recognized as a source of landscape heterogeneity in DNP and has contributed substantially in shaping the nesting habitat of the black vulture[1] (Poirazidis et al., 2004), wildfires are acknowledged among the major threats of the park (Catsadorakis et al., 2010). Quite paradoxically, it is the 'wild', unexpected, and uncontrollable character of such a fire that is particularly alarming, causing anxiety among local residents and protected area managers in a national park. Although fire has long been considered as a natural phenomenon that determined evolutionary traits of plant species in Mediterranean ecosystems (Arianoutsou et al., 2011; Robbins, 2004), it cannot be allowed to 'return' to DNP. However, fire suppression is not a risk-neutral solution itself, since it is expected to increase surface fuel load and, thereby, foster the ignition of a wildfire (Chuvieco et al., 2010; Piñol et al., 2007; Sletto, 2010, 2011). Fortunately, DNP has not experienced any devastating event, such as the fires in Greece during the summer of 2007. The last incident of fire that has been recorded in DNP dates back to September 2008, when there were three fires ignited by lightning within one day. These fires destroyed about 60 hectares of mixed forest (i.e., pine and oak) and were put out quickly due to the immediate intervention of the Fire Service supported by local people and members of WWF Greece.

2.2 Ecotourism development in DNP

Ecotourism development in DNP served initially an instrumental purpose in countering negative attitudes of local people towards the environmental management regime established after the designation of the protected area and creating support for nature conservation (Liarikos, 2005; Svoronou & Holden, 2005). Gradually, DNP has succeeded in becoming an exemplary case for ecotourism development in Greece (Hovardas & Korfiatis, 2008; Hovardas & Poirazidis, 2006). Recent trends show that annual visitor numbers tend to stabilize around 40000 visitors following a slight decrease after 2003 (Liarikos, 2010). Despite these numbers, the majority of visitors display soft ecotourism-type characteristics, namely, short duration of stay, relatively large visitor group size, and trip organization by travel agents (Hovardas & Poirazidis, 2006), which results in little income generated for local residents (Catsadorakis et al., 2010). Protected area managers urged for the revision of the ecotourism development strategy and an improvement of the ecotourism product so that high-value visitors are attracted and uncertainty among local entrepreneurs is relaxed (Catsadorakis et al., 2010).

High levels of overall visitor satisfaction have been recorded in Dadia (Arabatzis & Grigoroudis, 2010; Hovardas & Poirazidis, 2006). However, a satisfactory general impression from the visit can coexist with identification of problems signaling a possibility of being close to reach natural or social carrying capacities (Machairas & Hovardas, 2005). In an evaluation of ecotourism in Dadia (Hovardas & Poirazidis, 2006), watching birds at the Bird Observatory revealed relatively low levels of visitor satisfaction, which was attributed to visitors' frustration when they are not fortunate in watching vultures. There is a pressing need to diversify activities offered to visitors in DNP and reduce the dependence of the visit on the Bird Observatory Post, which will also assist considerably in avoiding any possible disturbance of vultures (Liarikos, 2005). Indeed, the primary concern voiced over

---

[1] The optimal nesting habitat of the black vulture in Dadia includes mature trees located in steep slopes, which are surrounded by openings or with low height vegetation. Fire was a crucial component in the formation of black vulture's nesting habitat (Poirazidis et al., 2004).

ecotourism development is the apparent vulnerability of fragile natural environments to degrading impacts stemming from excessive visitor numbers (Ólafsdóttir & Runnströ, 2009). Despite the potential it can offer for marginal areas (Hovik et al., 2010), tourism development has been characterized as a fickle endeavour, where precise forecasts are not possible (Cole & Razak, 2008). Ecotourism scholars have highlighted the tendency of tourism for accelerating growth (Butler, 1999; Fennell & Ebert, 2004; Eagles, 2002; Lawson et al., 2003). This opportunistic character of tourism is reinforced in protected areas under the withdrawal of primary sector activities (Puhakka et al., 2009; Sharpley & Pearce, 2007).

3. The emergence of new risk situations in DNP

3.1 The 'eviction' of fire from DNP and its come-back as a new risk situation

Much modern environmental thinking stems from the balance-of-nature metaphor (Hovardas & Korfiatis, 2011). According to this metaphor, natural systems tend towards a point of relative stability. Systems in equilibrium resist change and when balance of nature is upset by human intervention, self-regulatory forces work to return the system to the state of equilibrium. Nature conservation in the form of establishing and operating protected areas has been based on the balance-of-nature metaphor and on the separation of the natural realm from the social realm by the demarcation of boundaries (Diamond, 1975). Although the balance-of-nature metaphor has been challenged by developments in the field of ecology (Hovardas & Korfiatis, 2011), it still continues to determine perceptions and policies related to nature conservation (Clapp, 2004; Nygren, 2000; Durand & Vázquez, 2011).

With the balance-of-nature metaphor acting as an organizing principle, the ecological discourse refers to a political project which wishes to establish a 'state of lost harmony' (Stavrakakis, 1999). The illusory character of the harmonious image of nature is revealed by manifestations of elements which cannot be accommodated by the balance-of-nature metaphor and assume a symptomatic form[2] (Stavrakakis, 1999). In order for nature-in-balance to remain coherent, any evidence that is not compatible with this image and threatens to destabilize it has to be repressed and excluded from nature's symbolization. However, under this Lacanian frame, we have to accept that no repression can guarantee the disappearance of the undomesticated, excluded symptom, which should be expected to return to its place at an unsuspected time (Stavrakakis, 1999). It is in this way that fire is rendered a symptom exactly when it should be expected with certainty. A wildfire, which is a quite common phenomenon in forests all along the Mediterranean Basin, would have irreversible consequences for biodiversity conservation, ecotourism, and the local community in Dadia. Agee (1997) underlines that few managers plan for a devastating fire and they just hope that it will not occur on their watch. Yet, the event of an unexpected and unmanageable wildfire can never be excluded (Klenner et al., 2000).

3.2 The irreducible interrelatedness of society and nature

Instead of conceptualizing DNP as some kind of intact nature demarcated within park borders, we should perceive it as a new spatial configuration produced by a special forest management regime (Robbins, 2004). For instance, the reconfiguration of space through zoning in Dadia has been centered around the priority of black vulture conservation

---

[2] The 'symptom' is an element which is thought to introduce disharmony in a symbolization that would otherwise be coherent and harmonious (Stavrakakis, 1999).

(Poirazidis et al., 2004). In contrast to core zones which include black vulture nests, the buffer zone presents an overall higher diversity of habitats and taxa (Grill & Cleary, 2003; Kati & Sekercioglu, 2006; Kati et al., 2004; Schindler et al., 2008). DNP provides a case study for the challenges presented to protected area managers, which are result of delineation of boundaries (Fall, 2002). Managers have to maintain these boundaries that protect biodiversity and, at the same time, they have to strive to diminish the negative effects that are recognized as consequences of these same boundaries (Knight & Landres, 1998). Managers need to tightly control nature within park borders in order to maintain desired conditions (Wood, 2000). However, strict territorial control frequently leads to the production of a static, ahistorical space (Roth, 2008). This is exemplified in Dadia by the 'eviction' of fire from DNP. Although fire is a feature which has shaped Mediterranean ecosystems, both society and nature in Dadia cannot be reproduced unless fire is suppressed.

A static, ahistorical nature has to be matched against a society that shares analogous properties. This is exemplified in the measures proposed by protected area managers where activities such as grazing and agriculture are referred to as 'traditional'. One way of reading this adjective is that it addresses small-scale activities, which should be taken in contradistinction to large-scale mechanized agriculture or stock breeding. However, another way of conceptualizing the need of restoring 'traditional' activities is their supposed compatibility with protected area governance, since 'traditional' might denote a state of local community that can be integrated in the park. In this direction, analysis of the local press (Hovardas & Korfiatis, 2008) and of the textual material disseminated in the Information Center of Dadia (Stamou et al., 2009) has shown that production processes are hushed up, which obscures the competition between different land uses and social actors in the area. Quite interestingly, this eventuates in sealing off the production process that has resulted in the ecotourism product offered to visitors (Hovardas & Korfiatis, 2008). Such discursive practices that both produce and limit meaning foster the belief of an independent nature which is to be encountered in the Bird Observatory Post or while walking along trails in the forest (Stamou et al., 2009).

Rather than adopting a dualistic approach to address the relationship between society and nature, which opposes society to nature, or a monistic approach, which collapses the natural realm into the social or vice versa, we subscribe to a dialectical position. Our assumption is that society and nature are simultaneously shaping and being shaped by each other, while each maintains a measure of autonomy (Evanoff, 2005). In this dialectical perspective human societies are seen in relational terms as both constituting and being constituted by the environments they inhabit, which has important ontological and epistemological implications. 'Forests' do really exist and are to be found 'out there' independently of our existence and of our definitions (ontological assumption). However, there is no cognitive appropriation of any 'forest', which could provide an objective, socially and culturally unmediated access to some type of essence of this notion (Robbins, 2004). Namely, the cognitive appropriation of 'forests' is always socially and culturally embedded (epistemological assumption).

3.3 How many visitors are enough?

The calculation of the carrying capacity of DNP for ecotourism development has been considered as a crucial prerequisite for respecting core prescriptions of sustainability

(Svoronou, 2000). Most definitions of carrying capacity of destinations for tourism development involve two components: one relating to environmental impacts (natural carrying capacity) and a second one relating to the quality of the recreational experience (social carrying capacity) (Lawson et al., 2003; Seidl & Tisdell, 1999). The notion of carrying capacity implies a threshold of visitor numbers that corresponds to the maximum level of use that a destination can accommodate without unacceptable alteration in environmental aspects or in the quality of visitor experience (Sæþórsdóttir, 2010).

Despite the fact that carrying capacity indicators are considered as a crucial tool for determining levels of use and monitoring tourism development, there are many reservations expressed about the possibility of singling out consistent thresholds and clear benchmarks for quantifying carrying capacity (Fleishman & Feitelson, 2009; Manning et al., 2002). A number of studies attempted to incorporate stochastic and fuzzy decision rules in the estimation of carrying capacity to account for variability and uncertainty in observed indicators (Prato, 2001, 2009). Further, there can be substantial divergence between the natural carrying capacity and the social carrying capacity estimated for the same destination (Zacarias et al., 2011). Moreover, positive indicators (e.g., user satisfaction) might remain over minimum acceptable levels, while negative indicators (e.g., soil erosion from trails) are surpassing maximum acceptable rates. Additional difficulties in determining carrying capacity include uncorrelated crowding perceptions with actual visitor numbers and differences among various visitor groups in tolerating crowding (Leujak & Ormond, 2007). A final remark concerning the contribution carrying capacity in adaptive management relates to its reactive nature (Fleishman & Feitelson, 2009; Lawson et al., 2003): remediary action is initiated only after standards of quality have been violated or are close to be violated.

A question that arises at this point is how carrying capacity can still be regarded as an important management tool in the light of all these reservations concerning its estimation. Handling risk perception in the case of carrying capacities needed desperately for guaranteeing sustainable tourism development in Dadia would necessitate to rename uncertainty into a number of visitors or a range of acceptable natural and social conditions. However, the field of risk perception addresses any objective account of carrying capacities as naïve and maintains that thresholds and benchmarks are not simply derived by scientific measurements of natural attributes but result from motivations and interests of various competing stakeholders which are being projected to nature (Bradley & Morss, 2002). This critical reading recognizes the need of adaptive management but rejects any sharp distinction of 'brute' facts from values (Giessen et al., 2009; Ruppert-Winkel & Winkel, 2011; Warren, 2007). However, there is a possibility that the adoption of this critical perspective might end up in an absolute and unproductive relativism, which would ultimately question the background assumptions of adaptive management. Below we will see how the mixed-motive perspective of local residents in Dadia avoids such a relativistic outcome and how locals attempt to negotiate ecotourism development in the region.

4. From a 'Greens know best' perspective to a 'Nobody knows best' perspective

The history of forest management in Dadia can be reconstructed as a contradistinction of incompatible perspectives (Lebel et al., 2004), which comprise coherent sets of goals, beliefs and methods, guide decisions, and prioritize societal choices. The establishment of the

Dadia Forest Reserve in 1980 has been launched as an initiative of IUCN and WWF-International under a 'Greens know best' perspective, where scientific knowledge produced by environmental nongovernmental organizations was utilized as the knowledge base, which legitimized and guided forest management with a primary focus on biodiversity conservation. The fierce opposition of local people after the designation of the reserve can be understood within the frame of a 'Locals know best' perspective, where local knowledge based on local experiences and embedded in social practices was contrasted to scientific knowledge in order to delegitimize the privileged access of environmental nongovernmental organizations to decision making for forest management issues. Quite paradoxically, long-lasting weaknesses of the government apparatus in the Greek case and the recent financial crisis seem to have enrolled forest management in Dadia in institutional forms and arrangements that adhere to a negative version of a 'State knows best' perspective, where outdated legislation and shrinkage of public funding are recognized as core impediments to sustainable forest management. The establishment of Management Authorities for 27 protected areas in Greece in 2002, including Dadia, should be recognized as a move towards a 'Nobody knows best' perspective, where all interest groups are called to share the responsibility for the organization of social consensus in forest management. According to this perspective, a critical approach is necessary for building on both scientific knowledge systems and local knowledge systems. Further, uncertainties in forest management and trade-offs between different interests are acknowledged as ubiquitous and are handled by means of participatory institutional arrangements that need to become adaptive.

Under the 'Nobody knows best' perspective, contingency and 'not knowing' should be conceived as a precondition of freedom and choice of action, which provides a requested sense of controllability over the decision-making process (Keskitalo & Lundmark, 2010) and precludes powerful partners from determining the outcome unilaterally (Von Detten, 2011). Social partners are engaged in a procedure of mutual learning (Wynne, 2006), where experts are expected to question conventional wisdom, while lay people are expected to question expert assumptions (Parkins, 2006; Risse, 2000). However, 'sharing responsibilities' might be in itself far from adequate for facing the new risk situations which have been described previously. A new, affirmative conceptualization of sustainability in forest management is needed to guide decision-making under the urgency of risk and uncertainty and to guarantee public involvement and the contribution of all interest groups towards a common vision.

5. An attempt for an affirmative approach to sustainable forest management when 'nobody knows best'

5.1 Local participation in protected area governance and issues of accountability and trust

Protected area establishment and operation has frequently resulted in a clash between local communities and scientific experts (Hovik et al., 2010; Keskitalo & Lundmark, 2010; Laudati, 2010; Reser & Bentrupperbäumer, 2005). In this conflict, experts appear as the group that can guarantee scientific rationality, while local people point to their right to use natural resources (Clapp, 2004). Inclusionary models of environmental governance in contemporary planning share the premise that local inhabitants participation is a prerequisite of an effective management strategy (Aasetre, 2006; Ellis & Porter-Bolland, 2008; Kingsland, 2002;

Kleinschmit et al., 2009; Velázquez et al., 2009). At the same time, the inclusion of societal actors poses the question of accountability and legitimacy (Kleinschmit et al., 2009). Indeed, the lack of accountability and transparency between experts and the public has been highlighted as a considerable disadvantage of protected area governance (Apostolopoulou & Pantis, 2010; Dearden et al., 2005; Lund et al., 2009), which becomes increasingly important in situations of risk and uncertainty (Von Detten, 2011).

More often than not, initiatives to engage local stakeholders as partners in protected area governance have served as a means of countering local opposition to nature conservation and have not been organized as strategic institutional structures to mediate and reconcile opposing views on conservation (Hovik et al., 2010; Uddhammar, 2006). Environmental education interventions have commonly targeted local communities in protected areas with the central aim to transmit scientific knowledge under the anticipation that local attitudes and behaviors will be transformed by scientific accounts to align with nature conservation values and expectations (Durand & Vázquez, 2011; Nygren, 2004). Next to the contested approach to environmental education that these interventions follow (Hovardas, 2005), they reveal that equality in terms of participation and inclusion is not always accompanied by a corresponding equality in determining the outcome of participatory processes. This instrumental view of public involvement might acknowledge local communities as key actors in protected area management but local people are eventually treated as beneficiaries of projects that have been designed without their input and consent (Durand & Vázquez, 2011). The end result is that certain privileged groups still dominate decision-making practices despite the fact that the rhetoric of participatory governance has become a commonplace (Ojha et al., 2009).

Recent research has shown that the environmental discourse has diffused in rural areas (Hovardas et al., 2009). This is also valid in the case of Dadia (Hovardas & Stamou, 2006b). Although local communities in protected areas might have adopted core requirements of environmental conservation, this trend should not be perceived as an unproblematic adherence of local people to join the coalition of conservationists. What is needed for the promotion of effective local participation schemata and the organization of social consensus regarding natural resource management is power sharing between local people and experts (Arts & Buizer, 2009; Hovik et al., 2010). Unless this precondition is met, scientific experts will become so powerful in any type of deliberation process that local people or their interests will be excluded (Giessen et al., 2009; Lövbrand, 2009; Steffek, 2009) and the democratic legitimacy of the process will be severely undermined (Csurgó et al., 2008; Kleinschmit et al., 2009). An important barrier that has to be overcome in this respect is the restoration of trust between social actors with competing interests and powers. How can experts trust local people in issues that need to be based on the scientific background of forest management? On the other hand, how can rural residents in protected areas renounce their legitimate claims on the land and resources in their region?

5.2 The mixed-motive perspective of local people in DNP

The mainstream sustainable development discourse has been based on the predominant belief in 'win-win' options, namely on the potential reconciliation of economic and environmentalist interests (Arts & Buizer, 2009; Durand & Vázquez, 2011; Huttunen, 2009; Mulvihill & Milan, 2007; Veenman et al., 2009). However, 'win-win' conceptualizations fall short of addressing the issues of accountability and trust which have been described previously and point towards a 'win-lose' arrangement. Further, 'win-win' options cannot explain the dynamics of competition and negotiation between social actors before and after an agreement has been reached. The question is if there is any other possibility to conceptualize the interplay or contradistinction between nature conservation and economic development apart from naïve 'win-win' or tense 'win-lose' approaches. An alternative conceptualization might be indicated by the notion of distributive justice, which implies the allocation of benefits and burdens of an activity among affected social actors (Pelletier, 2010).

Local residents' dispositions (Table 1; for a detailed description of the methodology see Hovardas, 2010) can be interpreted as a mixed-motive perspective, which diverges from 'win-win' and 'win-lose' approaches in that it envisages gain solutions for both nature conservation and economic development while acknowledging that there will always be a distributive aspect (Hoffman et al., 1999; Hovardas & Korfiatis, 2008).

	Economic development	Nature conservation
Mutual gain	Creating forest clearings would provide significant additional income for local loggers	• Forest clearings in core areas would severely decrease the probability of a forest fire • Forest clearings would enable raptors to find their prey much easier
Distributive aspects	• Trees to be cut should be selected on biodiversity conservation criteria • Extensive reforestation programs should be banned	Investment in ecotourism should be enhanced as a supplementary source of revenue for locals

Table 1. Local residents' mixed-motive approach

As solutions to forest management disputes require the balancing of interests among a complex array of participants, and because this can only be achieved through negotiations inevitably associated with costs and benefits (frame of reference in win-lose models), the mixed-motive model offers a theoretical and empirical alternative to the opposing 'win-win' and 'win-lose' perspectives. In this regard, the range of players' interests does not bifurcate into simply economic development and nature conservation coalitions but there can be a mutual recognition and appraisal of interests, which is necessary for reaching an agreement. This confrontation of social actors might increase complexity considerably but it tends to involve greater opportunities to expand the scope of the debate, finding solutions that will improve the potential outcome simultaneously for both parties (integrative principle of win-win models).

Local people in Dadia claim that creating forest clearings would provide significant additional income for local loggers. At the same time, local people suggest that forest clearings in core areas would severely decrease the probability of a forest fire and enable raptors to find their prey much easier. In terms of distributive aspects, local residents accept that trees to be cut should be selected on biodiversity conservation criteria and that extensive reforestation programs should be banned. However, they expect investment in ecotourism to be enhanced so that ecotourism development will continue to comprise a supplementary source of revenue for the local community. Interestingly, the new risk situations arising in Dadia are grasped by locals in their argumentation (e.g., forest fire susceptibility as well as ecotourism development). This is a strong indication that the mixed-motive perspective might allow for the incorporation and negotiation of risk situations and potential responses among interest groups involved in forest management. It should also be noticed that nature conservation gains and distributive aspects of economic development were more prominent in local accounts than other elements in the presentation of their mixed-motive perspective. This might imply the attempt of local people to engage viewpoints which would be endorsed by park managers.

In terms of distributive aspects concerning economic development, local residents seem to acknowledge the need of multipurpose forest management and their suggestions follow closely the recommendations of park managers in Dadia (Catsadorakis et al., 2010; Gatzogiannis & Poirazidis, 2010). The economic gains they anticipate together with the gains expected for nature conservation can be inscribed in a 'productive forest' perception of forest health (Warren, 2007), which underlines the double objective of providing jobs in the forest industry while reducing the risk of wildfire. In that respect, local residents in Dadia present a position close to the emerging preventative paradigm in Mediterranean forested landscapes (Tàbara et al., 2003), which highlights the need to anticipate the intrinsic tendency of fire ignition as the highest priority. Finally, local people present ecotourism development as a distributive aspect related to nature conservation, since further investment in ecotourism would necessitate the reallocation of resources at the disposal of protected area managers for monitoring purposes and might initiate heated debates over natural and social carrying capacities. This position might indicate an implicit attempt to relate the risk concerning local unemployment rates attributed to the forest management regime to the risk of ecotourism development overriding carrying capacity thresholds. It has been reported that the existence of alternative branches of employment mediates views on the vulnerability of local forestry to nature conservation (Keskitalo & Lundmark, 2010; Stoffle & Minnis, 2008).

6. Conclusion

The Habermasian window of opportunity offered by the mixed-motive perspective is both timely and spatially delineated and guarantees the inclusion of all affected actors under a commitment of rational argumentative deliberation (Carvalho-Ribeiro et al., 2010; Durand & Vázquez, 2011; Kleinschmit et al., 2009; Ojha et al., 2009; Parkins & Davidson, 2008; Warren, 2007). However, conflict and negotiation should be acknowledged as indispensable constituents of a mixed-motive deliberation process. Foucauldian power differentials are not hidden between participating actors but have to be enacted to steer the negotiation or re-negotiation process. This enables hidden power structures to surface and be contested and provides instances for shifting power balances and multiple empowering effects (Berman Arévalo & Ros-Tonen, 2009; Winkel, in press). Social actors that participate will have to recognize both conflicting demands as well as the need to come to terms after negotiation. In this direction, claims of objective truth and of a single rationality have to be singled out as inadequate to serve the democratic mandate (Winkel, in press). Starting from the fact that uncertainty is irreducible by science (Borchers, 2005), social consensus is necessary to guide forest management decisions (Parkins, 2006). The possibility of any hegemonic attempt or discursive practice to frame the issue at stake will be counterweighted by declaring willingness to reach an agreement (Van Gossum et al., 2011) and being committed to support the final decision, against which negotiating social actors will be held accountable (Brechin et al., 2002).

'Final' in this case does not imply the possibility of any permanent solution but wishes to denote the outcome of a process, which will guide forest management for a certain period of time within a culture of experimentation. The final decision will always be marked by its temporary character, being subject to scrutiny and critical reappraisal. Within this frame, deliberation has to be considered as an incomplete process under the need of regular revision (Buizer & Van Herzele, in press), which is perfectly compatible with adaptive management. The mixed-motive perspective can be used for scenario analysis and planning, which is a tool for assessing alternative strategies under conditions of uncertainty and for guiding adaptive management (Swart et al., 2004; Von Detten, 2011). Different scenarios can be developed as narrative descriptions of alternative hypothetical futures especially when managers cannot anticipate future conditions by extrapolating from past trends (Daconto & Sherpa, 2010). In these scenarios, all affected social actors will have the opportunity to claim their participation.

7. Acknowledgments

I am grateful to Kostas Poirazidis for his multifarious help. I also wish to thank local residents in Dadia for their valuable input.

8. References

Aasetre, J. (2006). Perceptions of communication in Norwegian forest management. *Forest Policy and Economics*, Vol.8, No.1 (January 2006), pp. 81-92, ISSN 13899341

Adamakopoulos, T.; Gatzogiannis, S. & Poirazides, K. (1995). *Environmental Study of the Dadia Forest Reserve*. ACNAT Programme, WWF-Greece, EU DG XI, Ministry of Environment, Physical Planning and Public Works, Athens

Agee, J. K. (1997). The severe weather wildfire - Too hot to handle? *Northwest Science*, Vol.71, No.2 (May 1997), pp. 153-156, ISSN 0029344X

Álvarez Martínez, J.-M.; Suárez-Seoane, S. & De Luis Calabuiga, E. (2011). Modelling the risk of land cover change from environmental and socio-economic drivers in heterogeneous and changing landscapes: The role of uncertainty. *Landscape and Urban Planning*, Vol.101, No.2 (May 2011), pp. 108–119, ISSN 0169-2046

Apostolopoulou, E. & Pantis, J. D. (2010). Development plans versus conservation: explanation of emergent conflicts and state political handling. *Environment and Planning A*, Vol.42, No.4, pp. 982-1000, ISSN 0308518X

Arabatzis, G. & Grigoroudis, E. (2010). Visitors' satisfaction, perceptions and gap analysis: The case of Dadia–Lefkimi–Souflion National Park. *Forest Policy and Economics*, Vol. 12, No.3 (March 2010), pp. 163–172, ISSN 1389-9341

Arianoutsou, M.; Koukoulas, S. & Kazanis, D. (2011). Evaluating post-fire forest resilience using GIS and multi-criteria analysis: An example from Cape Sounion National Park, Greece. *Environmental Management*, Vol.27, No.3 (March 2011), pp. 384–397, ISSN 0364152X

Arts, B. & Buizer, M. (2009). Forests, discourses, institutions - A discursive-institutional analysis of global forest governance. *Forest Policy and Economics*, Vol.11, No.5-6 (October 2009), pp. 340–347, ISSN 1389-9341

Badia, A.; Saurí, D.; Cerdan, R. & Llurdés, J.-C. (2002). Causality and management of forest fires in Mediterranean environments: an example from Catalonia. *Environmental Hazards*, Vol.4, No.1 (March 2002), pp. 23–32, ISSN 1464-2867

Berman Arévalo, E. & Ros-Tonen, M. A. F. (2009). Discourses, Power Negotiations and Indigenous Political Organization in Forest Partnerships: The Case of Selva de Matavén, Colombia. *Human Ecology*, Vol.37, No.6 (December 2009), pp. 733–747, ISSN 03007839

Berninger, K., Adamowicz, W., Kneeshaw, D., & Messier, C. (2010). Sustainable forest management preferences of interest groups in three regions with different levels of industrial forestry: An exploratory attribute-based choice experiment. *Environmental Management*, Vol.46, No.1 (July 2010), pp. 117–133, ISSN 0364152X

Bernués, A.; Riedel J. L.; Asensio, M. A.; Blanco, M.; Sanz, A.; Revilla, R. & Casasús, I. (2005). An integrated approach to studying the role of grazing livestock systems in the conservation of rangelands in a protected natural park (Sierra de Guara, Spain). *Livestock Production Science*, Vol.96, No.1 (September 2005), pp. 75–85, ISSN 0301-6226

Borchers, J. G. (2005). Accepting uncertainty, assessing risk: Decision quality in managing wildfire, forest resource values, and new technology. *Forest Ecology and Management* Vol.211, No.1-2 (June 2005), pp. 36–46, ISSN 0378-1127

Bradley, B. S. & Morss, J. R. (2002). Social construction in a world at risk: Toward a psychology of experience. Theory and Psychology, Vol.12, No.4 (August 2002), pp. 509-531, ISSN 0959-3543

Brechin, S. R.; Wilshusen, P. R.; Fortwangler, C. L. & West, P. C. (2002). Beyond the square wheel: Toward a more comprehensive understanding of biodiversity conservation as social and political process. *Society and Natural Resources*, Vol.15, No.1, pp. 41-64, ISSN 0894-1920

Buizer, M. & Van Herzele, A. (in press). Combining deliberative governance theory and discourse analysis to understand the deliberative incompleteness of centrally formulated plans. *Forest Policy and Economics*, doi:10.1016/j.forpol.2010.02.012, ISSN 1389-9341

Butler, R. W. (1999). Sustainable tourism: A stateof-the-art review. *Tourism Geographies*, Vol.1, No.1 (February 1999), pp, 7 – 25, ISSN 1461-6688

Carvalho-Ribeiro, S. M.; Lovett, A. & O'Riordan, T. (2010). Multifunctional forest management in Northern Portugal: Moving from scenarios to governance for sustainable development. *Land Use Policy*, Vol.27, No.4 (October 2010), pp. 1111–1122, ISSN 0264-8377

Catsadorakis, G. (2010). The history of conservation efforts for the Dadia-Lefkimi-Soufli Forest National Park. In: *The Dadia-Lefkimi-Soufli Forest National Park, Greece: Biodiversity, Management and Conservation*, Catsadorakis, G., & Källander, H. (Eds.), pp. 241-252. WWF-Greece, ISBN 978-960-7506-10-8, Athens

Catsadorakis, G., Kati, V., Liarikos, C., Poirazidis, K., Skartsi, Th., Vasilakis, D., & Karavellas, D. (2010). Conservation and management issues for the Dadia-Lefkimi-Soufli Forest National Park. In: *The Dadia-Lefkimi-Soufli Forest National Park, Greece:*

Biodiversity, Management and Conservation, Catsadorakis, G., & Källander, H. (Eds.), pp. 265-279. WWF-Greece, ISBN 978-960-7506-10-8, Athens

Chuvieco, E.; Aguado, I.; Yebra, M.; Nieto, H.; Salas, J.; Martín, M. P.; Lara Vilar, L.; Martínez, J.; Martín, S.; Ibarra, P.; de la Riva, J.; Baeza, J.; Rodríguez, F.; Molina, J. R.; Herrera, M. A. & Zamora, R. (2010). Development of a framework for fire risk assessment using remote sensing and geographic information system technologies. *Ecological Modelling*, Vol.221, No.1 (January 2010), pp. 46–58, ISSN 03043800

Clapp, R. A. (2004). Wilderness ethics and political ecology: remapping the Great Bear Rainforest. *Political Geography*, Vol.23, No.7 (September 2004), pp. 839–862, ISSN 09626298

Cole, S. & Razak, V. (2008). How far, and how fast? Population, culture, and carrying capacity in Aruba. *Futures*, Vol.41, No.6 (August 2009), pp. 414–425, ISSN 0016-3287

Csurgó, B.; Kovách, I. & Kučerová, E. (2008). Knowledge, power and sustainability in contemporary rural Europe. Sociologia Ruralis, Vol.48, No.3 (June 2008), pp. 292-312, ISSN 00380199

Daconto, G. & Sherpa, L. N. (2010). Applying scenario planning to park and tourism management in Sagarmatha National Park, Khumbu, Nepal. *Mountain Research and Development*, Vol.30, No.2 (May 2010), pp. 103–112, ISSN 02764741

Dearden, P.; Bennett, M. & Johnston, J. (2005). Trends in global protected area governance, 1992-2002. *Environmental Management*, Vol.36, No.1 (July 2005), pp. 89–100, ISSN 0364152X

Diamond, J. (1975) The island dilemma: lessons of modern biogeographic studies for the design of natural reserves, *Biological Conservation*, Vol.7, No.2 (February 1975), pp. 129–146, ISSN 00063207

Durand, L. & Vázquez, L. B. (2011). Biodiversity conservation discourses. A case study on scientists and government authorities in Sierra de Huautla Biosphere Reserve, Mexico. *Land Use Policy*, Vol.28, No.1 (January 2011), pp. 76–82, ISSN 0264-8377

Eagles, P. (2002) Trends in park tourism: Economics, finance and management. *Journal of Sustainable Tourism*, Vol.10, No.2, pp. 132–153, ISSN 09669582

Ellis, E. A. & Porter-Bolland, L. (2008). Is community-based forest management more effective than protected areas? A comparison of land use/land cover change in two neighboring study areas of the Central Yucatan Peninsula, Mexico. *Forest Ecology and Management*, Vol.256, No.11 (November 2008), pp. 1971–1983, ISSN 0378-1127

Evanoff, R. J. (2005). Reconciling realism and constructivism in environmental ethics. *Environmental Values*, Vol.14, No.1 (February 2005), pp. 61-81, ISSN 09632719

Fall, J. (2002). Divide and rule: constructing human boundaries in 'boundless nature'. *GeoJournal*, Vol.58, No.4 (April 2002), pp. 243-251, ISSN 03432521

Fennell, D. A. & Ebert, K. (2004). Tourism and the precautionary principle. *Journal of Sustainable Tourism*, Vol.12, No.6, pp. 461 – 479, ISSN 09669582

Fleishman, L. & Feitelson, E. (2009). An application of the recreation level of service approach to forests in Israel. *Landscape and Urban Planning*, Vol.89, No.3-4 (February 2009), pp. 86–97, ISSN 0169-2046

Foster, B. C., Wang, D., Keeton, W. S., & Ashton, M. S. (2010). Implementing sustainable forest management using six concepts in an adaptive management framework. *Journal of Sustainable Forestry*, Vol.29, No.1 (February 2010), pp. 79–108, ISSN 10549811

Gatzogiannis, S., & Poirazidis, K. (2010). The Dadia Forest Complex: stand development and forest management. In: *The Dadia-Lefkimi-Soufli Forest National Park, Greece: Biodiversity, Management and Conservation*, Catsadorakis, G., & Källander, H. (Eds.), pp. 95-101. WWF-Greece, ISBN 978-960-7506-10-8, Athens

Gavashelishvili, A. & McGrady, M. J. (2006). Geographic information system-based modelling of vulture species to carcass appearance in the Caucasus. *Journal of Zoology*, Vol.269, No.3 (July 2006), pp. 365-372, ISSN 09528369

Giessen, L.; Kleinschmit, D. & Böcher, M. (2009). Between power and legitimacy – Discourse and expertise in forest and environmental governance. *Forest Policy and Economics*, Vol.11, No.5-6 (October 2009), pp. 452–453, ISSN 1389-9341

Grill, A. & Cleary, D. F. R. (2003). Diversity patterns in butterfly communities of the Greek nature reserve Dadia. *Biological Conservation*, Vol.114, No.3 (December 2003), pp. 427–436, ISSN 0006-3207

Hoffman, A. J., Gillespie, J. J., Moore, D. A., Wade-Benzoni, K. A., Thompson, L. L. and Bazerman, M. H. 1999. A mixed-motive perspective on the economics versus environment debate. *American Behavioral Scientist*, Vol.42, No.8 (May 1999), pp. 1254–1276, ISSN 00027642

Hovardas, T. (2005). *Social representations on ecotourism – Scheduling interventions in Protected areas*. Aristotle University, School of Biology, PhD-Thesis, Thessaloniki, Greece (in Greek with an English summary)

Hovardas, T. (2010). The contribution of social science research to the management of the Dadia Forest Reserve: nature's face in society's mirror. In: *The Dadia-Lefkimi-Soufli Forest National Park, Greece: Biodiversity, Management and Conservation*, Catsadorakis, G., & Källander, H. (Eds.), pp. 253-263. WWF-Greece, ISBN 978-960-7506-10-8, Athens

Hovardas, T. & Korfiatis, K. J. (2008). Framing environmental policy by the local press: case study from the Dadia Forest Reserve, Greece. *Forest Policy and Economic*, Vol.10, No.5 (April 2008), pp. 316-325, ISSN 13899341

Hovardas, T. & Korfiatis, K. J. (2011). Towards a critical re-appraisal of ecology education: Scheduling an educational intervention to revisit the 'balance of nature' metaphor. Science & Education, Vol.20, No.10 (October 2011), pp. 1039–1053

Hovardas, T. & Poirazidis, K. (2006) Evaluation of the Environmentalist Dimension of Ecotourism at the Dadia Forest Reserve (Greece). *Environmental Management*, Vol.38, No.5 (November 2006), pp. 810–822, ISSN 0364152X

Hovardas, T. & Stamou, G. P. (2006a) Structural and narrative reconstruction of representations on 'nature,' 'environment,' and 'ecotourism,' *Society and Natural Resources*, Vol.19, No.3 (March 2006), pp. 225–237, ISSN 08941920

Hovardas, T. & Stamou, G. P. (2006b) Structural and narrative reconstruction of rural residents' representations of 'nature', 'wildlife', and 'landscape'. *Biodiversity and Conservation*, Vol.15, No.5 (May 2006), pp. 1745–1770, ISSN 09603115

Hovardas, T., Korfiatis K.J., & Pantis, D.J. (2009). Environmental representations of local communities' spokespersons in protected areas. *Journal of Community and Applied Social Psychology*, Vol.19, No.6 (November 2009), pp. 459-472, ISSN 10529284

Hovik, S.; Sandström, C. & Zachrisson, A. (2010). Management of protected areas in Norway and Sweden: Challenges in combining central governance and local participation.

Journal of Environmental Policy and Planning, Vol.12, No.2 (June 2010), pp. 159–177, ISSN 1522-7200

Huttunen, S. (2009). Ecological modernisation and discourses on rural non-wood bioenergy production in Finland from 1980 to 2005. *Journal of Rural Studies*, Vol.25, No.2 (April 2009), pp. 239–247, ISSN 0743-0167

Kati, V. & Sekercioglu, C. H. (2006). Diversity, ecological structure, and conservation of the landbird community of Dadia reserve, Greece. *Diversity and Distributions*, Vol.12, No.5 (September 2006), pp. 620-629, ISSN 13669516

Kati, V.; Dufrêne, M.; Legakis, A.; Grill, A. & Lebrun, P. (2004). Conservation management for Orthoptera in the Dadia reserve, Greece. *Biological Conservation*, Vol.115, No.1 (January 2004), pp. 33–44, ISSN 0006-3207

Keskitalo, E. C. H. & Lundmark, L. (2010). The controversy over protected areas and forest-sector employment in Norrbotten, Sweden: Forest stakeholder perceptions and statistics. *Society and Natural Resources*, Vol.23, No.2 (February 2010), pp. 146–164, ISSN: 0894-1920

Kingsland, S. (2002). Designing nature reserves: adapting ecology to real-world problems. *Endeavor*, Vol.26, No.1, pp. 9-14, ISSN 0160-9327

Kleinschmit, D.; Böcher, M. & Giessen, L. (2009). Discourse and expertise in forest and environmental governance − An overview. *Forest Policy and Economics*, Vol.11, No.5-6 (October 2009), pp. 309–312, ISSN 1389-9341

Klenner, W.; Kurz, W. & Beukema, S. (2000). Habitat patterns in forested landscapes: management practices and the uncertainty associated with natural disturbances. *Computers and Electronics in Agriculture*, Vol.27, No.1-3 (June 2000), pp. 243–262, ISSN 0168-1699

Knight, R. L. & Landres, P. B. (1998). *Stewardship across boundaries*. Island Press, ISBN 1559635169, Washington, D.C.

Laudati, A. A. (2010). The Encroaching Forest: Struggles Over Land and Resources on the Boundary of Bwindi Impenetrable National Park, Uganda. *Society and Natural Resources*, Vol.23, No.8 (August 2010), pp. 776–789, ISSN 1521-0723

Lawson, S. R.; Manning, R. E.; Valliere, W. A. & Wang, B. (2003) Proactive monitoring and adaptive management of social carrying capacity in Arches National Park: an application of computer simulation modeling. *Journal of Environmental Management*, Vol.68, No.3 (July 2003), pp. 305–313, ISSN 0301-4797

Lebel, L., Contreras, A., Pasong, S., & Garden, P. (2004). Nobody Knows Best: Alternative perspectives on forest management and governance in Southeast Asia. *International Environmental Agreements: Politics, Law and Economics*, Vol.4, No.2, pp. 111–127, ISSN 15679764

Leujak, W. & Ormond, R. F. G. (2007). Visitor perceptions and the shifting social carrying capacity of South Sinai's coral reefs. *Environmental Management*, Vol.39, No.4 (April 2007), pp. 472-489, ISSN 0364152X

Liarikos, C. (2005). *The Dadia Forest Reserve: Conservation Plan for the after-LIFE period*. WWF Greece, Athens

Liarikos, C. (2010). Development trajectories and prospects in the Dadia-Lefkimi-Soufli Forest National Park. In: *The Dadia-Lefkimi-Soufli Forest National Park, Greece: Biodiversity, Management and Conservation*, Catsadorakis, G., & Källander, H. (Eds.), pp. 47-62. WWF-Greece, ISBN 978-960-7506-10-8, Athens

Lövbrand, E. (2009). Revisiting the politics of expertise in light of the Kyoto negotiations on land use change and forestry. *Forest Policy and Economics*, Vol.11, No.5-6 (October 2009), pp. 404–412, ISSN 1389-9341

Lund, D. H.; Boon, T. E. & Nathan, I. (2009). Accountability of experts in the Danish national park process. *Forest Policy and Economics*, Vol.11, No.5-6 (October 2009), pp. 437–445, ISSN 1389-9341

Machairas, I. & Hovardas, T. (2005). Determining visitors' dispositions towards the designation of a Greek National Park. *Environmental Management*, Vol.36, No.1 (July 2005), pp. 73–88, ISSN 0364152X

Maes, W. H., Fontaine, M., Rongé, K., Hermy, M., & Muys, B. (2011). A quantitative indicator framework for stand level evaluation and monitoring of environmentally sustainable forest management. *Ecological Indicators*, Vol.11, No.2 (March 2011), pp. 468–479, ISSN 1470160X

Manning, R. E.; Lawson, S.; Newman, P.; Laven, D. & Valliere, W. (2002). Methodological issues in measuring crowding-related norms. *Leisure Sciences*, Vol.24, No.3-4 (July 2002), pp. 339–348, ISSN 01490400

Mulvihill, P. R. & Milan, M. J. (2007). Subtle world: Beyond sustainability, beyond information. *Futures*, Vol.39, No.6 (August 2007), pp. 657–668, ISSN 0016-3287

Nygren, A. (2000). Environmental narratives on protection and production: nature-based conflicts in Rio San Juan, Nicaragua. *Development and Change*, Vol.31, No.4, pp. 807–830, ISSN 0012155X

Nygren, A. (2004). Contested lands and incompatible images: the political ecology of struggles over resources in Nicaragua's Indio-Maiz Reserve. *Society and Natural Resources*, Vol.17, No.3 (March 2004), pp. 189-205, ISSN 08941920

Ojha, H. M.; Cameron, J. & Kumar, C. (2009). Deliberation or symbolic violence? The governance of community forestry in Nepal. *Forest Policy and Economics*, Vol.11, No.5-6 (October 2009), pp. 365–374, ISSN 1389-9341

Ólafsdóttir, R. & Runnströ, M. (2009). A GIS approach to evaluating ecological sensitivity for tourism development in fragile environments. A case study from SE Iceland. *Scandinavian Journal of Hospitality and Tourism*, Vol.9, pp. 22–38, ISSN 1502-2269

Parkins, J. R. (2006). De-centering environmental governance: A short history and analysis of democratic processes in the forest sector of Alberta, Canada. *Policy Science*, Vol.39, No.2 (June 2006), pp. 183–203, ISSN 00322687

Parkins, J. R. & Davidson, D. J. (2008). Constructing the public sphere in compromised settings: Environmental governance in the Alberta forest sector. *Canadian Review of Sociology*, Vol.45, No.2 (May 2008), pp. 177-196, ISSN 17556171

Pelletier, N. (2010). Environmental sustainability as the first principle of distributive justice: Towards an ecological communitarian normative foundation for ecological economics. *Ecological Economics*, Vol.69, No.10 (August 2010), pp. 1887–1894, ISSN 0921-8009

Piñol, J.; Castellnou, M. & Beven, K. (2007). Conditioning uncertainty in ecological models: Assessing the impact of fire management strategies. *Ecological Modelling*, Vol.207, No.1 (September 2007), pp. 34–44, ISSN 0304-3800

Plummer, R. & Fennell, D. A. (2009). Managing protected areas for sustainable tourism: prospects for adaptive co-management. *Journal of Sustainable Tourism*, Vol.17, No.2 (March 2009), pp. 149–168, ISSN 1747-7646

Poirazidis, K., Skartsi, T., Catsadorakis, G. (2002). *Monotoring plan for the protected raea of Dadia-Lefkimi-Soufli Forest*. WWF-Greece, Athens

Poirazidis, K.; Goutner, V.; Skartsi, T. & Stamou, G. P. (2004). Nesting habitat modelling as a conservation tool for the Eurasian black vulture (*Aegypius monachus*) in Dadia Nature Reserve, northeastern Greece. *Biological Conservation*, Vol.118, No.2 (July 2004), pp. 235-248, ISSN 0006-3207

Prato, T. (2001). Modeling carrying capacity for national parks. *Ecological Economics*, Vol.39, No.3 (December 2001), pp. 321–331, ISSN 0921-8009

Prato, T. (2009). Fuzzy adaptive management of social and ecological carrying capacities for protected areas. *Journal of Environmental Management*, Vol.90, No.8 (June 2009), pp. 2551–2557, ISSN 0301-4797

Puhakka, R.; Sarkki, S.; Cottrell, S. P. & Siikamäki, P. (2009) Local discourses and international initiatives: sociocultural sustainability of tourism in Oulanka National Park, Finland. *Journal of Sustainable Tourism*, Vol.17, No.5 (September 2009), pp. 529–549, ISSN 09669582

Reser, J. P. & Bentrupperbäumer, J. M. (2005). What and where are environmental values? Assessing the impacts of current diversity of use of 'environmental' and 'World Heritage' values. *Journal of Environmental Psychology*, Vol.25, No.2 (June 2005), pp. 125-146, ISSN 02724944

Risse, T., 2000. Let's argue! Communicative action in world politics. *International Organization*, Vol.54, No.1, pp. 1–39, ISSN 00208183

Robbins, P. (2004). *Political Ecology*. Blackwell Puvblishing Ltd, ISSN 139781405102667, Oxford, UK

Roth, R. J. (2008). "Fixing" the Forest: The Spatiality of Conservation Conflict in Thailand. *Annals of the Association of American Geographers*, Vol.98, No.2 (April 2008), pp. 373–391, ISSN 00045608

Ruppert-Winkel, C. & Winkel, G. (2011). Hidden in the woods? Meaning, determining, and practicing of 'common welfare' in the case of the German public forests. *European Journal of Forest Research*, Vol. 130, No.3 (May 2011), pp. 421-434, ISSN 16124669

Sæþórsdóttir, A. D. (2010). Planning nature tourism in Iceland based on tourist attitudes. *Tourism Geographies*, Vol.12, No.1 (February 2010), pp. 25–52, ISSN 1470-1340

Schindler, S. (2010). *Dadia National Park, Greece – an integrated study on landscape, biodiversity, raptor populations and conservation management*. Doctoral Thesis, University of Vienna

Schindler, S.; Poirazidis, K. & Wrbka, T. (2008). Towards a core set of landscape metrics for biodiversity assessments: A case study from Dadia National Park, Greece. *Ecological Indicators*, Vol.8, No.5 (September 2008), pp. 502-514, ISSN 1470160X

Seidl, J. & Tisdell, C.A. (1999). Carrying capacity reconsidered: from Malthus' population theory to cultural carrying capacity. *Ecological Economics*, Vol.31, No.3 (December 1999), pp. 395–408, ISSN 0921-8009

Sharpley, R. & Pearce, T. (2007) Tourism, marketing and sustainable development in the English National Parks: The role of national park authorities. *Journal of Sustainable Tourism*, Vol.15, No.5, pp. 557-573, ISSN 09669582

Skartsi, Th.; Elorriaga, J. N.; Vasilakis, D. P. & Poirazidis, K. (2008). Population size, breeding rates and conservation status of Eurasian black vulture in the Dadia

National Park, Thrace, NE Greece. *Journal of Natural History*, Vol.42, No.5 (February
    2008), pp. 345-353, ISSN 0022-2933
Skartsi, Th., Elorriaga, J., & Vasilakis, D. (2010). Eurasian Black Vulture: the focal species of
    the Dadia-Lefkimi-Soufli Forest National Park. In: *The Dadia-Lefkimi-Soufli Forest
    National Park, Greece: Biodiversity, Management and Conservation*, Catsadorakis, G., &
    Källander, H. (Eds.), pp. 195-206. WWF-Greece, ISBN 978-960-7506-10-8, Athens
Sletto, B. (2010). The mythical forest, the becoming-desert: environmental knowledge
    production and the iconography of destruction in the Gran Sabana, Venezuela.
    Environment and Planning D: Society and Space, Vol.28, No.4, pp. 672-690, ISSN
    02637758
Sletto, B. (2011). Conservation planning, boundary-making and border terrains: The desire
    for forest and order in the Gran Sabana, Venezuela. *Geoforum*, Vol.42, No.2 (March
    2011), pp. 197–210, ISSN 0016-7185
Stamou, A. G.; Lefkaditou, A.; Schizas, D. & Stamou, G. P. (2009). The discourse of
    environmental information: Representations of nature and forms of rhetoric in the
    information center of a Greek reserve. *Science Communication*, Vol.31, No.2
    (December 2009), pp. 187-214, ISSN 10755470
Stavrakakis, Y. (1999). *Lacan and the political*. Routledge, ISSN 0415171873, London and New
    York
Steffek, J. (2009). Discursive legitimation in environmental governance. *Forest Policy and
    Economics*, Vol.11, No.5-6 (October 2009), pp. 313–318, ISSN 1389-9341
Stoffle, R. & Minnis, J. (2008). Resilience at risk: epistemological and social construction
    barriers to risk communication. *Journal of Risk Research*, Vol.11, No.1–2 (January–
    March 2008), pp. 55–68, ISSN 1366-9877
Sugimura, K. & Howard, T. E. (2008). Incorporating social factors to improve the Japanese
    forest zoning process. *Forest Policy and Economics*, Vol.10, No.3 (January 2008), pp.
    161–173, ISSN 1389-9341
Svoronou, E. (2000). *Planning pilot actions for the development of ecotourism*. Greek National
    Tourism Organization and WWF-Greece, Athens (in Greek)
Svoronou, E. & Holden, A. (2005). Ecotourism as a tool for nature conservation: The role of
    WWF Greece in the Dadia-Lefkimi-Soufli Forest Reserve in Greece. *Journal of
    Sustainable Tourism*, Vol.13, No.5 (November 2005), 2005, pp. 456-467, ISSN 0966-
    9582
Swart, R. J.; Raskin, P. & Robinson, J. (2004). The problem of the future: Sustainability
    science and scenario analysis. *Global Environmental Change*, Vol 14, No.2 (July 2004),
    pp. 137–146, ISSN 09593780
Tàbara, D.; Saurí, D. & Cerdan, R. (2003). Forest fire risk management and public
    participation in changing socioenvironmental conditions: A case study in a
    Mediterranean region. *Risk Analysis*, Vol. 23, No.2 (2003), pp. 249-260, ISSN 0272-
    4332
Triantakonstantis, D. P.; Kollias, V. J. & Kalivas, D. P. (2006). Forest re-growth since 1945 in
    the Dadia Forest nature reserve in northern Greece. *New Forests*, Vol.32, No.1 (July
    2006), pp. 51-69, ISSN 0169-4286
Uddhammar, E. (2006). Development, conservation and tourism: conflict or symbiosis?
    Review of International Political Economy, Vol.13, No.4 (October 2006), pp. 656–678,
    ISSN 1466-4526

Van Gossum, P., Arts, B., De Wulf, R., & Verheyen, K. (2011). An institutional evaluation of sustainable forest management in Flanders. *Land Use Policy*, Vol.28, No.1 (January 2011), pp. 110–123, ISSN 02648377

Vasilakis, P. D.; Poirazidis, S. K. & Elorriaga, N. J. (2008). Range use of a Eurasian black vulture (*Aegypius monachus*) population in the Dadia National Park and the adjacent areas, Thrace, NE Greece. *Journal of Natural History*, Vol.42, No.5 (February 2008), pp. 355–373, ISSN 0022-2933

Veenman, S.; Liefferink, D. & Arts, B. (2009). A short history of Dutch forest policy: The 'de-institutionalisation' of a policy arrangement. *Forest Policy and Economics*, Vol.11, No.3 (May 2009), pp. 202–208, ISSN 1389-9341

Velázquez, A.; Cué-Bär, E. M.; Larrazábal, A.; Sosa, N.; Villaseñor, J. L.; McCall, M. & Ibarra-Manríquez, G. (2009). Building participatory landscape-based conservation alternatives: A case study of Michoacán, Mexico. *Applied Geography*, Vol.29, No.4 (December 2009), pp. 513–526, ISSN 0143-6228

Vierikko, K.; Vehkamäki, S.; Niemelä, J.; Pellikka, J. & Lindén, H. (2008). Meeting the ecological, social and economic needs of sustainable forest management at a regional scale. *Scandinavian Journal of Forest Research*, Vol.23, No.5 (October 2008), pp. 431-444, ISSN 1651-1891

Von Detten, R. (2011). Sustainability as a guideline for strategic planning? The problem of long-term forest management in the face of uncertainty. *European Journal of Forest Research*, Vol. 130, No.3 (May 2011), pp. 451-465, ISSN 16124669

Warren, W. A. (2007). What Is a Healthy Forest? Definitions, Rationales, and the Lifeworld. *Society and Natural Resources*, Vol.20, No.2 (February 2007), pp. 99–117, ISSN 1521-0723

Winkel, G. (in press). Foucault in the forests — A review of the use of 'Foucauldian' concepts in forest policy analysis. *Forest Policy and Economics*, doi:10.1016/j.forpol.2010.11.009, ISSN 1389-9341

Wolfslehner, B., & Seidl, R. (2010). Harnessing ecosystem models and multi-criteria decision analysis for the support of forest management. *Environmental Management*, Vol.46, No.6 (December 20010), pp. 850–861, ISSN 0364152X

Wood, B. (2000). Room for nature? Conservation management of the Isle of Rum, UK and prospects for large protected areas in Europe. *Biological Conservation*, Vol.94, No.1 (June 2000), pp. 93-105, ISSN 00063207

Wynne, B., 2006. Public engagement as a means of restoring public trust in science–hitting the notes, but missing the music? *Community Genetics*, Vol.9, No.3 (May 2006), pp. 211–220, ISSN 14222795

Zacarias, D. A.; Williams, A. T. & Newton, A. (2011). Recreation carrying capacity estimations to support beach management at Praia de Faro, Portugal. *Applied Geography*, Vol.31, No.3 (July 2011), pp. 1075-1081, ISSN 0143-6228

Permissions

The contributors of this book come from diverse backgrounds, making this book a truly international effort. This book will bring forth new frontiers with its revolutionizing research information and detailed analysis of the nascent developments around the world.

We would like to thank Jorge Martín-García and Julio Javier Diez, for lending their expertise to make the book truly unique. They have played a crucial role in the development of this book. Without their invaluable contribution this book wouldn't have been possible. They have made vital efforts to compile up to date information on the varied aspects of this subject to make this book a valuable addition to the collection of many professionals and students.

This book was conceptualized with the vision of imparting up-to-date information and advanced data in this field. To ensure the same, a matchless editorial board was set up. Every individual on the board went through rigorous rounds of assessment to prove their worth. After which they invested a large part of their time researching and compiling the most relevant data for our readers. Conferences and sessions were held from time to time between the editorial board and the contributing authors to present the data in the most comprehensible form. The editorial team has worked tirelessly to provide valuable and valid information to help people across the globe.

Every chapter published in this book has been scrutinized by our experts. Their significance has been extensively debated. The topics covered herein carry significant findings which will fuel the growth of the discipline. They may even be implemented as practical applications or may be referred to as a beginning point for another development. Chapters in this book were first published by InTech; hereby published with permission under the Creative Commons Attribution License or equivalent.

The editorial board has been involved in producing this book since its inception. They have spent rigorous hours researching and exploring the diverse topics which have resulted in the successful publishing of this book. They have passed on their knowledge of decades through this book. To expedite this challenging task, the publisher supported the team at every step. A small team of assistant editors was also appointed to further simplify the editing procedure and attain best results for the readers.

Our editorial team has been hand-picked from every corner of the world. Their multi-ethnicity adds dynamic inputs to the discussions which result in innovative outcomes. These outcomes are then further discussed with the researchers and contributors who give their valuable feedback and opinion regarding the same. The feedback is then collaborated with the researches and they are edited in a comprehensive manner to aid the understanding of the subject.

Apart from the editorial board, the designing team has also invested a significant amount of their time in understanding the subject and creating the most relevant covers. They scrutinized every image to scout for the most suitable representation of the subject and create an appropriate cover for the book.

The publishing team has been involved in this book since its early stages. They were actively engaged in every process, be it collecting the data, connecting with the contributors or procuring relevant information. The team has been an ardent support to the editorial, designing and production team. Their endless efforts to recruit the best for this project, has resulted in the accomplishment of this book. They are a veteran in the field of academics and their pool of knowledge is as vast as their experience in printing. Their expertise and guidance has proved useful at every step. Their uncompromising quality standards have made this book an exceptional effort. Their encouragement from time to time has been an inspiration for everyone.

The publisher and the editorial board hope that this book will prove to be a valuable piece of knowledge for researchers, students, practitioners and scholars across the globe.

List of Contributors

Richard S. Mbatu
Environmental Sustainability, College of Science, Health and the Liberal Arts, Philadelphia University, Philadelphia, U.S.A

John Eilif Hermansen
Department of Industrial Economics and Technology Management, Norwegian University of Science and Technology, NTNU, Trondheim, Norway

Nelson Turyahabwe, Jacob Godfrey Agea, Mnason Tweheyo and Susan Balaba Tumwebaze
College of Agricultural and Environmental Sciences, Makerere University, Kampala, Uganda

Lucas Rezende Gomide
Federal University of Lavras, Brazil

Fausto Weimar Acerbi Junior, José Roberto Soares Scolforo, José Márcio de Mello, Antônio Donizette de Oliveira, Luis Marcelo Tavares de Carvalho, Natalino Calegário and Antônio Carlos Ferraz Filho
Federal University of Lavras, Brazil

X. Wei
Department of Earth and Environmental Sciences, University of British Columbia (Okanagan), Kelowna, British Columbia, Canada

J. P. Kimmins
Department of Forest Sciences, University of British Columbia, Vancouver, British Columbia, Canada

André Eduardo Biscaia de Lacerda
Embrapa Forestry, Paraná, Brazil

Maria Augusta Doetzer Rosot, Marilice Cordeiro Garrastazú, Betina Kellermann, Maria Izabel Radomski, Patricia Povoa de Mattos and Yeda Maria Malheiros de Oliveira
Embrapa Forestry, Paraná, Brazil

Afonso Figueiredo Filho
Midwest State University in Irati, Brazil

Evelyn Roberta Nimmo
University of Manitoba, Canada

Thorsten Beimgraben
Rottenburg University of Applied Forest Sciences, Germany

Kalyani Chatterjea
National Institute of Education, Nanyang Technological University, Singapore

Nur Muhammed, Roderich von Detten and Gerhard Oesten
Institute of Forestry Economics, University of Freiburg, Germany

Mohitul Hossain, Sheeladitya Chakma and Farhad Hossain Masum
Institute of Forestry and Environmental Sciences, University of Chittagong, Bangladesh

Koji Matsushita
Kyoto University, Japan

Anna Stier, Jutta Lax and Joachim Krug
Johann Heinrich von Thünen-Institute (vTI), Federal Research Institute for Rural Areas, Forestry and Fisheries, Institute for World Forestry, Germany

Edward Robak
University of New Brunswick, Canada

Jacobo Aboal
Dirección Xeral de Montes, Xunta de Galicia, Spain

Juan Picos
University of Vigo, Spain

Tasos Hovardas
University of Thessaly, Greece